Advances in Intelligent Systems and Computing

Volume 968

AF788477

The series "Advances in Intelligent Systems and Computing" contains publications on theory, applications, and design methods of Intelligent Systems and Intelligent Computing. Virtually all disciplines such as engineering, natural sciences, computer and information science, ICT, economics, business, e-commerce, environment, healthcare, life science are covered. The list of topics spans all the areas of modern intelligent systems and computing such as: computational intelligence, soft computing including neural networks, fuzzy systems, evolutionary computing and the fusion of these paradigms, social intelligence, ambient intelligence, computational neuroscience, artificial life, virtual worlds and society, cognitive science and systems, Perception and Vision, DNA and immune based systems, self-organizing and adaptive systems, e-Learning and teaching, human-centered and human-centric computing, recommender systems, intelligent control, robotics and mechatronics including human-machine teaming, knowledge-based paradigms, learning paradigms, machine ethics, intelligent data analysis, knowledge management, intelligent agents, intelligent decision making and support, intelligent network security, trust management, interactive entertainment, Web intelligence and multimedia.

The publications within "Advances in Intelligent Systems and Computing" are primarily proceedings of important conferences, symposia and congresses. They cover significant recent developments in the field, both of a foundational and applicable character. An important characteristic feature of the series is the short publication time and world-wide distribution. This permits a rapid and broad dissemination of research results.

**** Indexing: The books of this series are submitted to ISI Proceedings, EI-Compendex, DBLP, SCOPUS, Google Scholar and Springerlink ****

More information about this series at http://www.springer.com/series/11156

Cliff Sungsoo Shin
Editor

Advances in Interdisciplinary Practice in Industrial Design

Proceedings of the AHFE 2019 International Conference on Interdisciplinary Practice in Industrial Design, July 24–28, 2019, Washington D.C., USA

Springer

Editor
Cliff Sungsoo Shin
Department of Industrial Design
University of Illinois at Urbana Champai
Champaign, IL, USA

ISSN 2194-5357 ISSN 2194-5365 (electronic)
Advances in Intelligent Systems and Computing
ISBN 978-3-030-20469-3 ISBN 978-3-030-20470-9 (eBook)
https://doi.org/10.1007/978-3-030-20470-9

This Springer imprint is published by the registered company Springer Nature Switzerland AG
The registered company address is: Gewerbestrasse 11, 6330 Cham, Switzerland

Advances in Human Factors and Ergonomics 2019

AHFE 2019 Series Editors

Tareq Ahram, Florida, USA
Waldemar Karwowski, Florida, USA

10th International Conference on Applied Human Factors and Ergonomics and the Affiliated Conferences

Proceedings of the AHFE 2019 International Conference on Interdisciplinary Practice in Industrial Design, held on July 24–28, 2019, in Washington D.C., USA

Advances in Affective and Pleasurable Design	Shuichi Fukuda
Advances in Neuroergonomics and Cognitive Engineering	Hasan Ayaz
Advances in Design for Inclusion	Giuseppe Di Bucchianico
Advances in Ergonomics in Design	Francisco Rebelo and Marcelo M. Soares
Advances in Human Error, Reliability, Resilience, and Performance	Ronald L. Boring
Advances in Human Factors and Ergonomics in Healthcare and Medical Devices	Nancy J. Lightner and Jay Kalra
Advances in Human Factors and Simulation	Daniel N. Cassenti
Advances in Human Factors and Systems Interaction	Isabel L. Nunes
Advances in Human Factors in Cybersecurity	Tareq Ahram and Waldemar Karwowski
Advances in Human Factors, Business Management and Leadership	Jussi Ilari Kantola and Salman Nazir
Advances in Human Factors in Robots and Unmanned Systems	Jessie Chen
Advances in Human Factors in Training, Education, and Learning Sciences	Waldemar Karwowski, Tareq Ahram and Salman Nazir
Advances in Human Factors of Transportation	Neville Stanton

(continued)

(continued)

Advances in Artificial Intelligence, Software and Systems Engineering	Tareq Ahram
Advances in Human Factors in Architecture, Sustainable Urban Planning and Infrastructure	Jerzy Charytonowicz and Christianne Falcão
Advances in Physical Ergonomics and Human Factors	Ravindra S. Goonetilleke and Waldemar Karwowski
Advances in Interdisciplinary Practice in Industrial Design	Cliff Sungsoo Shin
Advances in Safety Management and Human Factors	Pedro M. Arezes
Advances in Social and Occupational Ergonomics	Richard H. M. Goossens and Atsuo Murata
Advances in Manufacturing, Production Management and Process Control	Waldemar Karwowski, Stefan Trzcielinski and Beata Mrugalska
Advances in Usability and User Experience	Tareq Ahram and Christianne Falcão
Advances in Human Factors in Wearable Technologies and Game Design	Tareq Ahram
Advances in Human Factors in Communication of Design	Amic G. Ho
Advances in Additive Manufacturing, Modeling Systems and 3D Prototyping	Massimo Di Nicolantonio, Emilio Rossi and Thomas Alexander

Preface

Industrial design has been one of the fastest growing and demanding areas which made significant contributions to our lives and societies. Traditionally, industrial designers focus on the aesthetics of products' form and style as well as ergonomics. This tradition has expanded its boundary to other subject matters such as service design, human–computer interaction, user experience design, sustainable design, and virtual and augment reality. This is a strong indication that interdisciplinary collaboration across disciplines became an essential and critical practice in design.

The International Conference on Interdisciplinary Practice in Industrial Design (IPID) aims at exploring and broadening the interdisciplinary practice in industrial design. The conference includes discussions on (1) a theoretical investigation as well as professional practice to foster interdisciplinary collaboration across disciplines, (2) design projects through interdisciplinary collaboration, (3) design process with external public and private sector partners with a solid record of interdisciplinary development experience, and (4) design methods and techniques to investigate productive and effective interdisciplinary collaboration in design. A total of seven sections presented in this book:

Section 1 Emotional Evidence
Section 2 Design for Health and Wellness
Section 3 Application of 3D Scanning to Product Design
Section 4 Interdisciplinary Practice for Automobile Design
Section 5 Interdisciplinary Design and Education
Section 6 Design Embracing Information Science, Intention, and Experience
Section 7 Research-Driven Design

Each section contains research papers that have been reviewed by members of the International Editorial Board. Our sincere thanks and appreciation to the board members as listed below:

Miso Kim, USA
YoungAe Kim, USA
Jiwon Park, USA
Scott Shim, USA

We hope that this book, which is the international state of the art in the Interdisciplinary Practice in Industrial Design, will be a valuable source of theoretical and applied knowledge enabling human-centered design of variety of products, services, and systems for global markets.

July 2019 Cliff Sungsoo Shin

Contents

Interdisciplinary Design and Education

Design Embracing Information Science, Intention and Experience

Research-Driven Design

Emotional Evidence

How Creative Mindset Is Involved in Positive Emotions and Attitude that Affects Creative Design Process

Joungyun Choi(✉)

College of Environmental Science and Forestry, State University of New York, 1 Forestry Dr., Syracuse, NY 13210, USA
jchoi@esf.edu

Abstract. Creativity is an important problem-solving tool on the design context. People who are in design education treat creativity as an important component of their self-identity. Thus, it is imperative to develop and be mindful of creativity in design education. Most traditional design fields focus on creative outcomes that impact revenue, trends, and popularity. However, this current research shows that it is even more important to focus on people's state of mind (mindsets) and attitude in building their creative ability. Depending on people's attitudes toward their projects, some design solutions are more creative than others. The primary research objective of this study is to identify the presence of design students' creative mindset and how it is manifested in their work. This research also seeks to determine how the creative mindset operates and develops with respect to students' creative performance by incorporating the concept of the fixed/growth mindset toward creativity. The qualitative data established the fact that: (1) The creative mindset involves students' attitude toward design projects and is manifested as a positive attitude, such as having an open mind and demonstrating readiness at the beginning of their design process; (2) A positive attitude from students' creative mindset is closely related to learning goals, which is one of the main concepts of the growth mindset; (3) The creative mindset greatly affects the outcome of the design, and more specifically improves one's design ability, which contributes to student success from a long-term perspective.

Keywords: Creative mindset · State of mind · Positive emotion · Growth mindset · Open mind · Learning goal · Design education · Higher education · Decision to be creative · Problem solving · Design ability

1 Introduction

Given that our society is becoming more complex, and people are undeniably facing a wider array of problems, design processes are considered to be an important tool for tackling complex social and cultural problems. Designers use design thinking, which provides us with a new way of tackling problems and encourages us to adopt divergent approaches and new solutions [1]. Currently, design is perceived as a crucial means of fostering creativity and innovation when employed as a problem-solving tool,

C. S. Shin (Ed.): AHFE 2019, AISC 968, pp. 3–12, 2020.
https://doi.org/10.1007/978-3-030-20470-9_1

insomuch as the solutions should not be banal or useless; rather, they should be novel and useful. Thus, design is a field that inherently involves a *creative* problem-solving activity and necessitates making decisions to fulfill certain objectives. From the definitions and conceptions of design, it is important to understand how creativity operates in the design process. Designers' ideas develop as their problem-solving progresses. Therefore, each designer constructs his or her own representation of the design problem, dealing with a problem that becomes specific to him or her [2]. In practice, different designers, supposedly solving the same design problem, reach different solutions [3]. In this process, some design solutions are more creative, while others are considered as less creative.

Design educators often see a similar process in the area of design education. Although students may have the same design project, they approach it with different design solutions. These various design solutions produce different processes and outcomes: some designs are more innovative, while others are less so. There are many reasons why students perform differently in their creative design processes. However, through my own experience and observation, there is a common reason – their attitudes – that explains how they approach design projects. However, their thinking habits precede their attitude formation.

This research originated from my interest regarding these two different attitudes and their subsequent outcomes. Why then do students act differently? Where do these attitudes come from? Do they adopt a certain mindset in terms of their new design project, like I do? If so, do these mindsets toward a design project affect their creativity? How do these mindsets influence students' creativity? The design area in higher education requires a high level of creativity for problem solving throughout the design process. Each design outcome tends to be unique and non-repetitive in its conception and development. People who are in design education treat creativity as an important component of their self-identity. If students' creativity can be affected by their mindsets, then investigating students' mindsets should be a first priority with respect to how they work differently.

The current research has the following perspective: the creative mindset is related to a growth mindset toward creativity. The concept of a growth mindset is based on the theory of implicit beliefs. Implicit belief theory is one of the most important concepts in educational psychology. This theory explains people's beliefs regarding their psychological traits and abilities, which plays an important role in influencing their own motivation and behavior [4]. According to research regarding implicit beliefs, people hold different beliefs that represent their state of mind and explain how and why they behave the way they do. Some people perceive their abilities as fixed (fixed mindset: unable to develop/change), while other people believe that their abilities are malleable (growth mindset: able to develop/change) [5]. This conceptualization shows how people develop beliefs about themselves; additionally, it shows how fixed and growth mindsets create their psychological worlds, influencing their thoughts, feelings, and behaviors. These mindsets explain why some people are motivated to work harder, and why others fall into patterns of passivity and self-defeating behavior. When people decide to do something, they believe that they can create, develop, change, and make a situation better. Thus, the creative mindset is related to a growth mindset toward creativity; in other words, the creative mindset can be learned and developed. While analyses of fixed and growth mindsets are rarely provided within the creative domain [6–8], a few studies regarding the fixed/growth mindset concept of creativity will be reviewed. This research will adopt specific methods to demonstrate the relationship between the concept of fixed/growth mindsets and the creative mindset.

2 Method

The current study is based on the personal experiences of design students, as well as design educators. The purpose aims to investigate whether design students decide to be creative (creative mindset) when they start a new design project. This study examines whether the creative mindset has a positive connection with the growth mindset in terms of creativity. Such a positive connection would establish the fact that students can develop their creative mindsets in their design projects. In seeking to understand these phenomena, methodological triangulation is used. There were three phases in the process of the study, which were as follows: an implicit belief of creativity (fixed versus growth mindsets toward creativity) survey, direct in-depth interviews, and observation. The survey used three different sections of questions: the mindset test [7–9], the Goals Inventory (GI) test [10], and the self-perception of creativity test [7–9] to identify students with a high growth mindset toward creativity. After the survey, direct in-depth interviews were conducted with the students who were identified as having growth mindsets toward creativity. The aim of the interviews focused on how to determine whether students had a creative mindset, as well as to identify the relationship between the creative growth mindset and the creative mindset. Participant observations were conducted throughout the research, between the survey and the interview, as well as during the interview. In the participant observations, the primary sources of data were what people said and how they behaved. Through their conscious and/or unconscious behaviors or conversations, students' attitudes toward creative design work could be identified. The purpose of these participant observations was to understand how design students' varied attitudes toward creative design work affect their creative design performance/outcomes by incorporating the experiences of growth mindset students and relating these experiences to their creative mindsets.

2.1 Participants and Setting

The sample for the current study consisted of 179 students who agreed to participate in this research since the first step of the study was used existing survey to the design students. The students were enrolled in a variety of design subject matter tracks, such as graphic design, architecture, interior design, product design, and apparel design in a couple of different major universities located in the US Midwest. These students were selected from two introductory-level design classes and two advanced-level design classes. After the participants took the fixed/growth mindset survey, the in-depth interviews were conducted with 12 students who demonstrated a growth mindset toward creativity. Throughout the process of the implicit belief tests regarding creativity and the interviews with the growth mindset students, participant observation was conducted to understand the relationships among the tests, interviews, and students' performance/design outcomes. Participation in this study was voluntary. The implicit belief tests regarding creativity took approximately 10 min in their classrooms, while the interviews took about an hour in a quiet area in the building. Each participant was given a $30 gift card as an incentive to participate in the direct interviews.

2.2 The Implicit Belief Test Regarding Creativity

In the initial phase of this study, participants were given the survey to measure their implicit beliefs about creativity, including the fixed/growth mindset test [7–9] the Goals Inventory (GI) test to determine students' goal orientations [10], and the self-perception of creativity test [7–9]. Participants were presented with a hard copy of the consent form before they took the tests. After reading and signing the informed consent forms, the test copies were distributed. The tests took about 10 min to complete. The results of the tests were used to select participants identified as having high growth mindsets toward creativity. The test consisted of 20 questions in total. These questions measured the mindset and goal orientations, along with three items to evaluate self-perceptions of creativity.

Dweck and Leggett [5] argued that there is a strong relationship between goal orientations and mindsets; thus, people who hold a growth mindset exert more effort and have more confidence, since they tend to form learning goals. Therefore, measuring students' goal orientations, particularly identifying students with learning goals is an important factor in researching fixed/growth mindsets [5, 8]. Overall, researchers have shown that fixed and growth mindsets are moderately negatively correlated, and growth mindsets have been shown to represent learning goals, which is a primary concept for developing self-efficacy. Based on the overall ideas of these previous studies, three measurements were implemented in this research: Fixed/Growth mindset toward Creativity, Goal Orientations and Self-efficacy of Creativity.

Participants indicated their responses to each item with a six-point Likert scale (1 = strongly disagree to 6 = strongly agree). The collected responses from the survey were analyzed and run to score and identify high growth mindset students (>4 growth mindset on a scale from 1 to 6). When scoring the data, negatively coded variables were reverse coded.

2.3 In-Depth Interview

When the survey was completed, 12 participants with an average score higher than or equal to 4.0 (>4 growth mindset on a scale from 1 to 6) were selected to continue to the interview phase. To determine the answers to the research questions, as well as the related questions (Do design students have creative mindsets? How do they describe their creative mindsets? Do creative mindsets have a similarity with the growth mindset regarding creativity?), a qualitative data collection method was used with in-depth interviews consisting of open-ended questions to identify participants' implicit beliefs about creativity and creative mindsets. Direct face-to face interviews were conducted because the observation of social cues such as facial expressions, body posture, or voice tone is an important part of the interview in determining participants' thinking habits, states of mind, and implicit beliefs. interviews were designed to collect descriptive data in the participants' own words and to develop an understanding with regard to participants' opinions and experiences. The interviews took approximately one hour each in the participants' design studios, where they usually spend their time in the creative design process on campus. Incentives were given to participants to elicit more active participation, and the interviews were audio recorded after participants signed the interview consent form. All interviews were recorded and transcribed; moreover, the interview situations were documented in field notes. The data were analyzed through an inductive content analysis.

2.4 Observations

Throughout the entire research process, participant observations were conducted to examine their behavior/performance and outcomes of the creative design process. The observation took place through nonparticipant observation: although the researcher entered the classroom and met with the students directly to ask about their feelings and thoughts about their work, the researcher did not become a member of the context, nor did the researcher participate in their activities. To enhance the validity of the observation data, the observation took four months (one semester) in two different design studio classes that were selected for phase 1 (the tests). Two classes were visited within two time periods (4 h) a week during the entire semester. The observations were made when the participants were given a new project and time to work on their own. In this way, the relationship between participants' creative mindsets (decision to be creative) and creativity (developing their project) in the design process could be observed through their behavior. Students' design processes were observed in detail, starting from the research to the production/presentation in order to record them objectively, without the researcher's personal bias. Throughout the design process, the participants were sometimes asked about their feelings and thought processes regarding their projects, and their responses were documented in written format. To analyze the field notes, (1) the collected data were organized into a narrative format of a day; (2) the narrative format of the information was organized, according to the outline of the research questions; and (3) a deductive content analysis was used, which "starts with the counting of words or manifest content, then extends the analysis to include latent meanings and themes" [11]. The outlined text information was analyzed to determine the frequency of the contents (students' habits of action). This information was later compared to the research questions and interview findings to match with students' performance and the creative mindset.

3 Results

Results from the observations: Two distinct groups of students with different attitudes toward their design projects were observed and selected from the broadly focused observations. There were students with neutral attitudes in between the two distinct groups; however, only the performance of the high growth mindset students and the high fixed mindset students were observed. Then, subsequently, only the high growth mindset students were selected and were invited to participate in the in-depth interviews. Their responses were then analyzed to address the research questions. At the beginning of the observation, two different groups were distinguished based on ten attributes that were selected from the frequency of content (students' habits of action). The ten attributes are as follows: (1) Active engagement; (2) Time management; (3) Relevant conversation; (4) Having an open mindset to critique; (5) A high attendance rate; (6) Mindfulness of the project; (7) Generating ideas; (8) Willingness to take challenges; (9) Preparing class materials and assignments; and (10) Sharing knowledge. These attributes were largely identified from students with high scores of implicit beliefs as having a growth mindset toward creativity. They usually engaged actively in class exercises, worked more than they were required, and managed their time well so that they always stayed on top of due dates. They tried to generate different ways of solving problems in given projects, and they were willing to share their knowledge with others. After the high creativity growth mindset students were identified from the survey, the narrowly focused observations took place in order to concentrate on high performance students' behaviors and outcomes. The most obvious characteristic of the growth mindset students was having an "open mind" among all of the ten attributes. All 12 participants (100% of the growth mindset students) had an open mind in terms of accepting others' critiques, accepting challenges, and collecting different ideas from various sources. However, not all participants with a high growth mindset did well on their project from the beginning. A total of 42% of the participants (5 out of 12 participants) showed a lack of both diverse ideation and development of their final outcomes on their first couple of projects. From the broadly focused observations, there was no significant difference in the first quarter of the semester with respect to their current design skills. However, when there was challenge or confusion in the assignment, the fixed mindset students clearly suffered from their confrontation with the challenge or the confusion. They ended up with the same level of work from the very beginning stage of the design process. There was no improvement throughout the project. They generally reacted with self-doubt and disruption, deciding quickly that they were not good at doing the project. That led them to give up easily. This would put them at a disadvantage to succeeding in the class. The growth mindset students were equivalent in the design skills that they brought to a task. However, they ended up displaying much different levels of performance and outcomes. From the narrowly focused observations, this appears to have occurred as a result of how they approached the project. Through their open-minded approach to a project, the challenge or confusion of an assignment allowed them to progress and improve in later projects. It turned out that being open minded was an essential attribute for students in influencing their success in their outcomes. An open mindset led students to enjoy their challenges or obstacles. They also remained very confident that they would succeed, saying things such as, "I almost got it now" or asking for a few more chances on the assignment because they were almost getting it. This group tended to maintain the positive mood they had displayed during a difficult assignment, but some of them become even happier about the assignment. The images below (Fig. 1) show the improvement in Student 3's (from Fig. 1) divergent thinking process in the same condition and criteria.

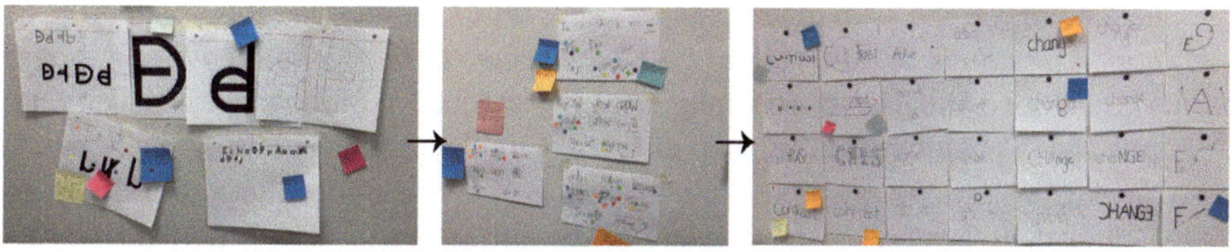

Fig. 1. Improvement in Student 3's divergent thinking process

Results from the Interviews: From the results of the narrowly focused observations and the growth mindset survey, 12 students with a high creative growth mindset were selected. From the interviews, (1) The students' thinking mode changes to being more open and looking for something more unique when embarking upon a new design project. Sixty-seven percent of the participants (8 out of 12) with high growth mindsets reported that they "changed" their way of thinking when starting a new project. This "change" means that the participants *got ready* to dive into their projects. In particular, their thinking mode changed to being open and looking for something more unique (creative mindset); (2) A creative mindset is described as a positive attitude, such as an open mind and readiness toward a design project. Participants were asked to describe what a creative mindset is in their perspective. The most common descriptions of a creative mindset reported by participants included a "starting point to be creative," "a mind that starts something positively," and "an activity of throwing questions to oneself that are raised from curiosity and interest in something." In this description, the creative mindset is usually apparent at the beginning of the design process. A creative starting point encourages students to have a positive attitude toward their projects. A positive attitude was described using several different words, which were repeated and stressed over time during the interviews: "open mind," "readiness," "decide," and "mode (approach)." Again, a creative mindset is a matter of people's state of mind and attitude on how they start a project.

4 Discussion

Two major findings contribute to defining the conclusions of the research objectives. Based on these findings, this study provides four important concepts, which involve: (1) the creative mindset; (2) the relationship between the creative mindset and the concept of the growth mindset; (3) the relationship between the creative mindset and students' performance/success in a project.

(1) The creative mindset: The creative mindset involves students' attitudes toward design projects and is manifested as a positive attitude, such as open mind and readiness, which is evident at the beginning of their design process: This concept of the creative mindset is highly meaningful in terms of dealing with the first research question, which examines whether such a creative mindset exists when students are given a design project, and how the creative mindset is manifested. Although the way in which the creative mindset was expressed differed,

depending on the participants' experience, *all of the participants articulated the notion that their "decision to be creative" occurs at the beginning of their project.* The creative mindset is generally seen in students' performance, which is indicated by a positive attitude, such as having an open mind and being ready to take on the project. As previously addressed in the finding chapter, all participants shared their experience regarding their design process. They reported that their design process is quite consistent with every type of project, and at the outset, their research, idea elaboration, and ideation sketches demand the most creativity in the design process. In this way, they consciously or unconsciously "change" their thinking mode to be open minded and to look for something more unique (the decision to be creative). This creative way of starting a project encourages students to have a positive attitude toward their projects. This positive attitude was described using several different words, such as "open mind," "readiness," "decide," and "mode (approach)." Also, these positive attitudes could be found from the following ten attributes: (1) Active engagement; (2) Time management; (3) Relevant conversation; (4) Having an open mind to critique; (5) A high attendance rate; (6) Mindfulness of the project; (7) Generating ideas; (8) Willingness to take challenges; (9) Preparing class materials and assignments; and (10) Sharing knowledge that was identified during the observations. Again, the creative mindset is a matter of people's state of mind and attitude in terms of how people start the design process.

(2) Relationship between the creative mindset and the concept of the growth mindset: The positive attitude (and emotion) from students' creative mindset is closely related to learning goals, which is one of the main concepts of the growth mindset. Fixed/Growth mindsets are involved in creativity. The growth mindset is manifested by a conscious effort to look for something new and develop creativity. For most of the time, participants keep looking for surprise associations in creating a new design. The participants reported that having a creative mindset always leads to gaining something new because of their open mind and readiness. An open mind allows people to have the desire to learn/try new things and master new projects successfully. Readiness for a project helps people overcome challenges. The creative mindset is a thinking pattern in which a person decides to approach solving a particular situation/problem creatively. This way of approaching a situation/problem leads a person to focus on more efforts to develop his or her creativity (Table 1).

(3) Relationship between the creative mindset and students' performance/success in a project: The creative mindset greatly affects the outcome of the design, more specifically, in terms of improving their design ability, which results in student success from a long-term perspective. This concept is mainly found from the observations. The growth mindset students, who start with creative mindsets, have a positive attitude toward design projects. As discussed in terms of the first concept, a positive attitude was associated with these behaviors: (1) Active engagement; (2) Time management; (3) Relevant conversation; (4) An open mind to critique; (5) A high attendance rate; (6) Mindfulness of the project; (7) Divergent thinking; (8) Willingness to take challenges; (9) Preparing class materials and assignments; and (10) Sharing knowledge. These positive behaviors significantly improve both

Table 1. Creative mindset vs. growth mindset toward creativity

"Creative mindset" vs. "Growth mindset toward creativity" in this study	
Creative mindset	*Decision to be creative* Thinking habits Attitudes Decision to generate new ideas, analyze these ideas and sell the ideas to others [12] The willingness to: 1. redefine problems in novel ways; 2. take sensible risks; 3. sell ideas that others might not initially accept; 4. persevere in the face of obstacles; and 5. examine whether their own preconceptions are interfering with their creative process [13]
Growth mindset toward creativity	*High growth mindset toward creativity* Implicit beliefs regarding creativity: self-perceptions of creativity How people think about their own creativity: They think that their creativity can be developed. Beliefs about the malleability of own creativity [4] The positive conceptions of psychological states th at are held by laypeople [12] Correlated with a learning goal orientation: The goal is to increase competence, and seek challenges that fostering learning Correlated with creative self-efficacy

the process and outcome, which ultimately affect students' grades in the class and their portfolio development. According to the findings of the study, half of the high creative growth mindset students did not engage in the creative design process, such as divergent thinking, rich ideation, and brainstorming during the first couple of projects. However, they likely possess self-esteem in their performance, which is a positive way of experiencing the creative process when they are fully engaged and are using their abilities to pursue the values of the class.

5 Conclusion

The explicit knowledge regarding the creative mindset from major finding of the study can support students' intrinsic motivation so that first, they believe that creative skills can be developed, and second, they can consciously make an effort to have positive emotions and attitudes toward their project to develop their creative mindset. Thus, by incorporating the concept of the growth mindset, an instructional element should include examples of how creative achievements are acquired after a process with multiple challenges to highlight the role of development and effort in the creative process. Also, knowledge about the relationship between the creative mindset and the concept of the growth mindset will help educators create an instructional environment and situation that can ignite students' creative mindsets. Particularly, creating a good atmosphere for the first impression of a project is important in leading students' creative mindsets, since the creative mindset involves students' attitudes toward the design project at the beginning stage. Also, the growth mindset is always involved in the process of creative thinking. As a result, educators can take advantage of situations in which students might express unexpected and surprising ideas. From this situation, educators praise the process and effort of creative thinking more than the outcome/creative ability, which is very important in encouraging students' intrinsic motivation.

Recommendation for future research is to examine the level of inequality between students with growth versus fixed mindsets. The current research dealt with finding the phenomena of influential aspects from the creative mindset by relating it to the concept of the growth mindset in design education. Thus, the finding concerning the relationship between the creative mindset and the creative growth mindset is still critical. However, there should be a comparison of mindsets, performance, and outcomes between students who have a high growth mindset and a high fixed mindset to examine how these different mindsets affect students' creative design process, outcomes, and their resulting success and happiness. Since there are unexpected findings regarding the factors that help ignite the creative mindset and suggestions to develop students' creative mindsets, this comparison process would provide a better understanding of fixed mindset students and could provide more specific solutions to them.

References

1. Brown, T., Katz, B.: Change by Design: How Design Thinking Transforms Organizations and Inspires Innovation, 1st edn. Harper Business, New York (2009)
2. Simon, H.A.: Problem forming, Problem finding and problem solving in design. In: Design & Systems, pp. 245–257 (1995)
3. Bonnardel, N., Marmèche, E.: Towards supporting evocation processes in creative design: a cognitive approach. Int. J. Hum.-Comput. Stud. **63**(4), 422–435 (2005)
4. Dweck, C.S.: Self-Theories: Their Role in Motivation, Personality, and Development. Essays in Social Psychology. Psychology Press, Philadelphia (1999)
5. Dweck, C.S., Laggett, E.L.: A social-cognitive approach to motivation and personality. Psychol. Rev. **95**, 256–273 (1988)
6. O'Conner, A.J., Akutsu, S.: Consequences of beliefs about the malleability of creativity. Creat. Res. J. **25**(2), 155–162 (2013)
7. Kawowski, M.: Creative mindsets: measurement, correlates, consequences. Psychol. Aesthet. Creat. Arts **8**(1), 62 (2014)
8. Hass, R., Katz-Buonincontro, J., Reiter-Palmon, R., Tinio, P.: Disentangling creative mindsets from creative self-efficacy and creative identity: do people hold fixed and growth theories of creativity? Psychol. Aesthet. Creat. Arts **10**(4), 436–446 (2016)
9. Hong, Y.Y., Chiu, C.Y., Dweck, C.S., Lin, D.M.S., Wan, W.: Implicit theories, attributions, and coping: a meaning system approach. J. Pers. Soc. Psychol. **77**(3), 588 (1999)
10. Roedel, T.D., Schraw, G., Plake, B.S.: Validation of a measure of learning and performance goal orientations. Educ. Psychol. Meas. **54**(4), 1013–1021 (1994)
11. Zhang, Y., Wildemuth, B.M.: Unstructured interviews. In: Wildenuth, B. (ed.) Applications of Social Research Methods to Questions in Information and Library Science, pp. 222–231. Libraries Unlimited. Westport (2009)
12. Sternberg, R.J.: The Nature of Creativity. Contemporary Psychological Perspectives. Cambridge University Press, Cambridge (1988)
13. Sternberg, R.J., Lubart, T.I.: An investment theory of creativity and its development. Hum. Dev. **34**(1), 1–31 (1991)

Happiness on Instagram – Content Analysis and Engagement Based on Attention Theory

Qiuwen Li[1(✉)] and Young Ae Kim[2]

[1] Department of Art, University of South Dakota, 414 E. Clark St., Vermillion, SD 57069, USA
Qiuwen.Li@usd.edu

[2] School of Design, South Dakota State University, 117 Grove Hall, Brookings, SD 57007, USA
YoungAe.Kim@sdstate.edu

Abstract. Instagram, a popular social networking platform, was ranked the number five most popular app by downloads in the first quarter of 2018 [1]. Instagram allows users to share happenings in their lives with other users by posting images and video content. Anyone with an Instagram account can also *like* and comment on others' posts. Social media generates attention-seeking behavior and obsessions with getting more *likes* and positive comments. According to psychologist John Robert Anderson, attention is the behavioral and cognitive process of selectively concentrating on specific information while editing out non-essential information [2]. Which photography posts get more *likes* and comments? How frequently do people feel truly happy when they use hashtag #happy and/or #happiness on Instagram? How often do people give sincere comments on Instagram posts? In this research, we conducted a content analysis of photography posts tagged with the hashtag #happy and/or #happiness. We downloaded the 200 most recent posts from individual accounts between January 9–20, 2019 from age ranges 10 to 55 years old. 91% of posts were from young adults or millennials (18–35 years old) and only 2% of posts were from adolescent and 7% of posts were from middle-aged adults. The study examined different value of life in happiness on Instagram and user engagement based with different types of photography posts on Attention Theory. The study found that the physical appearance (36%) was considered to be the aspect of happiness that is 15% more than happiness from the relationship and 27% more than happiness from the achievement regardless of biological sex. In addition, the study suggests that females post comments 2.3 times more than males and females click *like* 1.7 times more than males. Regardless of biological sex, both males and females received significantly larger numbers of insincere comments (85.7%) than sincere comments (13.1%) on their posts.

Keywords: Social media · Attention theory · Hashtag · Happiness · Photography posts

C. S. Shin (Ed.): AHFE 2019, AISC 968, pp. 13–20, 2020.
https://doi.org/10.1007/978-3-030-20470-9_2

1 Introduction

Instagram, one of the most popular social networks worldwide has been accessed by individuals and corporations for varying purposes. Since its launch in October 2010, it has attracted more than 1000 million active users as of June 2018. The United States have the highest number of active Instagram users with approximately more than 121 million on the same year [3]. Among all users, 62% of global active Instagram were young adults or millennials (18–35 years old) with 30% of female and 32% of male [4]. The average minutes spent per day for Instagram users increased from 27 min on July 2017 to 53 min on June 2018 [5]. These numbers demonstrate that Instagram has become a part of our everyday lives, revolutionizing the way we spend our time, the way we communicate and the way we think.

Hashtags, a type of metadata tag, have been used on Instagram in order to help users find specific contents of themes or topics. The function of a hashtag is to draw attention, organize, and promote contents to other users. The power of hashtags is seen in a 2014 study that found an Instagram post with at least one hashtag averages 12.6% more engagement than a post with no hashtags [6]. A study has also found that hashtag #happy was ranked as No. Six with 413.8 M usages for best Instagram hashtags for *likes* in 2016 [7]. What photography posts get the more *likes* and comments? How frequently do people feeling true happy when they use hashtag #happy and/or #happiness on Instagram? How often people give sincere comments on Instagram posts? The present study aims to improve our understanding by providing a content analysis of hashtag #happy and/or #happiness on Instagram.

Attention Theory. According to psychologist John Robert Anderson, attention is the behavioral and cognitive process of selectively concentrating on specific information while editing out non-essential information [8]. Social media generates attention-seeking behavior. The psychologist Emma Kenny found that it is a reward cycle for those who are obsessed. Every time a user get a *like* or a positive response on social media, the user get a squirt of dopamine [9]. There are many articles on the web to teach quick ways about how to get more *likes* and comment on Instagram, e.g. show your or someone else's faces. Research has found that photos with faces are 38% more likely to receive *likes* and 32% more likely to receive comments regardless of their age and gender [10]. But not all *likes* and comments are sincere, this study examined different user engagement by reviewing and categorizing all comments.

Formations Impression. People pay more attention to information that comes first because it influences how we perceive later information. Psychologist Solomon Asch's experiments on "*Formations of Impression Personality*" concluded that when adjectives describing a person appear in sequence, the order in which the words were presented is important for their impression. The first adjective is the most impactful one compared with the later ones. A person will get more positive ratings from others when given adjectives with more positive meaning were given first followed by words with less positive meaning, however, the participants tend to rate that person less positive if the order was reversed [11]. For example, a person will get a more positive impression when described with adjectives in Group A rather than words in Group B when group A and B contain the same traits but in reverse order.

Table 1. Asch's Formations Impression

Group	Adjective 01	Adjective 02	Adjective 03	Adjective 04	Adjective 05	Adjective 06
A	Intelligent	Compassionate	Skillful	Critical	Stubborn	Lazy
B	Lazy	Stubborn	Critical	Skillful	Compassionate	Intelligent

Instagram, as the primary visual social media using a mix of text, images or videos, offers ways of presenting information and opinion. In this paper's exploration of methodology, we focus on consideration of different types of photography posts with the hashtag #happy and/or #happiness to understand what value of life people are considering as their happiness. When the post presented a verbal statement with a contradictory visual image, the order in which the information was presented is important for our analysis.

Exchange Theory. Exchange Theory in sociological social psychology says individuals will tend to do what increases the outcomes they value and decreases outcomes they dislike. Social relationships formed and sustained exclusively as people develop mutually beneficial exchanges [12]. For example, as in factory owner- worker relation, owners need workers' labor and workers need the money they earn to buy goods and services. They all get what they need by giving their resources accordingly, not only in the matter of material things that we are to be content, such as food or shelters, but also social goods as status or social validation and attention when interactive with others. For instance, politicians will promise anything they think will get them votes and avoid any position that makes voters hostile. Action and reward become linked; the more an action is rewarded, the more it tends to be repeated [12].

Social networking platform, such as Instagram, changing the way we interact with each other. It provides an opportunity to keep social relation with a relatively low cost compared to offline relations [13]. Hashtags like #like4like, #follow4follow get popular for people wants more *likes* and followers. #like4like means they will *like* one of your posts if you *like* theirs. #follow4follow means if someone follows you, you follow them back. The present study compared the insincere comments with sincere ones in the relatively low-cost online environment.

Homophily Principle. As the old saying goes, "Birds of a feather flock together". Homophily occurs in a vast array of network studies. Each person has their own way of experience happiness and other emotions. Individuals in homogeneous relationships have common characteristics (beliefs, values, etc.) that make it easier to communicate and form relationships [14]. The present study aims to improve our understanding of different value of life in happiness on Instagram by analyzing different types of photography posts tagged with the hashtag #happy and/or #happiness.

Instagram and Value of Life. The use of social media platforms, such as Instagram, has increased in recent years [15]. Technology has created a new set of beliefs, fears, and aspirations and affected millennials' approach to their life. Today's young people are in no rush to get married and they invest in themselves more than any other generations. In addition, they do not remember their life without the internet; they are the first generation of *Digital Natives* [16]. Therefore, it is important that scholars, social media platform operators, designers, educators, and parents are aware of the value of life and emotion aspects of millennials' social media use.

2 Method

2.1 Procedure

This study examined the influences of attention theory on happiness and how an individual portrays one's happiness. We conducted a content analysis of posts tagged with hashtag #happy. We downloaded the 200 most recent posts January 20, 2019 and coded for the presence or absence of specific content elements. The Instagram posts were posted on Instagram and tagged with the hashtag #Happy and/or #Happiness between January 9 and January 20. In addition, we analyzed all comments of 200 posts into two group "sincere" and "insincere". All variables were coded dichotomously (0 = absent, 1 = present) by one coder. For the reliability analysis, we randomly selected 10% of the material. The analysis generally indicated reliable measurement. Krippendorff's α values were acceptable all (all values, $\alpha > .76$).

2.2 Variables

Reference to Happiness. This variable was coded as present when the post included an explicit reference to happy thoughts, plans, value, or expression. All posts were categorized by 9 types of life values that considered to be happy - Not Defined (no clear reference), Smile (no content reference), Physical Appearance, Relationship, Travel/Experience, Achievement, Wealth, and spiritual.

Sincerity. This variable was coded as present when the comments of post included an explicit reference to thoughtful expression. All comments were counted and categorized into two groups, sincere and insincere.

Emotions. The following emotional expressions were coded when they were visually or verbally present: Happiness/Happy (e.g., smiling faces, proud/confident faces and gesture).

Demographics. Atotal of 200 most recent Instagram post from both groups, 100 males and 100 females, were downloaded. We coded whether the post visually (e.g., pictures, visuals) or verbally (e.g., words, phrases) presented a female or male. We did not code for sex when the post ambiguously represents in terms of biological sex. Furthermore, the age of the individual was coded. We coded whether the post included young individuals, older adults, or held no age-related information.

Inconsistency. We coded whether the post explicitly addressed a conflict experience when internal cognitive and emotional content was inconsistent. Inconsistency was coded as present, for example, when the post presented a verbal statement with a contradictory visual image.

Data Analysis. We calculated descriptive statistics with percentage values and 95% confidence intervals based on bootstrapping techniques. The percentage values are significantly different when the confidence interval does not overlap. The confidence interval lower and upper limits are reported after the percentage value.

3 Results

There was an explicit reference to happiness in all reviewed posts. In Fig. 1, physical appearance was by far the most frequently depicted expression of happiness, followed by travel/experience, relationship, and achievement. Happiness posts included young adults (91%, 95% CI = 84.2–96.2), followed by youth (2%, 95% CI = 0.0–5.4) and adults (7%, 95% CI = 2.8−10.4). Regardless of biological sex, the physical appearance (36%) was considered to be the aspect of happiness that is 15% more than happiness from the relationship and 27% more than happiness from the achievement. Only a few posts included wealth (1.5%) or spiritual (1%) as consideration of their happiness. We learned that the physical appearance is by far importance factor for young adults' happiness than experience, relationship, or achievement. Not defined are smiles in only 8% of posts for which there are no content descriptions that reference the picture in the post. It was too difficult to identify what value of life they are considering as their happiness.

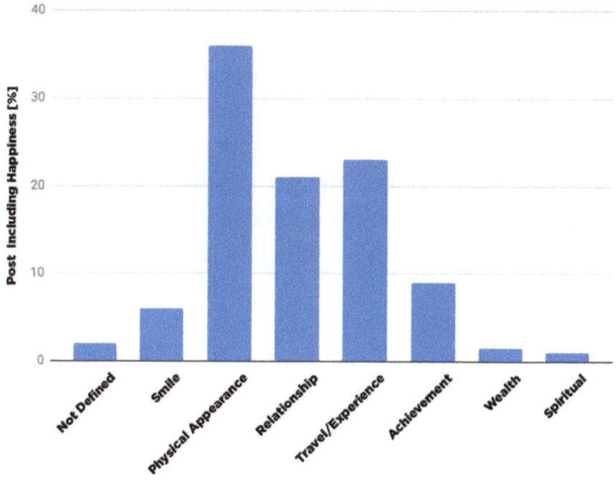

Fig. 1. Percentage of posts showing a specific type of happiness. N = 200.

Each post has an average of 5.98 (male) and 13.52 (female) comments. Females are 2.3 times likely post more comments and 1.7 times likely click *likes* the post than males. However, it is important to consider that there were variety numbers of *likes* for each post based on the popularity of an individual. Regardless of biological sex, each post is receiving only between 12.1% (female) and 15.5% (male) sincere comments among total number of comments. Majority comments between 86.2% (female) and 84.6% (male) were insincere comments. Moreover, the study found that females are 2.3 times more likely to leave comments than males, as seen in Table 1. Regardless of biological sex, both male and female received a significantly larger number of insincere comments (85.7%) than sincere comments (13.1%) of their posts (Table 2).

Table 2. Instagram post of the total number of Likes vs Comments.

Biological sex	Likes	Total comments	Sincere comments	Insincere comments
Male	19283	589	91 (15.5%)	498 (84.6%)
Female	32615	1352	163 (12.1%)	1165 (86.2%)
Total	51898	1941	254 (13.08%)	1663 (85.68%)

4 Discussion

The research found that millennials are a largely idealistic group. The way they communicate is vastly different than previous generations. The best way to reach out to them is through the internet and social medias, such as Instagram, Twitter, LinkedIn, and Snapchat because it does not require an interruption to their day, they can present them at their desired way, and they can access information at their leisure [17].

Ancient Greeks saw self-care as a factor of honest citizens who were more likely to care for others. Today, Findings suggest that Millennials are the most superficial, when it comes to physical appearance. Self-care sees [18] Millennials spending twice as much as boomers on self-care (e.g. diet plans, workout plans, etc.) and making personal improvement commitments, in contrast to the generation before [19]. Relationship, experience, and achievement are not as important to millennials as their physical appearance. Millennials want the positive responses from others by presenting their flawless images (or controlled images) of himself or herself. In fact, the majority of pictures in Instagram are low-quality images, focused on the individual, not their surroundings. In other words, Millennials do not seem to grasp the concept of anything but the visual presentation of themselves, which is heavily influenced by social medias and the communication methods.

5 Limitations

This study has several limitations. First, we only investigated the content of posts on American hashtags; decreasing the generalizability of our finding to cross culture. Second, the study used a descriptive and exploratory approach without a focus on a deductive test of hypotheses, but it enabled us to reveal an important phenomenon (subliminal value of happiness). Third, the sample of posts was small compared with the number of posts that are available on that hashtags. Fourth, we investigated only static image posts, however, the static image with no given written contexts is difficult to interpret clearly. We studied the importance role of moving images to recognize one's emotions comparison to static images and how the sound changes whole emotional status instantly regardless of given visual contexts [20]. Nonetheless, this limitation enabled us to find the individuals' interpretations of life value.

6 Future Directions

Further research is required to investigate more ranges of emotions (i.e., anger, fear, sadness) with more representative sample sizes. Social media communication is proven to be the most effective for millennials. Research has found that frequent positive visual exposure encourages positive emotions [21]. Future research can provide explanations of how these influences shift their viewpoints and to what they give value.

References

1. Nelson, R.: The top mobile apps, games, and publishers of Q1 2018: Sensor Tower's Data Digest (2018). https://sensortower.com/blog/top-apps-games-publishers-q1-2018
2. Anderson, J.R.: Cognitive Psychology and its Implications. Macmillan, London
3. Leading countries based on number of Instagram users as of October 2018 (2005). https://www.statista.com/statistics/578364/countries-with-most-instagram-users/
4. Distribution of Instagram users worldwide as of October 2018, by age and gender. https://www.statista.com/statistics/248769/age-distribution-of-worldwide-instagram-users/
5. People spend almost as much time on Instagram as they do on Facebook (2018). https://www.recode.net/2018/6/25/17501224/instagram-facebook-snapchat-time-spent-growth-data
6. Shively, K.: Simply measured Q3 2014 Instagram study (2014)
7. Top 100 HasTags on Instagram. https://top-hashtags.com/instagram/
8. Anderson, J.R.: Cognitive Psychology and its Implications. Macmillan, London (2005)
9. Harvey-Jenner. C.: The psychology of a like: how social media is really affect ing your brain (2017). https://www.cosmopolitan.com/uk/reports/a9931660/psychology-social-media-likes-mental-health-issues/
10. Bakhshi, S., Shamma, D.A., Gilbert, E.: Faces engage us: photos with faces attract more likes and comments on instagram. In: Proceedings of the SIGCHI Conference on Human Factors in Computing Systems, pp. 965–974. ACM (2014)
11. Asch, S.E.: Forming impressions of personality. J. Abnorm. Soc. Psychol. **41**(3), 258 (1946)
12. Hewitt, J.P., Shulman, D.: Self and Society: A Symbolic Interactionist Social Psychology. Allyn and Bacon, Boston (1979)
13. Surma, J.: Social exchange in online social networks. The reciprocity phenomenon on Facebook. Comput. Commun. **73**, 342–346 (2016)
14. McPherson, M., Smith-Lovin, L., Cook, J.M.: Birds of a feather: homophily in social networks. Annu. Rev. Sociol. **27**(1), 415–444 (2001)
15. Lenhart, A., Duggan, M., Perrin, A., Stepler, R., Rainie, H., Parker, K.: Teens, social media & technology overview 2015, pp. 04–09. Pew Research Center, Internet & American Life Project (2015)
16. Hutt, R.: What do young people value? (2016). https://www.weforum.org/agenda/2016/01/what-do-young-people-value/

17. Alton, L.: Phone Calls, Texts or Email? Here's how millennials prefer to communicate (2017). https://www.forbes.com/sites/larryalton/2017/05/11/how-do-millennials-prefer-to-communicate/
18. Silva, C.: The Millennial Obsession With Self-Care (2017).https://www.npr.org/2017/06/04/531051473/the-millennial-obsession-with-self-care
19. Millennial in Adulthood: Detached from Institutions, Networked with Friends (2014). http://www.pewsocialtrends.org/2014/03/07/millennials-in-adulthood/
20. Li, Q., Kim, Y.A.: Analysis of facial emotion recognition technology and its effectiveness in human interaction. In: International Conference on Applied Human Factors and Ergonomics, pp. 217–226. Springer, Cham (2018)
21. Kim, Y.A.: Emotional evidence: influences on happiness from the frequent positive visual exposure. In: International Conference on Applied Human Factors and Ergonomicsm, pp. 227–237. Springer, Cham (2018)

A Journey of Taking Industrial Design Students into Uncharted Territories. Designing Crowd-Sourced Playful Interactions for a Public Art Event

Magnus Feil(✉) and Milagros Zingoni

The Design School, Herberger Institute for Design and the Arts,
Arizona State University, Tempe, USA
{mfeil,mzingoni}@asu.edu

Abstract. The field of industrial design is finding itself in a highly transitional state. While at its core, the traditional values of teaching visualization techniques, form, design research, human factors, and manufacturing methods remain critical to the creation of innovative artifacts of mostly singular nature, the world around us has changed both in added complexity and increased digitalization. The scope of designing future designers therefore calls for reevaluating better strategies to address a big picture approach of solving complex problems at the intersection of users, behaviors, and interactions, with a plurality of possible artifacts or systems. Traditional models of the underlying design process tend to treat the "Design" as a noun, emphasizing the final outcome (the artifact) as the ultimate goal of the undertaken creative process. In contrast to this approach, there is an increasing recognition towards embracing "design" as a verb, by shifting the focus and design contributions onto the process itself to ultimately leverage the realm of results [1].

Either approach to the "big-D" or "small-d" design process models represents an important learning experience to the students. I tend to treat them as complementary rather than as exclusive to each other.

As an industrial design educator, the gamut of my responsibilities towards learners therefore ranges from nurturing the skillset and mindset to embracing design thinking, practice, and understanding user behaviors and interactions at the very core of solving complex problems. One of these challenges presented itself for the 2018 Scottsdale Canal Convergence Event – An art festival along the downtown waterfront of the Arizona Canal drawing scores of internationally renowned artists to exhibit installations featuring light and sound. A grant was extended to the Arizona State University's Design School to participate in an interdisciplinary collaboration between the Scottsdale Arts Commission, Interior Architecture, Industrial Design, and a local middle school, to envision interactive installations as reflections on the theme of community and water. While the field of Art is normally not considered to be a playground for industrial designers, this unique project collaboration served as an ideal model to exercise a fundamentally re-envisioned process-driven design studio course. This paper illustrates in-depth the classroom experience of an interdisciplinary design project centered around gathering crowd-sourced primary research, playful and explorative prototyping, form development, as well as the creation of functional

C. S. Shin (Ed.): AHFE 2019, AISC 968, pp. 21–29, 2020.
https://doi.org/10.1007/978-3-030-20470-9_3

interactive models. With an estimated 208.000 visitors after 10 days, the 2018 Scottsdale Canal Convergence event offered the students not only an opportunity to showcase their work to a broad audience, but also demonstrated the critical importance of observing and evaluating user-interactions of the created artifacts in the wild.

Keywords: Design education · Cross-disciplinary collaboration · Crowd sourcing · Playful interactions

1 Is Design a Noun or a Verb?

The field of industrial design finds itself in a highly transitional state. While the "digitalization" of artifacts has greatly increased both function and capabilities of goods that are part of our daily lives, the complexity of how people understand and interact with said artifacts has risen as well. Designers of the past predominantly focused on the aesthetics of an object itself, without paying greater attention to the larger surrounding environment. While advances in human factors and ergonomics challenged past designers to envision better artifacts with presumptive "users" in mind, gained insights mostly manifested themselves in changes to the physical form to increase compatibilities with the human body, and to a much lesser degree, human understanding. The digitalization of artifacts has brought another dimension of functionality to everyday goods – and most of the increases in capabilities have nothing to do with the physical form itself – they are intangible.

> *"The artefact designer has a complex role here for while the designer's intentions are in the artefacts, the functionality of the artefact often goes well beyond what the designer anticipated or envisaged. Both inputs from users and outputs of the artefacts can be unanticipated, unforeseen, and harmful." (Johnson 2004)* [2].

We are currently experiencing the onset of Internet of Things (IoT) devices that will soon affect all aspects of our daily lives. Yet the importance of the physical form of IoT devices, such as Amazon's Echo Dot "smart" assistant/hub, pales in comparison to the design of the complex interactions that take place between the object, a person, and a network of further connected IoT devices. Designers of the present and future hence face the challenge of shaping both the physical and intangible. They need to be the authority and agents for the design of the traditional physical form, shaping ongoing interactions, and deeply understand user behaviors.

The design education system has not fully understood, nor embraced, this ongoing development that affects both technology and our society at large. "It is necessary but not sufficient for the designer to have an understanding of the past and present, which is where most education has the luxury to end" [3].

Traditional models of teaching industrial designers are often based on challenges that treat the "Design" as a noun by overly emphasizing the importance of the final outcome (the artifact) as goal of the undertaken creative process [1]. Dennis Hambeukers acknowledges benefits to this traditional design approach, as it often leads to aesthetically crafted artifacts that are designed down to the last details. Following this process, students learn critical insights to the mastery of shapes, forms, proportions, materials, and manufacturing methods. Additionally, they develop an eye (or obsession) for details – a virtue that Jonathan Ive,

Apple's head of design, equates to "finishing the back of the drawer" [4].

While only very few would question the value and benefits of beautifully designed objects in our daily lives, this "big–D" approach to designing comes with a big caveat: It is undervaluing, or even neglecting, the path on how to get to a final outcome [1]. This path is by no means a linear road nor corridor of even width. The Double Diamond design process model, developed by the British Design Council (BDC) [5], conceptualizes the design process as a visual map from start towards a final outcome. Two adjacent diamonds, each vertically divided into two halves, are representing phases of Discovery, Definition, Development, and Delivery. Inspired by BDC's model, Dan Nessler provides us with an "extended" version of the Double Diamond design process (Fig. 1) [6] that further fills out the voids within the two diamonds with detailed steps to explore and tasks to do. For designers, navigating this very path results in shifting focus away from the yet undefined final outcome towards a trajectory towards a better final outcome. Hence the act of designing turns into a verb [1]. Embracing the individual phases of this "small-d" design philosophy rewards the designers willing to research and explore the yet unknown with the capability to gather deep insights and ultimately make informed decisions towards more relevant and innovative outcomes.

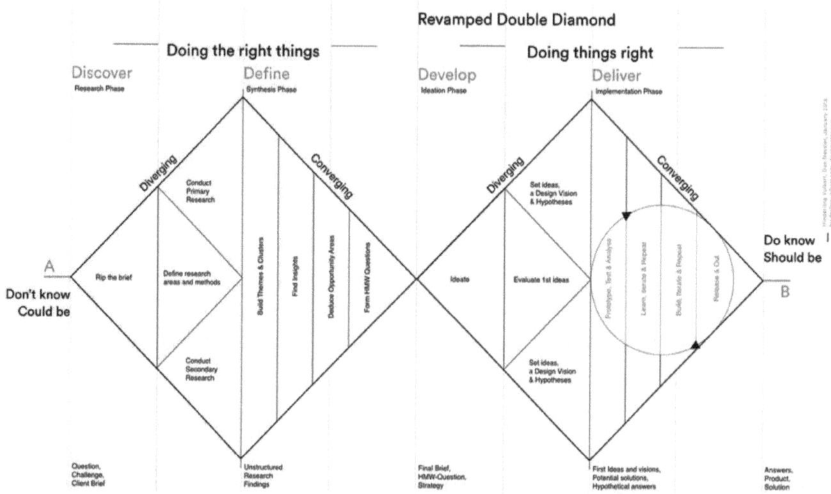

Fig. 1. Dan Nessler; The Double Diamond Re-envisioned (2018)

2 Designing for Process and Research

For educators in the field of Design it is paramount to seek real-world challenges for their studio-based learning environments to infuse relevance and distill significance into the students' creations. Acquired patterns and approaches of problem solving during this period of training will profoundly shape the skillset of design students and subsequently enable learners to address problems of increasing complexity. Bill Buxton reminds designers of artifacts of the significance to "make our best efforts to understand the larger social and physical context, within which it is intended to function" [7].

With the generation of a deep understanding at its core, shaping the conditions of observation for research therefore plays a critical role for instructors and students alike. David Woods highlights three classes of research methods available for designers to shape the conditions of observations listed in increasing order (Fig. 2) [8]: Natural history methods include "diverse collection of observation in situ" for discovery, "experiments in the field or field experiments (staged or scaled worlds)" for both discovery and verification, and "spartan lab experiments (experimenter created artificial tasks)" for a more targeted verification of emerging ideas and concepts [8].

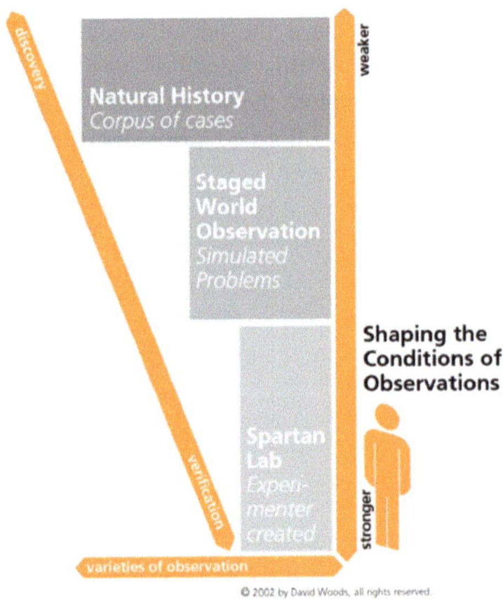

Fig. 2. Dave Woods; shaping the conditions of observation (2002)

All too often however, the work of design students is created under a vacuum of artificiality by lacking sufficient research, such as gathering of input from affected stakeholders, or observations at "the intersection of people, technology, and work" [9]. Consequently, the outcomes of design studio projects remain highly presumptive with an overly strong focus on the visual appearance as cover up for deficiencies in research. Furthermore, design studios often take place cloistered behind the walls of the design schools, with presentations only given to like-minded designers and peers, instead of seeking real feedback from the targeted stakeholders themselves. This situation creates both a subpar learning experience and evokes both skewed results and a false sense of proficiency among learners.

To address superficiality of designed outcomes, instructors need to enable a learning environment that encourages, demands, and facilitates research opportunities for shaping the students' conditions of observation as their foundation for further explorations of visual or physical/tangible nature.

3 Context: Scottsdale, Scottsdale Public Arts Commission, and the Canal Convergence Event

Situated north-east of Arizona's capital Phoenix, the City of Scottsdale is part of the greater Phoenix metropolitan area. Uniquely positioned in the Sonoran Desert with an average of 314 sunny days per year, the city is a desirable place to live for both permanent and seasonal residents. With an estimated population of 242,700, the city stretches out for 31 miles from north to south and covers an area of 184.5 square miles [10]. Fourteen years before Scottsdale was officially incorporated as a city in 1951, the architect Frank Lloyd Wright chose this area to establish his Taliesin West school for architecture and made it his winter residence.

Ten miles south of Taliesin West, Scottdale's vibrant "Old Town" city center is a pedestrian-friendly area with galleries, bars, restaurants, and retail facilities. Just north of "Old Town", the Arizona Canal crosses an area named the Scottsdale Waterfront in a park-like setting surrounded by upscale condominiums and flanked on the eastern side by the Soleri Bridge and Plaza by the heralded artist, architect, and philosopher Paolo Soleri [11]. Commissioned by the Scottsdale Public Arts Commission in the 1990s, and completed 2011, Soleri's public art installations includes a pedestrian bridge across the Arizona Canal that is suspended by two columns strategically oriented to cast a beam of light during the summer and winter solstices. These two pylons form the center of an open square south of the canal. This plaza is adjacent to Paolo Soleri's 1969 Goldwater Bell assembly, which echoes the artist's "life work in architecture and ecology" [11].

The Scottsdale Public Arts Commission also stands out as the organizers of the annual Canal Convergence Event. This 10-day long arts festival along the downtown waterfront of the Arizona Canal draws scores of internationally renowned artists to exhibit installations featuring light and sound. For the duration of 10 days in the month of November the Scottsdale Waterfront is being transformed into a public event space that draws spectators of all ages in excess of 200,000 visitors. In 2018, a grant was extended to the Arizona State University's Design School to participate with an interdisciplinary collaboration between the Scottsdale Arts Commission, Interior Architecture, Industrial Design, and a local middle school, to envision interactive installations as reflections on the theme of community and water.

4 Cross Disciplinary Collaboration

While the field of Art is normally not considered to be a playground for industrial designers, this unique project collaboration served as an ideal model to exercise a fundamentally re-envisioned process-driven design studio course. This unlikely pairing of approximately one-hundred 8th grade middle schoolers, thirty-four industrial design juniors, and ten interior architecture graduate students created at first a few obstacles that needed to be addressed in order for the project to succeed. The alignment of schedules was only a minor problem that could be overcome by the willingness of our college students to free up time to meet outside of their regular class schedules. The vast difference in age across all participants posed a greater challenge that had to be addressed with a series of "ice-breaker" activities during the initial gatherings. After being coached by a colleague in theater production, our design students felt emboldened to approach the 8th graders with a series of playful exercises to establish a sense of familiarity and mutual trust.

Under the supervision of their teachers, the design students were given access to individual classes at the middle school for one hour at a time. This brought the benefit of working with smaller groups of students once at a time, but created the necessity for the design students to repeat the same exercises multiple times in different classrooms.

4.1 Crowd Sourced Research

With the given topic of "Water and Community" in mind, the design students began to shape the conditions of observation by developing multimodal exercises for each of the scheduled gatherings. The careful enablement of voices from within the community of the participating middle schoolers was crucial for research in this initial phase.

To achieve this goal, the 8th graders were encouraged to unleash their creativity in a series of playful drawing, paper construction, and writing assignments (Fig. 3). The following examples reflect comments made by the high schoolers during a multi-modal exercise that included both construction and writing elements. Once prompted on the importance of their own community, students responded with: *"My community is always there for me."; "My community is filled with unique people."; "My community is important to me because they protect me. They have my back and I will always have theirs."; "My community motivates me."; "My community is full of water, people, nature, and greetings."*

Fig. 3. Multi-modal exercises conducted with the 8th graders (Feil, 2018)

After each meeting with the high schoolers, the design students carefully analyzed all collected material to identify emerging patterns. The students realized that the middle schoolers provided a rather broad perspective on their community – yet one commonality emerged related to water and weather. Rain, a rare occurrence in the hot climate of southern Arizona, was repeatedly identified as an event of particular significance. Rain can take different forms in Arizona – from light drizzles to powerful monsoons in the hottest times of the summer. The latter often evoke a chain event of events, including substantial amounts of downpours, flooded streets, and the sudden swelling of otherwise dry washes. Rain often results in a short, yet significant, drop in temperatures, and causes the ubiquitous creosote bushes to emit a distinct herbal smell.

Encouraged by the emerging commonalities in voices, the design students started to hone in deeper on the subject of rain. The following assignments fielded to the 8th graders now included various drawing challenges on the cyclical nature of monsoons. Furthermore, the middle schoolers were invited to re-enact the stages of a storm with the help of improvised "musical instruments". Each step in the process was carefully documented with photographs, scans of the children's work, and videos recorded of their performances. This resulted in a treasure trove of data that was shared among all design students.

After weeks of interacting with the middle schoolers, the design students started to feel confident enough to translate the gathered data and insights into opportunities for the upcoming public art event at the Scottsdale Waterfront. Following the voices of the high schoolers, the Arizona monsoon was chosen as a metaphor for a series of interactive installations to engage the visitors of the event. The decision was made to create a centerpiece, christened "Cumulus", as a cloud to showcase selected quotes from the middle schoolers. In close proximity, a series of fifteen interactive installation (named "Drops"), should be designed to introduce the audience to unique aspects of a rainstorm.

4.2 Research and Design of Playful Interactions

Moving along the Double Diamond design process model, a new challenge emerged – how to design interactions? Traditionally, designers use various sketching and drawing methods to visualize ideas. Once a promising form has been identified, mock-ups and prototypes are being created to evaluate form and proportions. A painted appearance model as a "realistic" stand-in for the envisioned concept marks in general the ceiling for the students' creations.

The problem however is, that all of these forms of creative expression are static and merely visually mimic given functions. They are intended to look at from a distance, and do not allow participation. The design of artifacts for interactions therefore necessitate a new approach for testing both the visual, functional, and interactive qualities of emerging ideas.

Inspired by Bill Buxton's guide to "Sketching User Experiences", design students started with building various representational forms of their concepts in full scale. Only objects in realistic scale enabled students to test interactions through reenactments. During presentations, invited guests were challenged to interact with the prototypes with only little instructions or guidance of the designers. This immersive form of prototyping often had more resemblance to stage play, and similar to an intermission at a theater, changes to the concepts could be accomplished quickly to alter impressions after the curtains go up again. None of these critical insights could have been achieved with conventional methods of designing as described above. Initially critical students soon discovered the power of playful prototyping.

> *"[...] in order to design a tool, we must make our best efforts to understand the larger social and physical context within it is intended to function."* [7]

4.3 Final Steps and Preparation for Public Display

With the evolution of the individual concepts and refined user experiences, the design students started to change pace and reverted to steps similar to the traditional "big-D" design process. An increased focus of the visual appearance of the individual artifacts was necessary to ensure that the visual aesthetics of the artifacts were of equal quality to the designed interactions. Communication across all students and concepts became critical in this phase to guarantee a unified identity across all interactive installation pieces. The students hereby developed a visual language that incorporated shared form elements such as colors and the application of a logo.

Since all models were to be displayed out in the open for a duration of ten days, all models had to be built and finished to withstand the elements.

5 The Finale: Canal Convergence 2018

With an estimated 208,000 visitors after 10 days, the 2018 Scottsdale Canal Convergence event offered the students not only an opportunity to showcase their work to a broad audience, but also demonstrated the critical importance of observing and evaluating user-interactions of the created artifacts in the wild. Most student projects will be presented at the end of a semester to a limited audience and will soon disappear into obscurity. For this particular project, the students were able to witness first-hand how their ideas were perceived by a larger audience of all ages. Having created something of substance that is being appreciated by a broad audience in an inspiring setting has proven to be of tremendous value for this class of future design professional. Most importantly, the design students were able to welcome the middle schoolers with their parents and teachers during the Canal Convergence event. Seeing co-created artifacts that carry the distinct handwriting of all involved designers, no matter how old they were, created a new community, and will hopefully inspire this upcoming generation of students to consider a career in design.

References

1. Hambeukers, D.: How the design world is being ripped apart right now (2018). https://medium.com/design-leadership-notebook/how-the-design-world-is-being-ripped-apart-right-now-efe9f01a33d8
2. Johnson, D.: Integrating ethics and technology. In: European Conference Computing and Philosophy (E-CAP2004_ITALY), Abstract, Pavia, Italy, 2–5 June 2004

3. Ruecker, S.: The perennial and the particular challenges of design education. Visible Lang. **46**(1/2), 122–131 (2012)
4. Richmond, S.: Jonathan Ive interview: simplicity isn't simple. https://www.telegraph.co.uk/technology/apple/9283706/Jonathan-Ive-interview-simplicity-isnt-simple.html
5. The British Design Council (2012). https://www.designcouncil.org.uk/news-opinion/design-process-what-double-diamond
6. Nessler, D.: Applying the Revamped Double Diamond to tackle the challenge in four steps. https://www.dannessler.com/intro-process
7. Buxton, W., ebrary, Inc. and MyiLibrary: Sketching User Experiences Getting the Design Right and the Right Design. Elsevier/Morgan Kaufmann, Amsterdam (2007)
8. Hollnagel, E.: Handbook of Cognitive Task Design. Chapman and Hall/CRC, Baton Rouge (2003)
9. Woods, D.D., Hollnagel, E.: Joint Cognitive Systems, 1st edn. CRC Press, Boca Raton
10. City of Scottsdale (2006). https://www.scottsdaleaz.gov/about
11. Scottsdale Public Arts. http://scottsdalepublicart.org/permanent-art/soleri-bridge-and-plaza

Teaching Industrial Design Through Real World Markets and Manufacturing

George K. Chow(✉)

University of Houston, Houston, USA
georgekchow@gmail.com

Abstract. Traditionally, Industrial Design studio classes consist of teaching Industrial Design through completing a series of projects at school. But what if there was a way to leverage existing real-world wholesale markets and manufacturing facilities outside of school to teach students about the realities of taking a design from "art to part"? This paper presents a case study in which students were challenged to design a product that could be sold at the Dallas Market, an international home goods marketplace in Dallas, TX for retailers from all over the world.

Keywords: Industrial design · Education · Markets · Manufacturing · Design entrepreneurship · Real-world · Teaching

1 Introduction

In the US, Industrial Design studio courses usually teach Industrial Design fundamentals by having students complete a project or series of projects. Fundamentals of Industrial Design include topics such as research, ideation, aesthetics, product semantics, affordances, ergonomics, sustainability and manufacturing methods. Industrial Design educators often look for ways to teach these fundamentals in more effective and efficient ways but may be limited in simulating real world issues such as manufacturing constraints, production costs, product viability and competitive markets in the classroom. What if there was a way to leverage a real world designing-building-selling experience to teach Industrial Design fundamentals as well as provide students with some design entrepreneurship experience? This paper presents a case study of utilizing the Dallas Market, an international wholesale home goods marketplace in Dallas, TX as an opportunity for students to learn ID fundamentals as well as experience real world issues associated with trying to take a design from "art to part".

2 The Dallas Market

The Dallas Market Center is "…a global business-to-business trade center and the leading wholesale marketplace in North America connecting retailers and interior designers with top manufacturers in home décor, furniture, gifts, lighting and fashion. Inside its dynamic, five million square foot campus near downtown Dallas, nearly 200,000 customers from all 50 states and 85 countries seek industry trends, business

C. S. Shin (Ed.): AHFE 2019, AISC 968, pp. 30–38, 2020.
https://doi.org/10.1007/978-3-030-20470-9_4

education and new products via open-daily showrooms and from exhibitors participating in trade events held throughout the year" [1]. The Dallas Total Home & Gift Market is an exhibit and marketplace held several times a year and is a "...premier destination for connecting retailers and manufacturers in home décor, gifts, gourmet, housewares, accessories and lighting" [2].

3 Project Brief

31 sophomore Industrial Design students at the University of Houston were tasked with designing and making an innovative product that would be appropriate to sell at the Dallas Total Home and Gift Market (Jan 17–20, 2019). The following lists of learning objectives, constraints and final deliverables were provided to guide them as they considered potential products to design:

Learning objectives

1. Practice designing functional, aesthetically pleasing, products with manufacturing in mind.
2. Learn how the disciplines of design, manufacturing and business work together in the formation of successful product designs.
3. Learn about local manufacturers' capabilities and appropriate utilization for prod uct manufacturing.
4. Learn, apply and practice new methods of manufacturing.

Constraints

1. Must be able to be mass-manufactured in low production quantities (at least 250).
2. Must utilize manufacturing within US.
3. Minimal finishing due to added cost and manufacturing time.
4. If the product requires assembly, assembly instructions must be provided.
5. If including hardware (i.e. screws), the bag that they are shipped in must be sourced.
6. If electric/battery powered, electrical sub-assembly must be off-the-shelf and UL pre-certified.
7. Must be branded and have simple, elegant, and inexpensive packaging to be shipped in.
8. Size limit: must fit in a 24" cube, disassembled.
9. Weight limit: 15 lbs. including packaging.

Final Deliverables

1. One functional prototype to bring to the Dallas Market.
2. Packaging for your product.
3. Bill of materials which will include each part, its source and cost.
4. ANSI D (22 x 34" portrait) Project Poster: product name, product photos, description.
5. Process Book PDF (research, ideation, prototyping, solution).

4 Process

The overall project development process was completed in five weeks and consisted of four main phases: research, ideation, validation and production. By the end of the 4th phase, prototypes were completed and presented in a final student exhibition held on Nov 30, 2018. At the exhibition, the top student designs were selected for the Dallas Market. Four Industrial Design faculty and five students attended the Market to promote and sell the products. The following sections detail each phase:

4.1 Research

The research phase was one week in length and began with the students familiarizing themselves with the Dallas Market via their website: dallasmarketcenter.com. The main goal was to better understand the retail buyers that attend the Market and the types of products that they may be interested in purchasing. The specific market event that the students were designing for was the "Total Home and Gift Market" so they researched and brainstormed various categories of potential products that would be appropriate for that specific showcase. Once the potential product categories were established, the students found photos of existing products in each of the categories and created mood boards with them. These mood boards helped the students to identify potential areas of innovation to pursue. The students were also taken on a field trip to a couple of local high-end home furnishing stores including Kuhl-Linscomb [3] and High Fashion Home [4] to see, feel and experience home furnishing products on the market. They were also encouraged to visit other stores in town including Crate and Barrel, Pottery Barn, IKEA, Sur La Table, At Home, Home Goods, Bassett Home Furnishings, Ashley Home Store, West Elm, and Design Within Reach.

A list of local manufacturers, fabricators and makerspaces was given to the students to research as a team. Each student was tasked to select at least one local manufacturer/fabricator or maker space, research their services, prices, location and present their findings to the rest of the class. The spreadsheet served as a reference for them throughout the project and could be updated as new information was found. Most students conducted their research online, however, some called and visited the locations in person. If the students found additional manufacturers that were not on the list, they were encouraged to add them to the spreadsheet.

The students also had the responsibility of researching specific manufacturing and finishing methods including laser, plasma and water jet cutting, stereolithography, 3D printing, CNC machining, vacuum-forming, injection molding, casting, metal bending and forming, anodizing, electroplating, power-coat painting, and clear coating. This was an on-going process that was done on an as-needed basis depending on the manufacturing and finishing methods required for specific product ideas. The studio class was also taught in conjunction with a materials and manufacturing class in which the students had the opportunity to experience and practice the manufacturing methods themselves.

4.2 Ideation

The ideation phase was one week in length and consisted of two sub-phases: brainstorming and refinement. In the brainstorm phase, students were tasked with sketching at least 20 different product concepts. These 20 concepts were reviewed together in class and the top five concepts for each student were selected for further refinement and exploration in the refinement phase. The top five concepts were visualized as refined and rendered 2D sketches as well as corresponding physical sketch models. From these five concepts, one final concept was selected to be further refined. The final concept was visualized with a rendered perspective drawing, design control (technical) orthographic drawing, refined physical sketch model and a fabrication plan. The fabrication plan was intended to help the students organize all the considerations for making their specific designs. The following list of questions was given to the students to help them in developing their fabrication plans:

1. Who will fabricate your product (at least 250 units)?
2. Who is supplying the raw materials for your product? Are pre-existing, sourced products required?
3. Who will finish your product (edge treatment, de-burring, grinding, sanding, electroplating, etc.)?
4. Who will paint (power coating, clear coat, etc.) your product, if applicable?
5. Does your product require assembly? If so, who will assemble it?
6. Who will make/print your packaging?
7. Who will pack your product?
8. Who will ship your product?
9. How much will 1 unit, 100 units, 250 units and 500 units cost? What is the cost difference?
10. What is the lead time for each of the different unit amounts?

From here, the students moved into the validation phase.

4.3 Validation

The validation phase consisted of building proof-of-concept prototypes in the actual manufactured production materials that could be tested by potential users. This required students to contact and obtain quotes from local manufacturers to build their parts and products. Some of the quotes were too expensive and/or would not be completed within the time frame so many of the students manufactured their proof-of-concept prototypes themselves in the school shop or at a local maker space. Once the prototypes were finished, they were put in front of users for testing and feedback. Based on their feedback and faculty guidance, the students further refined their prototypes and continued their refinement into the final production phase.

4.4 Production

The production phase included finishing their prototypes for the end-of-semester student exhibition. In parallel with this, students designed and made packaging prototypes that effectively communicated what the product was, highlighted its respective features, and represented how their products would be protected during shipping and displayed in stores.

The ID faculty developed a brand called "Texas by Design" (TXD) and provided students with a logo to place on their products and packaging. The intent was to provide a consistent, unified brand for all the student designs. In the same way that IKEA is a parent company with individually named products, Texas by Design is the name of the ID program's "company" with individually named student products.

If assembly of the products was required, the students developed assembly instructions to be shipped with the product. Some of the instructions were included on the packaging itself to save on printing costs and reduce the amount of paper consumption.

The students photographed their prototypes, created posters and digital process books, which documented their whole product development process from their beginning research to their final solutions.

The students were given a Bill of Materials spreadsheet to fill out for their concept. This included the part names, corresponding manufacturers/fabricators that would make the parts, the associated costs at different production rates (i.e. 100, 250, 500 units, etc.), material costs, packaging costs and total costs.

At the end of the production phase, the final concepts were showcased at a student exhibition where the top designs were selected.

4.5 Selection

At the end-of-semester student exhibition, a committee of 10 industry professionals, faculty and staff voted on the most appropriate products to showcase at the Dallas Total Home and Gift Market. The main criteria for selection included cost, manufacturing feasibility, functionality, marketability, longevity and aesthetics.

A 10 × 10 ft. booth at the Market was reserved by the college and financial support had been provided by a grant to cover the lodging, travel, food and event fees for five students. The ID faculty desired to take as many students as possible to the Dallas Market for the students' learning benefit, however, it was not possible or practical in terms of space and cost.

With these constraints in mind, a total of 15 products were selected by the committee: 5 to sell and 10 to display. The student designers of the five products being sold were invited to attend the Dallas Market with four ID Faculty.

The 10 products on display were chosen primarily to showcase the wide variety of products created by the class, to generate interest in the ID program and provide feedback as to the types of products Market buyers may be interested in for potential future sales opportunities (Table 1).

After the selection process was complete, the five selected students were informed and invited to attend the Dallas Market to represent the ID program and to promote and sell their products in person.

Table 1. List of the 15 selected student products for sale and display.

Product name	Description	Sell or display
1. Flight	Stainless steel wall hook	Sell
2. Happy whiskers	Cat dish	Sell
3. Two	Incense holder	Sell
4. Keeper	Recycling/trash bin	Sell
5. Tape simply	Stainless steel tape dispenser	Sell
6. Kaen	Tea light holder set	Display
7. Kle	Keychain and letter holder	Display
8. Keep stand	Cell phone holder	Display
9. Three	Stool	Display
10. Third coasters	Coaster set	Display
11. ManKala	Game set	Display
12. Vino	Stainless steel wine rack	Display
13. Uni	Wooden fruit tray	Display
14. Trio	Stool	Display
15. Caravan	Wooden animal toys	Display

5 Results

The five selected students and four ID faculty brought the 15 selected products to sell and display at the Dallas Total Home and Gift Market from January 17-20, 2019. Prior to departing for the Market, the students and college received media coverage in the school paper as well as from local television news stations. While at the Market, a substantial amount of invaluable feedback was received, many connections were made, a product was sold, several buyers expressed interest in licensing products and working further with the students and faculty to potentially bring some products to market. After the Market, the faculty and students followed up with interested buyers. The following sections provide the details of the Dallas Market results:

5.1 Media Coverage

Prior to departing for the Market, a news article was published in the school newspaper on Jan 8, 2019 about the students, their designs and their hopes for the Dallas Market. A couple of days later on Jan 10, a local television station, KHOU 11, interviewed the students and faculty and aired the interviews later the same day at 4 pm. On Jan 14, Fox Channel 26 conducted a live interview of the students and their products during their 9am morning segment.

5.2 At the Dallas Market

The ID faculty and students worked in two shifts to cover the four-day market. Every day the Market opened at 8:30am and closed at 6 pm (except for Sun, Jan 20, which closed at 4 pm). Over the four days, an abundant amount of invaluable feedback was given to the students and faculty regarding product viability and marketability, other manufacturing options, potential other shows/exhibits and websites to consider selling at and encouragement from many helpful and kind buyers. Over 30+ face-to-face connections were made with buyers and company owners, which included some people who were interested in potentially working on sponsored projects with the ID program. In addition to this, an order for six “Keeper” laser-cut aluminum recycling/trash bins was placed by Dandy Roll, a shop in Rogers, Arkansas with the goal of supplying “distinct goods for the modern home.” [5] The owners of Dandy Roll also expressed interest in potentially purchasing “Two” (aluminum and brass incense holder) as well as “Caravan” (wooden animal toy set) in the future if the manufacturing could be worked out so a viable price point could be reached.

Another company that sells pet products expressed interest in licensing “Happy Whiskers”, a low profile cat dish that allows cats to eat without disturbing their whiskers. In addition, an owner of a different company expressed interest in licensing “Tape Simply”, a one-piece bent stainless steel tape dispenser. One buyer expressed interest in “Third Coasters”, a laser-cut wooden coaster set that was on display but not for sale, with the hope of developing it further for production. The communication with Dandy Roll and the other interested buyers continued after the Market (Fig. 1).

Fig. 1. Five selected students and their respective products for sale at the Dallas market.

5.3 After the Dallas Market

After the Dallas Market, the ID faculty and students followed up with Dandy Roll and the other interested buyers. For the “Keeper” recycling/trash bin, a manufacturer was found that could cut the edges of the keeper bins much cleaner and smoother than what was presented at the Market while maintaining a similar price range, however, the anodizing process (for a matte finish) was much more expensive than anticipated. A proposal was made to Dandy Roll to clear coat the bins instead of anodizing to protect the raw aluminum from fingerprints and dirt. Dandy Roll agreed to the proposal and the bins were successfully completed and delivered.

The ID faculty and students are currently in discussions with the pet product company interested in licensing "Happy Whiskers" as well as the company interested in "Tape Simply". Protection of intellectual property is being pursued.

Discussions are also continuing with the connection who was interested in developing "Third Coasters" into a viable, marketable product.

The ID faculty are also continuing to work with the students who developed "Two" and "Caravan" to find less expensive manufacturers with the hope of reaching a viable price point for retailers as well as finding less expensive methods for achieving the desired matte finish on the "Keeper" bins.

6 Analysis of Results

After returning from the Market, the planning team held a debriefing meeting to discuss the results and gauge the potential of the Dallas Market in future design projects. The team unanimously agreed that it was an outstanding learning experience for both the students and faculty. For the students, the real world experience of taking their designs from concept to manufacturing, and even sales was invaluable. The realities of manufacturing constraints, costs, competitive products, markets and viable price points, packaging, shipping and selling cannot be achieved with text book learning alone.

For the faculty, the experience was invaluable to see what worked or did not work in reaching learning objectives for the class, motivating students, and the realities of a competitive product marketplace. The networking and new connections made at the Market have the potential to lead to other sponsored projects or new design work.

The Dallas Market was a great first starter venue to explore with selling and displaying student designs, however, many of the buyers gave the team feedback that the students' concepts were more contemporary and modern and may not be the right fit for all of the Dallas Market buyers. Instead, they suggested that the students' designs may have more traction and interest at the larger trade shows like AmericasMart in Atlanta, GA, NYNow Market in New York City, NY and the Licensing Expo in Las Vegas, NV.

Buyers also recommended exploring approaching online companies such as the American Design Club and The Grommet as possible venues for starting the students' design entrepreneurship careers.

Even though the sophomore students were able to complete their projects in only five weeks, the team agreed that in the future the project may be better suited for 3rd year students due to their more developed skill sets and design experience.

7 Conclusion

Dedicated Industrial Design educators continually search for more effective and engaging ways to teach students Industrial Design. Learning in the classroom is necessary but also limiting. Real world learning experiences such as the Dallas Market not only teach students Industrial Design, but they also expose them to the realities of manufacturing methods, constraints, costs, market viability, packaging, shipping, price points, selling, and all the details, considerations and effort involved in actually bringing a product to market. Even though the process is immensely challenging, the reward of real-life experience and the potential for sales or licensing can be well worth it.

The order that was placed for the "Keeper" bin may be first of many more orders and the beginning of a strong and lasting partnership with Dandy Roll and perhaps more stores interested in carrying the product. Licensing some of the students' designs may be the launching point for some of the students' careers. Further developing student designs with interested buyers and companies may yield more viable products that can reach the marketplace.

In conclusion, giving students a project that leverages real world markets such as the Dallas Market and the experience of working with real manufacturers to bring products to life can be a powerful, engaging and effective method for teaching Industrial Design.

References

1. The Dallas Market Center. http://www.dallasmarketcenter.com
2. Total Home and Gift Market. http://dallasmarketcenter.com/markets/markets.aspx?m=10936
3. Kuhl-Linscomb. http://www.kuhl-linscomb.com/
4. High Fashion Home., http://www.highfashionhome.com
5. Dandy Roll. http://www.dandyrollprints.com

How Color Coding Encourages Emotional Interactions While Using Smartphones

Young Ae Kim(✉)

School of Design, South Dakota State University,
117 Grove Hall, Brookings, SD 57007, USA
YoungAe.Kim@sdstate.edu

Abstract. Human cognition involves highly interrelated mental processes. The essential mental process is connected with the environmental influences that helps to store and remember. Color is the essential optical experience to user experiences as an influential communication channel to the human cognition, which is a meaningful position in improving memory performance. However, color coding can be detrimental under certain circumstances. Can color coding improve memory capacity? Can color coding influence emotional arousal? Can color coding help to create a mental map to navigate information in apps? This research discusses the relationship between colors, attention, memory, and emotional arousal. It addresses the color coding in similar and different settings, an exploration on the human memory, and the role of emotional arousal and memory while using smartphones.

Keywords: Attention · Color · Emotional intelligence · Mental recall · Positive emotion · Happiness · Zoom lens model of attention · Perceptual span

1 Introduction

For centuries artists and designers have manipulated color to evoke certain responses from their audience. Research on colors has been overflowing as a touchstone in one's memory and emotions in particular marketing, art, and design areas. Researchers have made observations about the psychology of color and how moods, feelings, and behaviors affected by colors; however, it is a relatively new science and determining the effects of color has been difficult.

Color is an influential communication instrument, which is used to influence mood and physiological reactions [1]. However, it is often greatly personal experiences in terms of how one is feeling about the color. For instance, black is used in many Western countries as a symbol of death and mourning while it represents life and rebirth in ancient Egypt. Color can be an extremely effective learning tool in educational settings, marketing, communications, sports, and more. Colors can increase brand identification by 80% according to the marking study. Color has been used as one of the influencing elements to get attention, shape attitudes towards products, and adds pressure to decision-making for consumers in advertisements [2]. It demonstrates the importance of color in marketing and how the information and message deliberations.

C. S. Shin (Ed.): AHFE 2019, AISC 968, pp. 39–49, 2020.
https://doi.org/10.1007/978-3-030-20470-9_5

Color can encourage one's the academic achievement and the cognitive abilities, including perception, attention, memory, and comprehension. For example, autistic patients reading without using a colored overlay, improves up to 35% of reading speed [3]. Since 1976 from Farley and Grant's research on the influence of color on attention, further experimental works on the influences of color on the human cognitive process were conducted [4].

It is vital to identify the emotional influences of color in this research, which focuses on how colors influence on emotion. Emotion occurs instantly upon the perceiving of advocative colors. This research proposes a view-point of how colors influence emotion with or without functional feedbacks, encouraging either positive or negative interactions while using products or services. Daniel H. Pink suggested the emotionally intelligent signage on words only to encourage one's respect towards others in need and understanding others' circumstances [5]; however, it has efficiency issues and not applicable in diverse platforms because it requires many words to explain the detail. We predict that colors influence one's desire to interact initiatively with low the stress and frustration from visual warnings of products and services we use.

This research is to deliver an enjoyable experience to an individual that is autotelic and worth doing. It must be a combination of internal and external experience to encourage an individual's motivation. However, the perception changes can be exceedingly influenced by current mood state more than general traits [6]. In other words, it changes an individual's perceptions and motivations by emotion that is influenced colors.

2 Attention, Color and Memory

Human Memory. William James and Waugh and Norman introduced the nature of memory that human memory is involved in short-term and long-term memory storage. Atkinson and Shiffrin proposed the model of memory that consist of sensory register store, short-term store, and long-term store. The huge amount of information can be registered in the sensory register store when the environmental influence comes in; however, the information quality is very brief [7]. The certain amount of attention determines whether it is transferred to the short-term store or long-term store. Thus, the information can be transferred to long-term store with the attention control processes, such as heuristics, memorization strategies, and rehearsal [8]. Also, Atkinson and Shiffrin discussed the importance of the memory process and the process to a deeper semantic memory process that the stimuli can be remembered better when stimuli are accessible in the environment to activate the attention. Color can enhance the memory performance and attention by arising the emotional arousal.

Selective Visual Attention. Every moment, people are exposed to constant sensory information. An individual encounter in the traffic noises from the street, the tinkering noise from a dog's collar, the click of the computer keyboard at the library, the chatter of people at the coffee shop, and more. Fortunately, people ignore those sensory experience every one of them. Instead, many things blend into the background and unnoticeable level while people give attention to only important things in a given environment. For instance, an individual can focus on the story that your friend tells while all of the irrelevant sounds can be toned down, which is an example of selective attention. Due to the limitation on one's ability to attend things in terms of capacity and duration, people have to decide on what they want to pay attention to. It is similar to the process of visual hierarchy – giving a highlight to a certain element and pushing inapt or less significant information to a margin in our perception. Because the attention is involved in a highly selective process on information based on one's subjective or objective goals, it only directs to the most relevant information and ignoring everything else [9].

The emotional factors were observed as a device to control the orientation to visual information in recent studies. The findings suggest that people are influenced by visual or motivational significance to their attention. There are two stages of visual attention from parallel fashion and sequential fashion. The parallel performance is involved in attention with the distribution consistency on visual and information processing. The sequential performance is involved state is involved in attention with a particular concentration [10]. William James, psychologist, described the spotlight model, which works as sounds (Fig. 1 shows an example). The spotlight or focus is an information extraction point from the visual information where visual attention is detected. The area surrounding this focus or spotlight is the fringe where low-resolution or low-attention happens [11].

Fig. 1. The spotlight model

The zoom-lens model [12] has been also used in many researches as a visual search. Depending on an individual's control of increasing or decreasing the size of the focus, such as zoom-lens of camera, the size of visual attention can change from sharply narrow area to widely distributed area. However, the widely distributed area takes longer time to process due to the amount of information that is included within the area.

Attention and Color. Attention is captured subconsciously before conscious attention happens. Color influences customers' purchase and behaviors, because it attracts one's attention to a service and a product. Customers understands the brand colors to recognize their products and services or search for different options. A study proved that visual color cues can predict one's purchasing behavior for well-known and un-known brands, which validated the voluntary and involuntary attentions and attractions are on colors when people choose packages [13]. For instance, the basic package colors of Crayola crayons stayed the same while it has changed their logo numerous times since

1934. In other words, a company generally keeps products the same color scheme, even if they change the look of the product, because consumers use color to search for brands. Research found that people give attentions to color they like unconsciously, which indicates that one's attention to products or services does not need conscious search and it is entirely based on color [14]. In addition, environment and the product or service colors interact [15]. In particular, warm colored displays gets more drawn attention and timeless – slow feel – than cool colored displays – limited and fast feel, but the cool colored displays are more favorable to public (Fig. 2 shows an example).

Fig. 2. Warm colored display design and cool colored displays design (*left and right*).

Color and Memory. The human observers rely on a wide range of experiences to understand objects. Color is one of experiences that defined as hue of an object. For example, we know that sky is blue, banana is yellow, apple is red, and strawberry is red. An individual observer who has seen and tasted an apple previously knows that an apple appears in a red hue typically. The awareness of the color represents the memory color in memory constitutes [16].

Memory color effect is the experience of colors directly rendering the appearance of the object with the authentic color. Research discovered that normal human trichromats identify memory color of an object rather than the color is presented. For example, people see the banana's memory color first regardless of different color of picture of banana that artificially changed. In other words, an individual regulates the color of the banana towards the opposite color of yellow in color wheel – blue (Fig. 3 shows an example). Following studies also indicated that memory color effect on man-made objects is exclusively noticeable on blue and yellow objects. Researchers have argued that blue and yellow objects display at a higher degree in the visual system for balancing for the variability in illumination to deliver a solid memory color effect because natural daylight changes from blue hue to yellow hue during the day [17].

Fig. 3. Banana image with gray and yellow (*left and right*).

Emotional Arousal and Words. MacKay and Ahmetzanov suggested that taboo words are more emotionally arousing than neutral words and can be remembered much longer. They found that the participants accomplished much more when they were associated with emotionally arousing words or taboo words [18]. MacKay and Ahmetzanov's research result support Heuer and Reisberg's study. They found that the higher level of emotional arousal leads to occur the better retention in long-term recall as well as short-term and long-term memory. Another research from Corteen proves that there is a higher recall after twenty minutes and two-weeks delay after aurally words presented as well as a single arousing word [19]. In other words, arousal words are more effective on recall performance than non-arousal words; however, short-term retention gets the negative effect of arousal. Words can generate higher arousal effect with higher memorability in long-term than short-term learning process. Therefore, there is a significant effect on memory performance in the short-term and long-term memory storage when the arousal elicited by environmental stimuli [20].

Emotional Arousal and Color. People are affected by various color stimuli differently depending on individual, because color can influence one's feelings of products, services, and environments, such as coziness, spaciousness, freshness, cleanliness, brightness, etc. In addition, an individual's preference on color differs depending on various influence. For example, an individual feeling comfortable with cold temperature prefers warm color spectrum, such as red, orange, and yellow when an individual feeling comfortable with hot temperature prefers cool color spectrum, such as blue, green, and purple [21].

Culturally-learned meanings of colors are quite powerful. They can be used to subtly affect mood and behavior in some individuals. A few studies have shown that cultural backgrounds have strong influences on color preference. The study results proved that people prefer the same colors when they are from the same region regardless of race. In addition, depending on one's locational background – nationality and different area of the same country, an individual's color preference is different [21].

Different culture sees differently in terms of common connotations of a color to a certain mood. A study inspected color and moods relationship with people from Poland, Germany, Mexico, Russia, and the United States. Researchers found some consistency in color red and black cross-culturally; however, Germans related jealousy with yellow while Polish linked purple both anger and jealousy [22]. The results demonstrate how perception of color and relationship between color and mood can be influenced by one's cultural differences. Nonetheless, there were cross-cultural resemblances in how people associate with certain colors while there were cross-cultural differences present. Table 1 is the example of how colors perceived differently by cross-culture [23].

Table 1. Color psychology by different culture.

Culture	Colors					
	Black	White	Red	Yellow	Green	Purple
Western	Death	Purity	Love	Jealousy	Wealth	Loyal
Eastern	Rebirth	Harmony	Happiness	Wealth	Fertility	Immortality

3 Emotional Design and Effective Color Coding

Whether it is aesthetically pleasing or functionally practical, everything around us has been designed in a certain way. When users interact with products, experiences occur and emotion sparks. We experience emotional interaction to a given context every moment whether we like or not. It becomes very personal experience. In other words, any end products will provoke an emotion from the audience regardless of the positive or negative type of emotions. Products facilitate social interaction and the efficiency of its use. Psycho-pleasure, one of the four pleases in emotional design, is increased from the finishing of a task. It helps in task accomplishment, which makes a pleasing experience. In addition, psycho-pleasure includes the productivity in a task can be done – i.e. a grid structure of app navigation organization increasing the efficiency of the app for each purpose.

Emotional Design Pyramid. A theory of human motivation was proposed by Abraham Maslow in 1943 [24], which suggests that the motivation is based on one's pursuing satisfaction and transformation through personal growth. Emotional design is a very similar process as Maslow's pyramid (Fig. 4 shows an example). For example, physiological attractiveness in a product influence how one's perception of the usability of a product. A pleasing aesthetic of a product gives a leading degree of one's satisfaction, which gives forgiveness in any imperfections with it – i.e. people who buy Apple products continue to buy it due to its pleasing slick designs. Emotions influence the brain operation process. Depending on what experience occurs the brain focuses on the diverse narrative thought process, which makes people feeling accordingly. For instance, the negative emotions grow if a website and app is poorly designed and does not meet the users' expectation while the positive emotions grow with good emotional design.

Fig. 4. Emotional design pyramid

Color Contrast. Color contrast encourages attention to a particular color from another. It diminishes eyestrain and brings users' attention to a screen (Fig. 5 shows an example).

Fig. 5. Impact of color contrast

Color theory is vital to the usability in order to create not only aesthetically pleasing design but also provide readability. When designing mobile screens, designers must take an account users possible situations and locations, because it influences colors due to brightness changes and screen glare. Thus, it is important to check the contrast ratios, which demonstrate the difference in two colors. Table 2 shows the W3C suggestions in terms of the contrast ratios in terms of the relationship between body and image text, which helps users with color blindness, low vision, and legibility of text on the screen. The same contrast ratios are recommended on icons and important elements.

Table 2. Contrast ratios

Small text	4.5:1
Large text	3:1 with 14pt bold/18pt regular and up or 18.5px/24pt
Color to another color	1:1 or 1:2

Color Scheme. Commonly, monochromatic colors (single color) have a comforting effect and work well with anything. It is comfortable to look and looks clean and classy. On the other hands, analogous color schemes use neighbor colors that one of colors becomes a focal point and rest colors supports the whole color scheme (Fig. 6 shows an example).

Fig. 6. Monochromatic and analogous color scheme (*left and right*).

Color contrast with complementary relationship provides the high level of user's attention while using a device; however, it works when the dominant colors with its opposite color were used. Figure 7 displays how complementary color relationship works in app information – red is used to alert the number of unread or missed, which differentiate from app icons.

Fig. 7. Unread email and missed phone calls

Creating a customized color scheme can accelerate attention at the most outstanding level. One bright accent color with monochromatic color can produce the most visually remarkable presentation. For example, Instagram app uses cool gray with white background and yellow and red to purple as an accent, which creates the Instagram color scheme (Fig. 8 shows an example).

Fig. 8. Instagram color scheme

Accessible Design for Color Blindness. Color blindness is one of the most common impairments (less than 10% of population) and affects individual using smart devices. Red and green are the most common challenging combination. Designers must apply several visual signals to give a visual attention due to diverse forms of colorblindness. In other words, there are several benefits and important role of color for attention, it is critical to create a design system using strokes, patterns, action text, texture, etc. that is color independent. Designers must simulate color blindness in helpful instruments in order to test of the look of design output with different types of color blindness.

Emotion and Perception. When an individual need to express their reactions to a given situation or an object, emotions are the tool they use. Emotions help to make decisions in a variety of short-term and long-term situations. On the other hands, the perception study has been relatively different from the emotion study. Due to the isolation of perception and emotion study, researchers have experimented the processes in perceptions and emotions indirectly. The unique interaction between emotion and perception is discussed in this research. The research predicts that emotion influence perception and distinct levels of visual perception encourage emotion also. Thus, emotional stimulations give the value and importance to information about a given situation or an object. Moreover, the study that is reviewed propose that information is formed with the collaboration between visual perception and environment, because people give higher level of attention to objects that have emotional and motivational relevance to an environment. Perception encourages immediate results without thinking about the costs of latent action and a given emotional objects or situations. It is very goal-oriented. In other words, the negative emotions, such as fear support a change in order to see potential treats while the positive emotions support to keep the present way to see things.

Emotional Intelligence Exceeds the Moment. Because human and animal have emotional wiring, the effects of emotion are extremely impulsive and primitive. Emotions contain subjective and objective information of an individual. The "affect-as-information" hypothesis [25] talks about the influence of mood and emotion on attention, judgment, and thought. The study suggests that the information is represented in encouraging individuals. People always ask themselves about how they feel about something before they are thinking about the current types of condition. Often, emotional experiences are exemplified as two-sided – i.e. pleasant vs. unpleasant and excited vs. calm. While moods or feelings provide information with authoritarian action, emotions are the temporary states manipulating perceptions. A biologist Randy

Nesse proposed that we spend time on things that work and less time on things that do not because moods regulate investment strategies [26].

The emotional expressions and reactions are very individual that are coded in one's genes and extremely comparable across all humanity – i.e. human smiles when a baby is laughing. The amygdala has a significant function in emotional arousal. It also controls the release of neurotransmitter essential for memory association that encourages strong and long-lasting emotional memories. Perceptions are regulated by an individual's energy that an individual can manage presently visible hindrances when people make decisions about actions. Emotions are reactions to people, circumstances, and objects while mood statuses return present assets in decision making. Emotions exhibit the imagination, the past, the present, the future, as well as current implications.

Research by Antonio Damasio suggested that feeling encourages heighten emotions into the physical states. Feelings are associated with the particular emotion about the collection of thoughts and memories at the subconscious level. Each emotion is temporary while feelings from the emotion endure over a lifetime. Thus, emotions and feelings have a critical function in how an individual experience with their environments. The process of one's decision making is always intuitively.

4 Discussion

Current information reminder system in smart devices are one color-coded (red) with a number as well as notification boxes, which encourage attention due to the color contrast, but not intuitive to differentiate what is less or more important. It creates a negative perception of given information and discourages the desire for interaction with given contents. Information we are given through smart devices almost look dangerous and scary due to the amount of information we receive every day, because information is treated the same whether it is urgent or not. Thus, we need to consider how to tell the information story to a user that it is safe, and in fact it is fun to explore, opening the full reflective force while the visceral level is operating at full force. Because the appeals are apt when the reflective system collapses.

The research found that objective with emotional and motivational relevance influence attention and the finding proposes that emotional stimuli deliver the involuntary effects without providing on the meaning of an individual's emotional change [27]. Therefore, the color coding can escalate attention to visual and content information in apps to be stored and strived positively. Color coding can influence an individual's interaction to human memory performance in a given context; however, both short-term and long-term memory retention can be maximized by fewer color combination (i.e. monochromatic or Analogous colors) with the higher level of contrast with saturated hue and the luminance of the color (i.e. complimentary colors). In addition, color impairment and any implication related to colors must be considered and other visual elements must be explored to support this particular impairment.

Future Directions. The research only mentioned the fundamentals of how color coding enhances the experience of using smartphones. It is important to examine what form of iconic stimuli with a particular color encourages one's positive or negative emotional moments in the future. A limited selection of iconic stimuli will be tested, then a wider collection of iconic stimuli will be examined in future studies.

References

1. Adams, F.M., Osgood, C.E.: A cross-cultural study of the affective meaning of color. J. Cross Cult. Psychol. **4**(2), 135–156 (1973)
2. Moore, R.S., Stammerjohan, C.A., Coulter, R.A.: Banner advertiser-web site context congruity and color effects on attention and attitude. J Advert. **34**(2), 71–84 (2005)
3. Ludlow, A.K., Wilkins, A.J.: Colour as a therapeutic intervention: the case of JG. J. Autism Dev. Disord. **39**(5), 815–818 (2009)
4. Farley, F.H., Grant, A.P.: Arousal and cognition: memory for color versus black and white multimedia presentation. J. Psychol. **94**(1), 147–150 (1976)
5. Pink, D.H.: Drive: The Surprising Truth About What Motivates Us. Penguin, London (2011)
6. Riener, C.R., Stefanucci, J.K., Proffitt, D.R., Clore, G.: An effect of mood on the perception of geographical slant. Cogn. Emot. **25**(1), 174–182 (2011)
7. Atkinson, R.C., Shiffrin, R.M.: Human memory: a proposed system and its control processes. In: Spence, K.W., Spence, J.T. (eds.) The Psychology of Learning and Motivation, pp. 89–195. Academic Press, London (1968)
8. Greene, T.C., Bell, P.A., Boyer, W.N.: Coloring the environment: hue, arousal, and boredom. Bull. Psychon. Soc. **21**(4), 253–254 (1983)
9. Posner, M.I.: Chronometric explorations of mind. Lawrence Erlbaum, Mahwah (1978)
10. Jonides, J.: Further toward a model of the mind's eye's movement. Bull. Psychon. Soc. **21** (4), 247–250 (1983)
11. Eriksen, C.W., Hoffman, J.E.: Temporal and spatial characteristics of selective encoding from visual displays. Atten. Percept. Psychophys. **12**(2), 201–204 (1972)
12. Eriksen, C.W., James, J.D.S.: Visual attention within and around the field of focal attention: a zoom lens model. Percept. & Psychophys. **40**(4), 225–240 (1986)
13. Kawasaki, M., Yamaguchi, Y.: Effects of subjective preference of colors on attention-related occipital theta oscillations. NeuroImage **59**(1), 808–814 (2012)
14. Babin, B.J., Hardesty, D.M., Suter, T.A.: Color and shopping intentions: the intervening effect of price fairness and perceived affect. J. Bus. Res. **56**(7), 541–551 (2003)
15. Hering, E.: Outlines of a theory of the light sense (1964)
16. Hansen, T., Olkkonen, M., Walter, S., Gegenfurtner, K.R.: Memory modulates color appearance. Nat. Neurosci. **9**(11), 1367 (2006)
17. Witzel, C., Valkova, H., Hansen, T., Gegenfurtner, K.R.: Object knowledge modulates colour appearance. i-Perception **2**(1), 13–49 (2011)
18. MacKay, D.G., Ahmetzanov, M.V.: Emotion, memory, and attention in the taboo Stroop paradigm: an experimental analogue of flashbulb memories. Psychol. Sci. **16**(1), 25–32 (2005)
19. Heuer, F., Reisberg, D.: Emotion, arousal and memory for detail. In: Christianson, S.A. (ed.) Handbook of Emotion and Memory: Research and Theory [Internet]. Lawrence Erlbaum Associates, New Jersey (1992). http://books.google.com.my/books?id=dXMs_dloSEcC&pg=PA39&source=gbs_toc_r&cad=4#v=onepage&q&f=false
20. Kleinsmith, L.J., Kaplan, S.: Paired-associate learning as a function of arousal and interpolated interval. J. Exp. Psychol. **65**(2), 190–193 (1963)
21. Whitfield, T.W., Whiltshire, T.J.: Color psychology: a critical review. Genet. Soc. Gen. Psychol. Monogr. **116**, 387–412 (1990)

22. Hupka, R.B., Zaleski, Z., Otto, J., Reidl, L., Tarabrina, N.V.: The colors of anger, envy, fear, and jealousy: a cross-cultural study. J. Cross-Cult. Psychol. **28**(2), 156–171 (1997)
23. SmarterTravel: What colors mean in other cultures, 26 January 2016. https://www.huffpost.com/entry/what-colors-mean-in-other_n_9078674?ec_carp=5740158985548787547. Accessed 15 May 2010
24. Maslow, A.H.: A theory of human motivation. Psychol. Rev. **50**(4), 370 (1943)
25. Schwarz, N., Clore, G.L.: Mood, misattribution, and judgments of well-being: informative and directive functions of affective states. J. Pers. Soc. Psychol. **45**(3), 513 (1983)
26. Goode, E.: Viewing Depression as Tool for Survival, 7. The New York Times F (2000)
27. Kim, Y.A.: Emotional evidence: influences on happiness from the frequent positive visual exposure. In: International Conference on Applied Human Factors and Ergonomics, pp. 227–237. Springer, Cham (2018)

Design for Health and Wellness

Experiencing Momentum Through an Effective Use of Technology in Museums

Anna Kabjin Kim(✉) and Eileen Harris(✉)

Carleton University School of Industrial Design,
1125 Colonel by Drive, Ottawa, ON K1S 5B6, Canada
annakim4@cmail.carleton.ca, eileen@cunet.carleton.ca

Abstract. Technology can serve as a medium for visitors to feel connected throughout their museum experience. This creates an opportunity for both educators and museum facilitators to promote interactive viewing experience. This paper examines examples of the ways museum spaces have incorporated effective technology that has assisted with movement and identifies design considerations that provided positive user experience. These projects reveal that technology in museums can be used either as stationary devices to generate movement, or movement tracked by mobile technology. The literature includes studies in movement yet there is a dearth of research on the concept of *momentum* - continuous movement or *energy* - in museum spaces. Momentum is hard to maintain given the myriad of factors that may interrupt museum visitors from experiencing the *momentum* in their user experience. How do current museum spaces incorporate *momentum* in the visitor experience through an appropriate use of technology?

Keywords: Movement · Momentum · Museum · Visitor experience · User experience · Interactive design · Technology · Efficiency · Literature review · Continuity · Uninterrupted

1 Introduction

Unlike the conventional ways to "collect and display objects" [10] in museum spaces, many facilitators are now gearing towards innovative ways to redesign their purpose to encourage visiting experience [3]. As museum facilitators are adopting new strategies to capture visitors' interest and curiosity, the pilot projects involving physical movement and technology reported to have positive visitor experience, which also enabled the transformation of the extant museum programs [2, 13]. By examining this important connection between physical movement and technology mediums used in specific museum studies, this paper aims to suggest innovative visitor experience through the concept of *momentum* - the continuous *movement*, to allow visitors to stay focused while viewing displayed artwork.

Often, museums are faced with unexpected challenges when new design strategies are incorporated into specific exhibition projects [3, 5]. As a result, visitors can experience unpleasant situations within the shared space while trying to enjoy artwork [12]. In order for museums to maximize *momentum* in visitors' experience, spatial

C. S. Shin (Ed.): AHFE 2019, AISC 968, pp. 53–58, 2020.
https://doi.org/10.1007/978-3-030-20470-9_6

considerations should also be acknowledged for unanticipated factors that can interrupt or hinder *momentum*.

2 Background

According to Chan and Cope [2], the role of *momentum* is identified as a source that "carries the user through an interface and creates experience to move forward" [2]. Moreover, Womack [15] defines the *momentum* as the "*progress, things that are happening and moving*" [15]. Tilburg [14] also employs the term, *momentum*, as a reminder for today's *UI (User Interface)* designers to be more engaged in their interactive product solutions for their users [14]. According to these definitions, *momentum* can translate into positive energy or force that can keep individuals to continue or move forward with their activities and goals [15].

Based on the recent *museum workout* participation by Morris [10] at the *New York Metropolitan Museum of Art*, continuous movement from this dance experiment seemed closely associated with keeping the *momentum* while the process of "*doing*" [10] occurred by the participant. Several museum studies reported successful visitor experience where *movement* was evident through various technology mediums [7] and physical activities [1]. In addition, recent research experiment in a particular museum reported that continuous movement during the scheduled exhibition has provided rewarding experience to the participants [10].

Based on the evidence through these examples from the extant literature on movement, the concept of *momentum* can relate to the user experience where the visitors can feel connected to the experience by continuous movement. Although this concept has not been explored in museum studies, the future research should be in pursuit of *momentum* where an effective use of technology can enhance the visitor experience and develop preventive strategies to overcome any challenges.

3 Current Studies

As demonstrated in the examples from the literature, the focus on physical movement is evident through diverse presentation of technology in specific museum spaces and the continuous body movement may apply to the concept of *momentum*. The following is one of the prominent examples of how physical movement can be encouraged through an effective use of technology in a museum space to the visitors. With six weeks of strategically planning and testing the project, the *Gallery One* successfully launched this interactive museum project in 2012 [1]. The gallery is situated in the ground floor of the main museum, which displays over 4,100 art collections through the use of digital technologies, such as the *Sculpture Lens* that allows the visitors to mirror a historical character's gestures in *x-ray* mode [1].

The study findings also revealed that this hands-on and transformative experience through technology mediums allowed the visitors to sense personal connection to the museum collections [1]. In addition, the results from an in-depth interview with museum visitors showed that the museum attendance had increased by 39 percent after the first year of opening the gallery [1].

Other examples of using physical activity to create movement can be examined by the following dance event. Morris [10], who participated in a choreographed dance at the New York Metropolitan Museum, wrote an article called *The Museum Workout* [10]. He informs the readers about the feelings of "expansive and empowerment" [10] experienced by other participants as they moved freely through the museum space that was available to them exclusively after the closed hours [10]. Morris also noted that, by focusing on the physical activity in the museum, choreographed dance in this case, the participants felt a sense of connection with the displayed artifacts and achieved a sense of purpose when they were moving forward with the flow [10].

This iterative process of choreographed movement with physical activities, such as *jumping jacks* and *stretching*, seemed to influence the participants to collect different viewpoints on the displayed artifacts like the statues, as they danced through the organized route [10]. This continuous movement may have also caused the participants to form their own perception based on personal experiences and later allowing them to share emotions which possibly have stemmed from the dance performance exercise [10]. In the doing of the physical exercise in the Metropolitan Museum, there seemed to have an increase in creative thinking and self-awareness in their lives [6].

Moreover, the examples below use both technology and physical activity to create movement. In *Stages of Experience* article, Scheibe [13] introduces narrative techniques that were originally created by Disney design teams for the development of attraction sites [13]. The *7 Stages of Experience* describe an individual's journey through theatrical performance which is a narrative method that can be applied to historic museums, such as the *Titanic Museum* in Missouri [13]. The *7 Stages of Experience* method is a structured process which guides the visitors visually through the exhibition until the last stage [13]. This process of moving through themed stages seemed to allow the visitors to feel physically and psychologically connected to the experience as interacted with the stage actors on board and the historic artifacts, such as the Grand Staircase where special events happened in the *Titanic* movie [13].

One of the *7 Stages of Experience* demonstrates an effective use of technology in design. The Interactive Room in *Titanic Museum* reveals the computer simulation technology which encourages the visitors to try sending telegraph signals through *Marconi Machine* and to walk at different inclines of the deck of the sinking ship, providing psychological connection to the actual sinking experience [13]. In addition, the author emphasizes the importance of "human contact" [13] as the visitors are greeted at the entrance, which seems to be the key element that adds value to providing an interactive experience to the museum guests [13].

Another good example that engages technology and physical movement is the *Reflection Companion*. It is a software application developed for mobile phones, which tracks daily physical activity and generates a dialogue for the user to maintain self-reflection [6]. This software application collects the data on the physical activities from other existing mobile applications and allows the interaction to happen through "mini-dialogues" [6] about any activities completed by the user.

By showing the fitness results through graphic charts and exchanging dialogues with the user, the user is more likely to engage in this reflective exercise [6]. In the long run, this could be an opportunity to motivate the user in a positive direction where physical movement can be increased in one's daily life and gain a sense of purpose through the "doing" [10]. According to this example, it is evident that the effective use of technology contributed to forming a connection with the physical activity.

4 Discussion

Extant literature focused on continuous movement in museum spaces where the visitor experience had common themes of incorporating physical activity or technology. During these studies, the relationship between the physical interaction with the displayed artifacts and innovative technology uses was evident through successful visitor experience [5, 10]. While the physical activity during the museum experience was uninterrupted, each visitor felt sense of connection with the displayed artwork and reported to have positive results on the visitor experience [10]. Since this user experience involved a physical activity that required a continuous movement in a controlled environment, the *Museum Workout* had to be scheduled after the museum's closed hours to avoid the possible incidents, such as running into other visitors [10]. There were elements of *momentum* discovered in the literature which also encompass unexpected challenges which the future research should factor in.

Other studies have facilitated movement through mobile technology as guidance systems that the museum visitors can either install into their own mobile phones or camera devices to carry with them [8]. Having a mobile device with a tracking technology allows the user to be connected with the museum activities [8]. This can allow the user to be readily accessible to museum programs through one's personal technology device and feel engaged [9]. A similar study has discovered movement through digital technology, where the museum visitors were physically interacting with the historic characters featured in the *x-ray screen* [1]. Many literature on movement demonstrated positive visitor experience in museums when the activities encouraged the users to be active through the use of technology mediums.

Since there was no previous discussion on the relationship between movement and *momentum* in the museum spaces, further research should investigate experiencing *momentum* through effective technology mediums. Due to potential limitations concerning the specific museum's special operation hours, maintenance of technology mediums in museum spaces, and issues in the accessible design, the visitors' experience may vary depending on the goals of the museum educators and the project characteristics. Therefore, future research requires meeting specific demands of the target museum visitors by conducting qualitative and quantitative measures, along with the investigation of an appropriate technology available for current and future museum events in the related fields.

5 Conclusion

Based on the examples from literature, previous research supported the evidence of movement in specific museum spaces and enhanced user experience by fostering innovative technology [5, 7, 9]. Through various uses of technology mediums and devices involving physical movement, participants responded in the surveys and in-depth interviews that they developed a sense of purpose or connection by being part of the theatrical performances or physical activities in museums [1]. The relationship between movement and *momentum* shows that the continuous movement during the physical activities allowed for *momentum* to be maintained. As technology emerges in today's world, design seems to play an important role in making conscious decisions in everyday life.

The findings in the extant literature suggest the future research to continue to develop visitor experience in specific museums by investigating ways technology can enhance physical movement. By being in connection with interactive experience examined by specific museum studies [11], the visitors will be able to stay more focused during their visits through continuous movement. Consequently, the museum business can also benefit from the increase in clientele through "inspiring" design of the museum space through an effective use of technology [9, 13]. In order to allow the museum visitors to feel engaged in their experience, experiencing the concept of *momentum* through physical activities and incorporating an appropriate technology expect to influence the emerging design practice.

This interdisciplinary connection between design and technology will make an important contribution to the proceedings of the Applied Human Factors Engineering (AHFE) 2019 International Conference on Interdisciplinary Practice in Industrial Design.

References

1. Alexander, J.: Gallery one at the cleveland museum of art. Curator: Mus. J. **57**(3), 347–362 (2014)
2. Chan, S., Cope, A.: Digital: strategies against architecture: interactive media and transformative technology at the Cooper Hewitt, Smithsonian Design Museum. Curator: Mus. J. **58**(3), 352–368 (2015)
3. George, A.: The Curator's Handbook. Thames & Hudson, New York (2015)
4. Hawkey, R.: All the (natural) world's a stage: museum theater as an educational tool. Curator: Mus. J. **46**(1), 42–59 (2003)
5. Jarrier, E., Bourgeon-Renault, D.: Impact of mediation devices on the museum visit experience and on visitor's behavioural intentions. Int. J. Arts Manag. **15**(1), 18–29 (2012)
6. Kocielnik, R., Xiao, L., Avrahami, D., Hsieh, G.: Reflection companion: a conversational system for engaging users in reflection on physical activity. Interact. Mob. Wearable Ubiquitous Technol. **2**(2), 1–26 (2018). ACM Press
7. Lee, D., Park, J.: Augmented reality based museum guidance system for selective viewings. In: Second Workshop on Digital Media and its Application in Museum & Heritage, pp. 379–382 (2007)

8. Linzer, D.: Learning by doing: experiments in accessible technology at the Whitney museum of American art. Curator: Mus. J. **56**(3), 363–367 (2013)
9. Montgomery, A.: Science museum launches £2.5m Interactive Gallery Project. Design Week (Online Edition), p. 5 (2014)
10. Morris, G.: Dance in the museum redux: Monica Bill Barnes and the museum workout. Danc. Res. J. 50(1), 17–19 (2018)
11. Ntalla, I.: The interactive museum experience: investigating experiential tendencies and audience focus in the galleries of modern London and the high arctic exhibition. The City University, London (2017)
12. Pelowski, M., Liu, T., Palacios, V., Akiba, F.: When a body meets a body: an exploration of the negative impact of social interactions on museum experiences of art. Int. J. Educ. Arts. **15**(14), 1–47 (2014)
13. Scheibe, E.: Stages of experience: theatrical connections between the seven stages of experience and historical museums. Theatr. Symp. **22**, 99–109 (2014)
14. Tilburg, M.: Maintaining Momentum Through Efficient Mobile Design. Smashing Magazine (2014). https://www.smashingmagazine.com/2014/06/maintaining-momentum-through-efficient-mobile-design/
15. Womack, J.: Get Momentum: How to Start When You're Stuck. Wiley, New Jersey (2016)

Better Mental Healthcare Through the Built Environment

Alanna Bamber(✉)

School of Industrial Design, 1125 Colonel by Drive, Ottawa, ON, Canada
alanna.bamber@carleton.ca

Abstract. This paper addresses the issue of the built environment's effect on mental health. It begins with an overview of the state of mental health in Canada, and what factors cause mental illness. The subject of healthcare facility design is then addressed. How a mental healthcare facility may affect one's mental health is discussed in regard to various factors. The subject of interdisciplinarity is introduced, and why healthcare facility design should be tackled interdisciplinarily is discussed. This is illustrated with two examples of interdisciplinary teams that design therapeutic environments.

Keywords: Interdisciplinary design · Mental healthcare · Healthcare facility design

1 Introduction

The built environment that people inhabit every day is deliberately designed and affects both human experience and behaviour. The built environment also influences the emotions and mental health of its users through interactions with the physical surroundings and social realm. In her book *Space, Place and Mental Health*, Sarah Curtis notes "there are interactions amongst the material and the social aspects of places, and the characteristics of individual inhabitants, which are important for many aspects of individual experience" [1]. It is critical that mental health facilities are built in a way that is therapeutic and improves the mental health of those seeking treatment [2]. Therefore, design has an integral part to play in the planning of these spaces. In order to improve the design of mental health facilities, however, designers need to work in cooperation with different disciplines, due to the "wickedness" of the issue of mental health.

Mental health is of major concern to Canadians. In any given year, one in five people in Canada will personally experience a mental health problem or illness [3]. By age 40, about 50% of the population will have or have had a mental illness [3]. Mental illness affects virtually everyone, either directly or through a family member or friend. Due to its multifaceted nature, however, it is a difficult issue to address.

Compounding the complexity of the issue of mental health is the stigma and the reluctance to seek treatment that accompanies it. Almost one half of people who feel they have suffered from depression or anxiety have never gone to see a doctor about

C. S. Shin (Ed.): AHFE 2019, AISC 968, pp. 59–66, 2020.
https://doi.org/10.1007/978-3-030-20470-9_7

this problem [3]. This can present a barrier to diagnosis, treatment, and community acceptance [3].

2 Causes of Mental Illness

Traditionally dichotomous models of mental illness have been drawn from the philosophy of Descartes, organic - functional neuropathology models of nineteenth century Europe, and the concept of computer functionalism [4]. Kenneth Kendler rejects the traditional dichotomous model of the human mind/brain system which considers the root of illness originating either from the mind (software), or the brain (hardware), stating that "the causes of psychiatric illness are dappled, distributed widely across multiple categories." [4]. These causes include: molecular genetics, molecular neuroscience, systems neuroscience, aggregate genetic effects, other biological risk factors (such as obstetric complications, season of birth, exposure to drugs, intrauterine viral exposure, famine, abnormalities in neuroendocrine levels, abnormalities in immune function, brain-derived neurotrophic factor levels), neuropsychology, personality and cognitive/attitudinal patterns, trauma exposure, and social, political, and cultural factors [4].

Bishop and Dzidic also recognize that mental health issues do not just arise within the individual and emphasize the importance of context. They state that the field of community psychology was "born from the recognition that mental health issues were not purely intrapsychic", going on to state that "the discipline needed to conceptualize mental health in terms of the social, political, economic, historical and psychological context in which problems occurred" [5].

3 Healthcare Facility Design

Healthcare facility design poses complex challenges. The built environment and especially the design of hospitals and mental wellness centers have multiple facets (ambient, architectural, design, and social) that must be considered [2]. This is due to the complicated nature of mental health as well as the enormity of the task of planning and designing a building that will act as both a place of treatment and a workplace for the staff. There is currently a dearth of research on the links between built environments and mental health outcomes that can help guide planners, designers, and architects in designing mental health settings [6]. Additionally, as more research about mental illness and the impacts of design on mental health are uncovered, the nature of healthcare design is complex and evolving [7].

The built environment exerts many effects on people, through different avenues; "personal control, socially supportive relationships, and restoration from stress and fatigue are all affected by properties of the built environment" [8]. Therefore, thoughtful construction of the built environment is essential in order to deliver these experiences.

Agency is imperative for patients in mental health facilities. Autobiographical narratives written by patients with severe mental illnesses often stress the necessity for "a humane and empowering mental health system that encourages patient choice, agency, and autonomy" [9]. Personal control over the environment and the social landscape is not easily achieved in multi-bed rooms in healthcare facilities. Construction of single patient rooms in hospitals reduces stress associated with having a roommate, increases communication between patients and staff, and when combined with decentralized charting does not require higher nurse staffing than multi-bed rooms [10].

Other effects caused by the built environment arise from the experience of sensory stimuli. In poorly designed or regulated environments, "bright lights, incessant sounds, uncomfortable textures and temperature, strange smells and stranger tastes abound" [8]. This is a critical problem in healthcare settings, which often assault the senses [7]. Nanda refers to Rapoport's concept that as monotony stems from sensory deprivation and chaos from sensory overload, the ideal sensory environment falls somewhere in the middle between these two extremes [7]. Similarly, Lawton's Environmental Press theory suggests that people are constantly balancing their ability to cope with the level of environmental demands placed on them, and the ideal point occurs where the person's capacity is equally balanced with the environmental demands [7].

When these demands are balanced optimally, they can provide an ideal setting for healing and rehabilitation. "Careful use of color, light, texture and sound combine to create a healing environment" [1]. Curtis discusses the relationship between mental health, psychological healing, and aspects of the material environment through the frameworks of Biophilia, Topophilia, Attention Restoration Theory, the significance of green and blue landscapes for mental health, and the characteristics of the architectures of happiness [1]. From a perspective of user-centered design, these frameworks can help create better user experiences for a variety of stakeholders through the design of the built environment [11].

4 Qualities of the Built Environment that Affect Mental Health

4.1 Light and Circadian Rhythm

Time of light exposure, light quality, and exposure to natural sunlight are all factors that are critical to control in the built environment. These factors have a large effect on both physical and mental health, through loss of sleep and circadian rhythm. Circadian rhythm is a series of changes (physical, mental, and behavioral) that occur in organisms over a 24-hour time period, responding primarily to light and dark in the environment [12]. These patterns regulate when sleep occurs, and dysregulation leads to detrimental outcomes. When the circadian rhythm becomes dysregulated, regulatory hormones become unbalanced and negative health effects can occur [13]. These can be physical, such as cancer, high blood pressure, and cardiovascular disease, or mental, correlating directly to manic-depressive disorder and seasonal affective disorder as well as clinical depression, especially in adults, the elderly, and hospitalized patients [13].

This indicates that patient areas should be properly lit during the day, and dark at night. One way that this can be achieved is through exposure to daylight. Higher daylight exposure in patient rooms has added benefits, directly reducing depression and pain [10]. To achieve this in a healthcare environment would require a coordinated effort by different actors: architects (the building would have to be built so that patient rooms are located adjacent to exterior walls), lighting designers, psychologists, and possibly software designers (to address the blue light issue associated with screens).

4.2 Interior Design and Visual Aspects

Research indicates that what psychiatric patients see in their environment affects their mental health and therapeutic outcomes [8]. Visual elements within the environment include wall color, artwork, and house plants. Research indicates that warm and bright colors helps patients with depression and cool blues and greens help patients with anxiety and stress disorders [13]. Monochromatic colors can have a negative effect on people in mental health settings, disturbing and overwhelming patients with mental and attentional disorders [13]. There is also a benefit to using green, due to its association with nature [13].

A connection to nature is also beneficial regarding artwork and plants within patient rooms. "Artwork tends to be most healthful when it is of restful nature scenes, especially in rooms without exterior windows" [13]. However, caution must be used when choosing artwork; "color patterns should not be able to be misperceived as threatening or confusing in any way – such as bugs, animals, or figures – by medicated or distressed patients." [13]. Interior plants also have recorded beneficial effects, such as improved productivity and attentiveness and reduced stress levels and physical discomfort [13].

Exterior windows, besides aiding in regulating circadian rhythm, have additional benefits when they depict natural scenes. These views promote faster healing, lowered stress levels, and better attention, focus, and mood [13]. Actual access to green spaces is also highly beneficial. "Recent research has shown that time in natural environments is associated with reduced negative emotions and better energy levels, attention span and feelings of tranquility compared with being in synthetic settings" [14]. This is also associated with improved psychological states, such as higher self-esteem and better mood, decreased stress and anxiety, and regulated mood and emotional disorders [13]. Like access to daylight, creating these improved therapeutic environments require a collaborative effort between fields in order to put these theoretical benefits into practice.

4.3 Noise

Excess noise is associated with negative effects on patients, especially those in psychiatric facilities [13]. Patients in mental healthcare institutions are especially vulnerable to excess noise, as they may not be able to escape the noise, noise sensitivity could be a predisposing factor for mental illness, and "there exists a demonstrable correlation between noise (its effects and manifestations) and the psychopathology of the individual affected, eliciting greater 'stress' or arousal responses in those with psychological or psychiatric problems" [11].

However, creating a silent environment should not be the goal; as mentioned above, sensory deprivation can lead to monotony. Both the level and clearness of noise are important; the optimal soundscape is one in which the sounds are intelligible and legible [11]. Nature sounds and music controlled by patients can act in a soothing capacity [13]. Appropriate noise may also have a role in social regulation, "to secure a degree of privacy for oneself, to exclude others or as a source of solidarity among friends and colleagues" [11]. The role of planners, designers, and architects is to ensure that patient areas are in quiet, calm environments [11].

5 Interdisciplinarity

Given the complexity that arises from designing a mental health facility, there is a growing demand for interdisciplinary expertise [15]. The scope of this task includes: understanding how the built environment affects mental health, urban planning, building design, interior design, service design, lighting and sound design, ergonomics and human factors, et cetera. The social, cultural, political, and environmental contexts must be considered. These factors simply cannot be accounted for within one discipline. "An interdisciplinary approach often arises because of the grounded nature of problems and also because a single discipline does not have the range of expertise to deal with all aspect of the issues" [5]. In fact, interdisciplinarity is unavoidable in design; "design is by necessity interdisciplinary because it has to consider so many issues" [16].

The psychological impact of the design of a healthcare facility is essentially an issue grounded in context. Psychologists cannot take a clinical approach when contributing to such a project. Bishop and Dzidic regard a significant amount of research conducted in a lab as unhelpful and irrelevant due to its lack of context, and lack of procedural information, and state the need for psychology to step out of purely clinical settings [5]. To be an agent of change, researchers and psychologists must understand that their actions are embedded in a broader social context [5].

Considering context is important to ensure the knowledge and insights gained from research reflect reality. This involves different layers of analysis from separate disciplinary lenses to create a single viewpoint, creating a messier, but more accurate picture of the object of study; "this level of complexity is essential if we are to develop a discipline that reflects the realities of people's lived experiences and not some abstracted version that suits [psychology's] disciplinary view of the world" [5]. Clark and Wallace also consider that looking through the viewpoint of different disciplines is necessary to achieve what they term "knowledge integration" [17]. Integration is a "big picture" view; it "refers to the combining of disparate things, events, or processes with the goal of fitting them together in such a way as to better understand both the parts and the emergent whole" [17]. Understanding the emergent whole is only possible through interdisciplinarity.

6 Interdisciplinary Design Team

This section illustrates two examples of projects focused on therapeutic built environments. The examples are disparate in scope and goal, but both take an interdisciplinary approach.

6.1 Behavioral-Healthcare Design

CannonDesign is a design firm that, in their own words "unites a dynamic team of architects, engineers, industry experts and builders driven by a singular goal—to help solve our clients' and society's greatest challenges" [18]. They also recognize the value of research that is grounded in context, blending their expertise with empirical and ethnographic research [18]. CannonDesign works in an interdisciplinary team with 27 members, including architects, planners, designers, engineers, programmers, and advisers, functioning as a collection of experts selected to lead behavioral-healthcare design projects [19].

Part of the work that CannonDesign does is to create crisis-center intervention units, the aim of which is to provide a calmer environment for assessing patients exhibiting both mental and physical distress, and ultimately avoid hospitalization for those who may not need it [19]. These spaces are becoming more common in children's hospitals; Diane Osan, the leader of CannonDesign's pediatric practice, states that "good design can alleviate the strain on staff and young patients" [19].

As stated earlier, agency and personal control are important in stress reduction and healing. CannonDesign is aware of this. They promote "therapeutic platforms" that combine several different environments of healing, ranging from private to semi-public. "Key to a therapeutic platform's success is patient choice. Private rooms, which help preserve patient dignity and a sense of normalcy, are now favored over dormitory-style rooms with multiple beds" [19]. They also design for patient independence, aiming to provide increasing amounts of freedom wherever possible [19].

This mode of design also promotes safety for patients and staff and reduction of physical and chemical restraints. In modern healthcare facilities, "clients are turning to design-driven ways to de-escalate high risk situations or avoid them entirely" [19]. Creating a peaceful, soothing environment can help avoid physical violence [19].

This example illustrates how interdisciplinary design can account for more aspects than a single disciplinary approach. By working with many different disciplines this approach can provide a better experience for service-users, more effective therapeutic outcomes, and a reduction of traditional practices of restraint and seclusion.

6.2 Smart Environment Architecture for Emotion Detection and Regulation

In this more theoretical example, academics from computer science, audiovisual technology, robotics, medicine, and education are collaborating in order to create a smart environment that can detect and react to a user's emotions. "The aim of the proposal is to detect the patient's emotional state by analyzing his/her physiological signals, facial expression and behavior. Then, the system provides the best-tailored actions in the environment to regulate these emotions towards a positive mood when possible" [20]. The stimuli that are employed to regulate emotion are music, color, and light. The goal of this project is to enhance quality of life and care delivery to the user [20].

The group involved in this project recognizes the importance of working with multiple disciplines, stating that "the success of the project is highly dependent of the involvement and participation of subject matter experts" [20]. Involved in the team are psychologists, neurobiologists, physicians and musicians [20].

Although this type of smart environment is not likely to be deployed in a therapeutic setting any time soon, this example further illustrates what kind of solutions are possible when working in an interdisciplinary team. This project involves aspects such as machine vision, engineering, psychology, and fine arts, and these aspects must each be represented by domain experts in order to function together.

7 Conclusions

Working interdisciplinarily in healthcare research is not simple. Working in an interdisciplinary team requires that one must "navigate and negotiate with differences at many levels" and can remove one from his or her "paradigmatic/theoretical comfort zones" [21]. Building an interdisciplinary team requires much time and funding and can be impeded by disciplinary "turf wars" [21]. However, the complexity of the field indicates that it is critical that teams designing healthcare facilities must work interdisciplinarily in order to create the best outcomes for patients, workers, and visitors.

It is imperative that designers become more comfortable working on interdisciplinary teams. Central factors in successful interdisciplinary healthcare teams are: "listening to others, asking questions, persuasive rethinking, respect, a supportive environment, leadership with autonomy, and trust" [21]. If design researchers and practitioners can put these concepts into practice, working interdisciplinarily has the potential to create better designs and ultimately better mental health outcomes for anyone seeking treatment.

References

1. Curtis, S.: Space, Place and Mental Health. Ashgate Publishing Limited, Farnham (2010)
2. Karlin, B.E., Zeiss, R.A.: Best practices: environmental and therapeutic issues in psychiatric hospital design: toward best practices. Psychiatr. Serv. **57**(10), 1376–1378 (2006)
3. Mental Health Commission of Canada: Making the case for Investing in mental health in Canada (2013)
4. Kendler, K.S.: The dappled nature of causes of psychiatric illness: replacing the organic-functional/hardware-software dichotomy with empirically based pluralism. Mol. Psychiatry **17**(4), 377–388 (2012)
5. Bishop, B.J., Dzidic, P.L.: Dealing with wicked problems: conducting a causal layered analysis of complex social psychological issues. Am. J. Community Psychol. **53**(1–2), 13–24 (2014)
6. Platt, L.S., Bosch, S.J., Kim, D.: Toward a framework for designing person-centered mental health interiors for veterans. J. Interior Des. **42**(2), 27–48 (2017)
7. Nanda, U.: A sensthetic approach to designing for health. J. Interior Des. **42**(2), 7–12 (2017)
8. Evans, G.W.: The built environment and mental health. J. Urban Health **80**(4), 536–555 (2003)
9. Drake, R.E., Whitley, R.: Recovery and severe mental illness: description and analysis. Can. J. Psychiatry **59**(5), 236–242 (2014)
10. Ulrich, R.S.: Essay: evidence-based health-care architecture. Lancet **368**, S38–S39 (2006)

11. Brown, B., Rutherford, P., Crawford, P.: The role of noise in clinical environments with particular reference to mental health care: a narrative review. Int. J. Nurs. Stud. **54**, 1514–1524 (2015)
12. National Institute of General Medical Sciences. https://www.nigms.nih.gov/Education/Pages/Factsheet_CircadianRhythms.aspx
13. Dao, A.E.: This is Your Brain on Architecture: An Exploration in Designing a Neurologically Healthy Mental Wellness Center (2018)
14. Lachowycz, K., Jones, A.P.: Towards a better understanding of the relationship between greenspace and health: development of a theoretical framework. Landscape Urban Plann. **118**, 62–69 (2013)
15. Kasali, A., Nersessian, N.J.: Architects in interdisciplinary contexts: representational practices in healthcare design. Des. Stud. **41**, 205–223 (2015)
16. Boradkar, P.: Taming wickedness by interdisciplinary design. In: The Oxford Handbook of Interdisciplinarity, pp. 456–460 (2016)
17. Clark, S.G., Wallace, R.L.: Integration and interdisciplinarity: concepts, frameworks, and education. Policy Sci. **48**(2), 233–255 (2015)
18. CannonDesign. https://www.cannondesign.com/about/our-firm/
19. Okamoto, K.: Mindful spaces. Metropolis, 95–101 (2018)
20. Fernandez-Caballero, A., Martínez-Rodrigo, A., Pastor, J.M., Castillo, J.C., Lozano-Monasor, E., López, M.T., Fernández-Sotos, A.: Smart environment architecture for emotion detection and regulation. J. Biomed. Inform. **64**, 55–73 (2016)
21. Hesse-Biber, S.: Doing interdisciplinary mixed methods health care research: working the boundaries, tensions, and synergistic potential of team-based research. Qual. Health Res. **26**(5), 649–658 (2016)

Leveraging Disciplinary and Cultural Diversity in the Conceptualization Stages of Design

Alex Close(✉) and Eileen Harris

Carleton University, 1125 Colonel by Dr, Ottawa, ON K1S 5B6, Canada
alexclose@cmail.carleton.ca

Abstract. This paper explores the involvement of increasingly diverse viewpoints in design and makes a case for how using low fidelity 3D prototyping in conceptualization stages might be able to leverage these viewpoints. It compares traditional uni-disciplinary methods with interdisciplinary and cross-cultural collaboration and explores why this might be relevant at this time. It also examines how participants can be embodied in the early design process to allow divergent cultural and disciplinary perspectives to emerge and develop more sophisticated and innovative designs with a global scope. The paper also explains some challenges with collaborative cross-cultural and cross-disciplinary approaches, and offers recommendations while suggesting areas of influence that could be looked at in further research.

Keywords: Interdisciplinary · Cross-cultural · Prototyping · Embodiment · Collaboration · Globalization

1 Introduction

As globalization increases, the resulting interactions between humans and the issues that arise become more complex. To have a comprehensive understanding of these issues, problems increasingly need to be looked at from multiple perspectives [1]. This is the case in the field of design. In the face of globalization, design companies face an increasing need to communicate with stakeholders in multiple locations, some of which could be in emerging countries [1]. Since more diverse players will interact, more attention needs to be paid to how people interact with objects and physical systems.

Many different types of expertise might be needed to come up with an effective solution in the case of increasingly complex design situations. This means that different disciplines must often work together. Divergent thinking can yield creativity, and could be a great opportunity when taken advantage of in the early stages of conceptualization [2]. The involvement of increasingly diverse viewpoints in design could lead to physical prototyping playing a key role in making innovative physical systems and objects in contemporary design.

C. S. Shin (Ed.): AHFE 2019, AISC 968, pp. 67–74, 2020.
https://doi.org/10.1007/978-3-030-20470-9_8

2 Background

When searching for insights on user experience, Burdek [3] notes that the degree of sociocultural differences between users means that assessments conducted from afar might not be able to produce adequate conclusions for product policy and design. He writes that in the contemporary world, it is clear that national frontiers and disciplinary boundaries are now blurred because of globalization. This leads to the need for a "more extensive scope and depth of expertise to stay in the game, work in distributed real and hyperreal decentralized environments and co-design solutions," [3]. The need for increased scope and depth suggests that it is no longer enough to design with simply one perspective in mind. More platforms for design exist (such as software or service design) and more diverse locations are playing a role in the market for designed objects and systems.

One possible way to cope with these changes is to involve a more diverse group of people in the conceptualization stages of design. In some cases, this could mean using codesign and allowing users in the countries the design will be made for to have a say in the final product, as well as working on interdisciplinary teams to have a broader perspective of the problem. To develop a common language in this setting, participants could be embodied in the design process by making quick and cheap prototypes (low-fidelity 3D prototyping) to come up with ideas. Embodiment would mean that their ideas become linked to material.

3 Communication Challenges

There are numerous examples of companies failing to understand the target market, leading to loss in revenue. One such case is Whirlpool's attempted sales of the World Washer in India that lost the company money due to a failure to understand the fabrics that would commonly be used in the region. It "had been given a single, generalized, emerging-market reference point by designers with limited understanding and direct experience of the customs and modes of dress in South India. They did not ask the right questions of target users—if they talked to them at all," [4]. To learn more about these groups, prototyping could be used in the early stages of brainstorming while interacting with these communities to find the easiest language with which to communicate.

Even though a 3D low-fidelity prototyping approach might lead to more fluidity and useful prompts to elicit ideas, there are still many other variables that could cause collaboration to be challenging, one of which is the idea of interdisciplinary collaboration in general. Some people might place higher value on their discipline and may be less willing to take others' ideas seriously [1]. This could inhibit brainstorming or ideation (when multiple disciplines are needed at this stage). Also, clients might sometimes not be able to understand the language of design (e.g. technical drawings), and thus feel that they are not a part of the conversation – in this case, using 3D prototypes might be able to reduce communication issues.

3.1 Diversity Within Design Teams

There are often people on design teams who hold more than one cultural background, or choose to interact with it in different ways. Some theories suggest that when people hold multiple cultural meaning systems (perhaps with conflicting knowledge bases), "which of these is activated depends on what situational constraints are salient." [5]. In other words, the setting/context can affect a participant's contribution. In addition, each culture (and in fact each person) comes with a different set of philosophies (or reactions to philosophies) that may have been influenced by the culture in which a person was brought up. For example, certain cultures have different views of individualism vs. collectivism, and feminism vs. masculinity. This might affect the value people place on group activities and in turn the amount they are willing to contribute.

No person truly represents a pure national culture - instead people are subject to and identify with culturally constructed categories [6]. These include gender, ethnicity, nationality, race, institutional affiliation, or subculture. These categories interact in complex ways that shape their interpretations and actions in the world [6]. For example, there might be a person on a design team that comes from a mixed background (e.g. American and Chinese) that possesses multiple cultural codes, switching back and forth depending on the context.

There are various ways that someone on a design team could access cultural information. If one is brought up with a specific type of cultural knowledge, it might be effortless to activate this, whereas someone from a different culture might need explicit activation cues from design material/props or other participants and designers to access relevant cultural knowledge about the emotional value of the product or system in a different cultural context than their primary one [5]. This is another reason that material prototypes in the early stages of design ideation could be useful.

Contrary to some traditional uni-disciplinary design approaches, having different groups participate in this early stage could mean that later mistakes might be prevented. It might be important to have a diverse team at this stage because they can "[cover] a broader territory of information, [tap] into a broader range of networks and perspectives, and can have enhanced problem-solving, creativity, innovation, and adaptability." [2]. Though homogenous groups might have improved communication, "creativity is clearly a divergent process, and the creative benefits of heterogenous team compositions have been supported by numerous studies" [2]. Because low-fidelity physical prototypes are less mediated and therefore seem less coded in disciplinary languages, they could be an immediately changeable way for more people to effectively share their ideas. If multiple perspectives are respected and considered at the beginning of the design process, more innovative and suitable design solutions could be made at this stage, preventing mistakes that could lead to financial losses for a company.

4 Design Approaches

4.1 Traditional Design Approaches

The expanded scope of design due to globalization may call for an examination of traditional design approaches. Because of the complexity of issues and variety of platforms of design that interact today (e.g., software design, product design, interaction design) interdisciplinarity is sometimes necessary to achieve goals. When working on interdisciplinary design teams in a context aware of global scope, the design process could be important to consider when aiming for innovation.

In many cases in traditional uni-disciplinary design approaches, some mistakes are not realized until it is too late. This might have to do with the ego of some designers, a push to get a product to market, or just failure to do enough research. Like the previous example of Whirlpool's design failure, in the case of environmental design, "a typical product design process does not incorporate environmental assessments until late into the design process, if at all" [7]. The preliminary conceptualization stage can be where many crucial decisions are made, so one way to make sure challenges are met and problems become innovative is to encourage divergent thinking in this early process. This can be done by making low-fidelity physical prototypes as ideation methods in a group brainstorming session.

4.2 Collaboration

Collaboration is a complex topic and is alone a subject in need of further exploration in a time of increasing globalization [8]. Visualisation can be a medium for researchers and community members to engage in dialogue, but visual expression in some places is not graphical, and the values we associate with visual qualities could be culturally dependant. [9] In a way this could suggest that 3D prototypes could be useful in communicating with a broader scope of communities since there is less mediation of the design concept.

Being aware of cultural differences can lead to a more holistic approach to future design by allowing designers to learn about how people treat physical systems (object/space). Creating objects early on in the design process with people from different backgrounds can allow information systems to be interacted with and synthesized by different groups, inspiring a broad scope of challenging questions from each different perspective.

Though using 3D prototypes might make for easier communication in many cases, there could still be situations where individual or cultural attitudes could affect a person's view of collaboration (i.e. individualism vs. collectivism) and therefore the results of each brainstorming/ideation session. Teamwork could also be influenced by how much value each discipline might place on other areas. Additionally, an individual's personality could play a role in their contribution to the collaborative process. Though it will likely be difficult to achieve, it seems that successful collaboration is essential to consider in such a complex era of globally oriented design.

When expanding a company to an emerging country, it might be useful to consider how the people there can become empowered, how to have a conversation with these communities "so that they can form a robust relationship with the world[.] Otherwise, it becomes a kind of colonialism" [9]. Instead of companies from developed countries imposing their cultural trends, a more nuanced relationship and strengthened cultural understanding could be achieved through conversation in the early prototyping stages. It could be possible to empower people in emerging countries, and learn from them by providing them with a stronger voice in the products to be distributed within their location.

5 Designer and User Interaction

5.1 User Experience

The reality of the user experience in the contemporary world is clearly complex. In some cases, this might require an increased focus on human factors and collaborative approaches compared to a more conventional top-down model. This focus can extract stories about the users/designers, that could improve the sensitivity of designs. The surface in the final objects can subsequently be an arena "through which these narratives can be publicly understood and articulated." [10]. If the user is not paid attention to, the opposite could be true. The sentiment a thing is designed with in conceptualization could carry through to production and finally influence the interaction of people and their environment or others.

Sensitivity to users and the complex ideologies that construct how they perceive their world today can be seen in the stories told by the objects that construct peoples' lives. The mentality of integration can be present from the selection of participants in the design process to early ideation stages, all the way to final products. When diversity in disciplinary expertise and cultural backgrounds is leveraged, connections between people and places that might not be able to be made through language can be made through design. In a time where global issues might need to be tackled, this could be one way of making collaboration easier.

5.2 Embodiment and Storytelling

Embodiment and use of physical props might help people link their ideas and stories to material. This in turn can lead to more comprehensive perspectives being integrated in the early design process. Embodiment of an idea (thoughts transferred to material) "may act as a trigger that [leads] people to think of something relevant but unexpected or [serves] as an inspiration for further design activities," [8]. As ideas go from abstract invisible concepts in the mind to solid visible shapes, a designer (or participant in the brainstorming session) can clearly envision ideas while simultaneously editing, and engaging with other team members. This iterative circular process "would be a critical driving force to turn an individual's early vague thoughts and ideas into a clear design solution and to transform controversies and conflicts into mutual agreement" [8]. This method would focus on the explicit visual elements through fluid critique and play during brainstorming.

By developing a common language and stage for people from diverse backgrounds to ideate, the "important dimensions of human experience and understanding [that] lie unexplored in the spaces between disciplinary boundaries or the places where they cross, overlap, divide, or dissolve" [11] might come bubbling to the surface. Also, when interacting with local cultures (e.g. co-design) the user might be able to easier communicate their specific needs.

Using 3D prototyping as an ideation method is one way that people could find common ground regardless of their background, since visual communication has a long history in most places in the world [12]. Since working with prototypes in ideation lends itself to intuition and experimentation, divergent thinking at this stage can lead to contradictions, but also harmonious moments where people could find commonalities and agree on what works in a design.

6 Discussion

As globalization increases, cross-cultural collaboration is becoming necessary in many cases. This might be an excellent opportunity for divergent ideas and attitudes to support creative designs. This is also true for interdisciplinary collaboration. People from different disciplines could ask different questions. Through bold integration, achieving a broad scope might eliminate blind-spots that would have been created in the alleys between each discipline's tunnel vision.

Communicating through visual and tactile means could be a way to eliminate language and disciplinary boundaries in ideation stages leading to more fluid ideation. Though other forms of research are likely also needed, by using a tactile and visual method of communication, the lack of language barriers could prevent mistakes by being concerned with the views of the users from the beginning. By embodying people in this process, it empowers them to have a voice, and thus creates a more comprehensive platform out of which a design can emerge.

Physical materials might be able to better explain inherent feelings or thoughts when cultural barriers inhibit communication, since they are explicit and require less translation than words or graphic representation (when used as an ideation method, not an artistic product). As such, it might be easier for participants to visualize the end product in a 3D model because they occupy physical space just like a final design would, leading to practical insights. Though 3D prototyping might be one of the most accessible options (since it lacks mediation such as computers or sketching), it might still be necessary to look at how people perceive collaboration in group settings and in what situations they would be least inhibited.

With the ideas of people from diverse locations, it is possible to tap into the underlying philosophies, methods of spatial understanding, graphic translation, and interpretation of artifacts in these different areas. This might lead to more creative and intelligent designs. More than one method would clearly be needed to achieve this, but involving communities in early prototyping stages could be one way of reaching a deeper understanding. Though many challenges exist, if globalization continues, finding more sophisticated methods of communicating within diverse groups (locally or globally) could become increasingly important in design.

Though it seems like a great opportunity, there are also significant challenges that stand in the way of cross-cultural and cross-disciplinary interaction. For example, the difference between developed and emerging markets can be a cause for confusion and misunderstanding due to language and cultural differences, or even residual colonial attitudes and naïve assumptions. Although using a 3D prototyping method can help in several ways, there will likely continue to be friction and challenges that can sometimes lead to invention, and sometimes lead to failure.

6.1 Future Research

Though diversity could clearly be a source of innovation, it might also be necessary to note that there are ways that a uni-disciplinary approach could be sufficient, if not more effective. There are questions to be asked about rigour and expertise and whether there are ways this would be sacrificed by involving so many people in design. There are also great practical challenges in collaboration when groups are diverse. A participant/stakeholder/user could have the knowledge of different cultural philosophies and customs, as well as individual views which may or may not align with these. Further research would need to be done about collaboration methods as well as the role of philosophy in cross-cultural collaboration. It might also be interesting to see how changing philosophies (such as modernism evolving to postmodernism and what exists after that) might have an effect on how people interact with objects in design. Changes in interaction with objects could be present in ideation as well as artifacts.

Further research could also be done to examine the effects of technology on creativity. Though there seem to be notable benefits of tactile interaction in prototyping stages in design, it might be useful to see whether there are creative benefits to ideating through a digital platform. This would also be relevant because as companies expand and need to talk to partners who might not be in the same country, it might not always be easy to travel abroad to converse with them so a digital platform might have to be used. It could be important to research the effects of this kind of collaboration.

7 Conclusion

Low fidelity 3D prototyping in the conceptualization stages of design could play a key role in making innovative physical systems and objects while the involvement of diverse viewpoints increases. There will likely be more interaction between stakeholders in different countries as companies expand, including interaction between developed and emerging countries (or cultures with different types of colonial history). This can happen at the global and local level (interactions within companies as well as interactions with users abroad). In addition to cross-cultural collaboration, there will likely also be interdisciplinary interaction on design teams due to complex problems needing to be solved that are too large for one discipline to handle alone. This paper explored how being embodied through making low fidelity prototypes could help people communicate their ideas and stories through material. By involving users from different cultural and disciplinary backgrounds, divergent thinking could be achieved that will lead to more innovative design for the future.

References

1. Moreno, L., Rogel, E.: Transdisciplinary design: tamed complexity through new collaboration. Strateg. Des. Res. J. **1**, 42–50 (2018). https://doi.org/10.4013/sdrj.2018.111.07
2. Stahl, G.K., Maznevski, M.L., Voigt, A., Jonsen, K.: Unravelling the effects of cultural diversity in teams: a meta-analysis of research on multicultural work groups. J. Int. Bus. Stud. **41**(4), 690–709 (2009). https://doi.org/10.1057/jibs.2009.85

3. Burdek, B.: Design and globalization. Des. Birkhäuser Basel, 71–224 (2005). https://doi.org/10.1007/3-7643-7681-3_3
4. Chavan, A.L., Gorney, D., Prabhu, B., Arora, S.: The washing machine that ate my sari. Interact. January + February **16**, 26–31 (2009)
5. Clemmensen, T., Ranjan, A., Bodker, M.: How cultural knowledge shapes core design thinking – a situation specific analysis. CoDesign **14**(2), 115–132 (2018). https://doi.org/10.1080/15710882.2017.1399146
6. Irani, L., Dourish, P.: Postcolonial interculturality. In: Proceedings International Workshop on Intercultural Collaboration, pp. 249–252 (2009)
7. Aguas, S.: The Design Process Paradox. Traditional Design Processes vs. Eco-Design Processes. Caleidoscopio Revista De Comunicacao E Cultura, Caleidoscopio Revista De Comunicacao E Cultura (2006)
8. Chung, W., Lacerda, L.: Jazz improvisation for effective design collaboration. In: Proceedings IASDR 2013, pp. 1–9 (2013)
9. Crichton-Miller, E.: Found in Translation. Crafts. (0306610X) (2006)
10. Jacob, S.: Beyond the flatline. Archit. Des. **81**(5), 24–31 (2011). https://doi.org/10.1002/ad.1292
11. Roy, T.: Did globalisation aid industrial development in colonial India? a study of knowledge transfer in the iron industry. Indian Econ. Soc. Hist. Rev. **46**(4), 579–613 (2009). https://doi.org/10.1177/001946460904600404
12. Rodil, K.: Reflection on Visualization in Cross-Cultural Design. Aalborg University Denmark (2015)

Interdisciplinary Approach of the Design Process for the Application of New Materials in Wheelchair Design

Paula Chacon-Cifuentes[1,3(✉)], Alejandro Zuleta-Gil[1], Gustavo Sevilla Cadavid[1], Andres Valencia-Escobar[1], Esteban Correa-Bedoya[2], and Felix Echeverria-Echeverria[3]

[1] Grupo de Investigación de Estudios en Diseño, Facultad de Diseño Industrial, Universidad Pontificia Bolivariana, Sede Central Medellín, Circular 1 No. 70-01, Medellín, Colombia
{paula.chacon, alejandro.zuleta, gustavo.sevilla, andres.valencia}@upb.edu.co

[2] Grupo de Investigación Materiales con Impacto – MAT&MPAC, Facultad de Ingenierías, Universidad de Medellín, Carrera 87 No 30 – 65, Medellín, Colombia
escorrea@udem.edu.co

[3] Centro de Investigación, Innovación y Desarrollo de Materiales (CIDEMAT), Facultad de Ingeniería, Universidad de Antioquia UdeA, Calle 70 No. 52-21, Medellín, Colombia
felix.echeverria@udea.edu.co

Abstract. The appropriation of new technology requires the design and development of products that are accepted by the market. Developing such products is a complex task that usually requires the integration of knowledge from multiple disciplines. In the field of new materials, developments that can be applied to assistive technology are constantly made. This work studies the collaboration between users and professionals from industrial design, engineering, human factors, and disability for the design of an all-terrain wheelchair. This product incorporates the use of magnesium alloys, a material with an excellent strength-to-weight ratio and good damping properties. With a human-centered design approach, tools such as interviews, focus groups, prototyping, and testing were used in several stages of the process. It allowed achieving a design that takes advantage of the properties of the new material and, at the same time, meets the mobility, stability and comfort requirements of the users.

Keywords: User-centered design · Light alloys · Product development · Assistive technology · Design practice

1 Introduction

Innovation involves, not only the development of new technology but its appropriation by society. Usually, a new technology finds its way to the market through the design of products aimed to solve people needs. For this to happen, these products must satisfy

C. S. Shin (Ed.): AHFE 2019, AISC 968, pp. 75–82, 2020.
https://doi.org/10.1007/978-3-030-20470-9_9

those needs taking advantage of the benefits of the new technology and ensuring its technical and economic viability [1]. Developing such products is a complex task that usually requires the integration of knowledge from multiple disciplines. This diverse knowledge allows to recognize the needs and deeply understand them, to identify and apply the knowledge to satisfy them, to develop the products in a detailed way and to propose the productive models that make possible their materialization and transference to the market.

In the field of new materials, developments that can be applied to the industrial design are constantly made. In the case of light alloys, recent advances include new ways of obtaining magnesium alloys and the development of surface treatments that allow expanding their field of application [2, 3]. Magnesium alloys are structural metals that have properties such as an excellent strength-to-weight ratio and good damping capability, which makes them attractive for applications in mobility [4]. In this context, reducing the weight of the products contributes to energy savings, and damping capabilities contributes to absorb vibrations, which gives greater comfort to the users.

However, to exploit the properties of these new materials in mobility products, an interdisciplinary approach is required. The collaboration between diverse disciplines will allow to achieve a product that, in addition to having a reduced weight, has an adequate mechanical behavior and resolves the needs of people in terms of mobility, safety and comfort. With this in mind, a user-centered design approach was incorporated. This approach is oriented to the design of more useful, usable and desirable products through the use of tools that allow a better understanding of the needs of the user and a constant evaluation of the results during the design process of a product [5]. This involves a process of constant learning from the users, which results in guidelines for the designers and engineers work.

2 Methodology

Three aspects were crucial for a successful development process: the participation of people with different backgrounds and disciplines to allow knowledge integration, a design process considering the close collaboration with users and the application of management tools and techniques to deal with the uncertainty related to research and development activities.

2.1 Interdisciplinary Team

The design of an all-terrain wheelchair applying a new material requires knowledge from fields such as industrial design, engineering, human factors, disability, and physiotherapy. The main team was conformed by materials engineers, a mechanical engineer, an industrial designer and a design engineer. Physiotherapists and manufacturing experts were consulted in several stages of the project.

Materials engineers were in charge of the alloy selection, the development of processes of forming and surface treatment to improve the magnesium alloy properties, and collaboration with designers and manufacturing suppliers to work with the new material. The mechanical engineer defined the structural criterion to develop and test the frame of the wheelchair, and the design of the propulsion and brake system. The industrial designer, with an expertise in human factors and disability, give the guidelines of the ergonomic, safety and comfort requirements, and defined the protocols to develop interviews, focus group and usability test with the users involved in the project. The product design engineer was in charge of the recompilation, analysis and synthesis of the information for the design process, the detailed design of the wheelchair and the management of the material and manufacturing process suppliers. All the team was involved in the ideation process and the development of activities were the users participated, like the definition and validation of requirements and the tests developed.

2.2 Design Process

Based on previous experiences in product design of several of the team members, it was decided to divide the design process into three phases named information, formalization, and conformation [6]. As mentioned above, the user-centered design approach involved the participation of users in several activities of the process.

In the *information phase*, researchers gathered information about wheelchair design and magnesium alloys considering academic research, science databases, commercial information, regulations and organizations dedicated to the fields mentioned. Potential users were identified and interviewed about their current wheelchairs and daily activities. In this phase, it was necessary to perform a test with functional models to define the position of the caster wheel, in which users were involved too. Every subject tested two models with different positioning of the caster wheel, driving the two models through an established path and answered a questionnaire about their perception of each model. All the information gathered were synthetized in the design requirements, using the product design specification tool, proposed by Ulrich and Eppinger [7].

The *formalization phase* began with an ideation process were all the team participated. Design alternatives were evaluated and selected by the researchers based on the requirements. Digital models of two final design alternatives of the wheelchair frame were constructed and evaluated with finite element analysis (FEA). The structure with the best behavior in terms of strength, stiffness, and stability were selected as the final design. The wheelchair was sized according to anthropometric measurements of three potential users, and components like the seat, backrest, footrests, propulsion, and brake system were designed in detail. Before generating all the information for manufacturing, a rudimentary prototype constructed in polyvinylchloride (PVC) tubing was made, beginning the *conformation phase*. A *rudimentary prototype* is a low cost and rapid elaboration prototype used to communicate in a physical model the main characteristics of the design proposed and learning from the user perception as early, cheap and fast as possible [8]. This prototype was used in a focus group with researchers and users, to validate their perception about the wheelchair frame. Also, the digital model was checked with the manufacturer to evaluate the fabrication processes alternatives. After the final adjustments suggested by users and suppliers, all the technical drawings required to fabrication were finished.

The *conformation phase* also consists of the fabrication of functional prototypes and their validation. The first functional prototype is being manufactured in aluminum alloy, due to its ease of access in the local context. This prototype will be used to develop technical and usability tests. The technical tests will be made according to the standard ISO 7176 [9]. For the usability tests potential users will be involved. After this testing, final adjustments are expected. The resultant design will be prototyped in AZ31 magnesium alloy and the same tests mentioned will be applied. With this information it will be possible to establish the improvements made and evaluate the advantages of using a magnesium alloy in this type of products.

On the other hand, the processes for the treatment of the surfaces and the formation of pieces by powder metallurgy, were developed by materials engineers in parallel subprojects. The results of these subprojects will be integrated into the design process when said manufacturing processes are scaled for their application in the magnesium prototype.

2.3 Project Management

The project was planned considering schedules, budget and human resources needed to achieve the project´s scope. During the project development, a close monitoring was considered, with meetings every two weeks, where every sub-team showed its advances, discussed the obstacles and possible solutions. Changes to the original schedule and budget were evaluated and approved by all the team. One partial report was presented by the end of the first year and one final report will be presented at the end of the second year.

2.4 Ethics

As the project involved user participation, informed consent was signed with each of them. In the document, each subject was informed about the activities in which he would participate, the purpose of these activities and how the results would be used. For usability tests, they were informed about the safety precautions implemented and the risks to which they could be exposed during the tests. Permission for the use of images and video recorded during the project activities was also requested. Finally, the document established the commitments of the participating institutions with each subject.

3 Results

3.1 Knowledge Integration

The technical knowledge about the properties of the materials and their processing, the design of structures, anthropometry, ergonomics, and disability, allowed the identification of a problem in a real context and the conception of an integral solution, considering the technological aspects and human factors. The practical experience with the application of the methodologies from each field of knowledge allowed realistic planning, as well as the identification of necessary adjustments during the progress of the project.

Maintaining constant communication among all project members was crucial, not only to keep track of the development of project activities but also to create a positive environment and teamwork dynamic, understanding the impact that the work of each individual has for the achievement of common objectives. Additionally, as mentioned by Toft et al., the collaborative work between designers, engineers and human factors experts allowed to develop new skills and gain understanding of the different languages and methods of each discipline, which reduces cross disciplinary conflicts in the design process [10]. All of the above increases the success probabilities of the project.

3.2 Product Design

The participation of the users at the beginning of the design process allowed to recognize their needs and contrast it with the existing products to identify design opportunities. Through the users and experts, the local wheelchair provision system could be understood and taken into account during the design process. The interviews provided in-depth information on the needs of each user, and from this it was possible to identify patterns to define the design requirements. The first tests with functional models allowed to establish the best position of the caster wheel, but also the adaptation of an existing design resulted from a prior research project [11]. Figure 1 shows the functional models used during the test, made of plywood sheets. The use of this material, together with the CNC cutting process and slots assemblies, allowed to obtain a low cost object, fast manufactured and with all the possibilities of personalization according to the anthropometric measurements of each user. These models allowed to establish, according to the perception of the users, that the caster wheel should be placed in the front position.

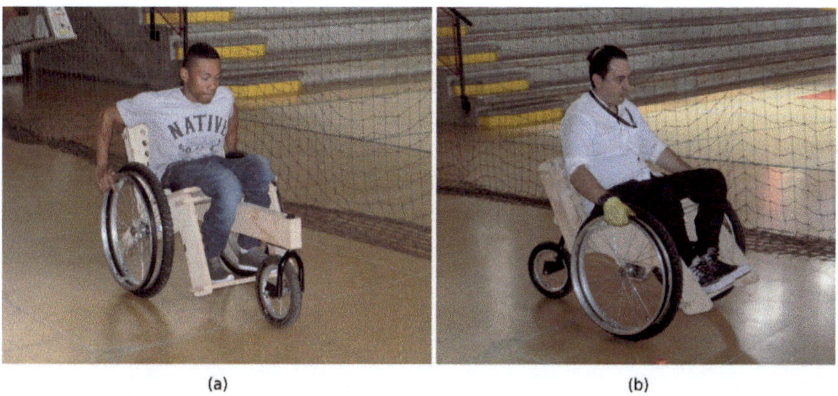

Fig. 1. Plywood functional models used to test the position of the caster wheel in (a) front position and (b) rear position.

The use of functional models and rudimentary prototypes in the design process was crucial for testing fast and with a low cost in order to learn from the design decisions early and make the validations and adjustments on time. In Fig. 2 can be observed the process of construction of the PVC prototype and Fig. 3 shows the focus group session where the wheelchair design was discussed with users and the project team. The PVC tubing prototype as a way to communicate to the team and to the users the characteristics of the design conceived was more efficient than a digital model, due to the different capabilities of imaging a real object from an image. It allowed to identify concerns about the use and fabrication of the wheelchair before manufacturing an aluminum prototype, with a much lower investment, both in time and money.

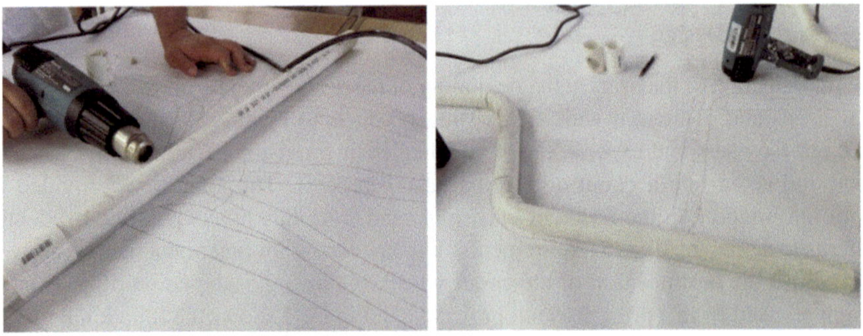

(a) (b)

Fig. 2. Construction of a rudimentary prototype of the wheelchair using PVC tubing, heat guns and technical drawings of the frame. Detail (a) shows the heating of the PVC tubing and detail (b) shows the bent tubing over the technical drawing of the frame.

The final design of the all-terrain wheelchair incorporates the use of AZ31 magnesium alloy and the new manufacturing processes and surface treatments to achieve a light product. In addition, it proposes a new design for the wheelchair frame that allows adaptation to the anthropometric and anatomical conditions of users, with a formal configuration that ensures an adequate structural behavior. The frame features a frontal caster wheel without obstacles to the transfer of the user and positioning of their legs and foots. Finally, it incorporates a manual lever-propelled mechanism and the use of three wheels, to reduce the physical effort needed to move and maneuver the wheelchair.

(a) (b)

Fig. 3. The PVC tubing prototype and the digital model of the wheelchair (a) used in the focus group session with users and the project team (b) to discuss the design of the wheelchair.

3.3 Project Impact

With the development of the project, a local appropriation of knowledge about the use of magnesium alloys as a structural material was achieved, with potential applications in products aimed at mobility. Although the project is responsible for the design of a wheelchair, this knowledge can be extended to other assistive technologies.

On the other hand, the participation of multiple institutions in the project, including the Universidad de Antioquia, with its Centro de Investigación, Innovación y Desarrollo de Materiales, CIDEMAT, the Universidad Pontificia Bolivariana with its Grupo de Investigación de Estudios en Diseño, GED, and the Universidad de Medellín with its Grupo de Investigación de Materiales de Impacto, MAT&MPAC, fosters relationships between three research groups that provide diverse knowledge to create new developments with high added value. With the close collaboration of these three groups, it is possible to continue developing projects aimed to integrate knowledge from engineering and industrial design. This integration allows the scientific knowledge developed to be transferred to final products that can be used by the population and solve needs from the local context.

In addition, this project is aimed at resolving a lack of products aimed at the rural population in Colombia. A wheelchair suitable for this environment will provide greater independence to its user, since it will require less assistance to overcome the obstacles found in irregular terrain. By improving efficiency in mobility, the risks of upper limbs injuries due to excessive efforts will be reduced. This will help the user to enjoy a higher level of independence for a longer time by maintaining the good health of their hands, arms, and shoulders, which are their main tool to develop the activities of daily life. In this way, new technologies will contribute to improving the quality of life of people with disabilities.

4 Conclusion

The above shows the importance of collaboration between professionals and researchers of diverse disciplines, users and manufacturers in the process of designing a product that aims to solve a problem through the incorporation of a new technology. In the case of this wheelchair, satisfying the needs of the users was only possible with a design that takes advantage of the properties of the new material and at the same time, meets the mobility, stability and comfort requirements.

Achieving this result required a design process where, in addition to the collaboration of the interdisciplinary team, users were involved in the definition of design requirements and the evaluation of possible solutions. With a human-centered design approach, tools such as interviews, focus groups, prototyping and testing were used in several stages of the design process. Those tools were crucial for the deep understanding of the particular needs of the users in relation to their daily activities and the characteristics of the local context. Also were useful to evaluate the design decisions and their impact in the usability of the wheelchair. On the other hand, collaboration with manufacturers who were willing to adapt their processes for the use of magnesium alloys, contributed to make viable the construction of the prototypes and latter to propose a productive model for the implementation of the product in the market.

Acknowledgments. The authors are grateful to "Departamento Administrativo de Ciencia, Tecnología e Innovación (COLCIENCIAS)" of the Colombian Government (Grant 1115745-58322), "Centro de Investigación para el Desarrollo y la Innovación (CIDI)" from the Universidad Pontificia Bolivariana through UPB-Innova Rad: 747B-03/17-35, Universidad de Antioquia, SENA and Universidad de Medellín.

Funding. Departamento Administrativo de Ciencia, Tecnología e Innovación en Colombia (COLCIENCIAS), grant 1115745-58322.

References

1. Tidd, J., Bessant, J.R.: Managing Innovation Integrating Technological. Market and Organizational Change. Wiley, Hoboken (2013)
2. You, S., Huang, Y., Kainer, K.U., Hort, N.: Recent research and developments on wrought magnesium alloys. J. Magnes. Alloy. **5**, 239–253 (2017)
3. Berrio-betancur, L.F., Echeverry-rendón, M., Correa-bedoya, E., Zuleta, A.A., Robledo-restrepo, S.M., Castaño-gonzalez, J.G.: Development of the magnesium alloy industry in Colombia - an opportunity. DYNA **84**, 55–64 (2017)
4. Polmear, I., StJohn, D., Nie, J.-F., Qian, M.: Light Alloys: Metalurgy of the Light Metals. Butterworth-Heinemann, Oxford (2017)
5. Trujillo Suárez, M., Aguilar, J.J., Neira, C.: Los métodos más característicos del diseño centrado en el usuario -DCU-, adaptados para el desarrollo de productos materiales. Iconofacto **12**, 215–236 (2016)
6. Sevilla Cadavid, G.A., Valencia-Escobar, A., Velázquez Gómez, J.: Estudio de caso sobre el diseño de Proaid E: silla de ruedas neurológica para niños. Iconofacto **13**, 128–143 (2017)
7. Ulrich, K.T., Eppinger, S.D.: Product Design and Development. McGraw-Hill Education, New York (2012)
8. Brown, T.: Design Thinking. Harv. Bus. Rev. **86**, 85–92 (2008)
9. International Organization for Standardization: International Standard ISO 7176-8 (2015)
10. Toft, Y., Howard, P., Jorgensen, D.: Human-centred engineers - a model for holistic interdisciplinary communication and professional practice. Int. J. Ind. Ergon. **31**, 195–202 (2003)
11. Cadavid, G.S., Valencia-Escobar, A., Gómez, J.V.: Proaid E. low cost neurological wheelchair design. In: Rebelo, F., Soares, M. (eds.) Advances in Ergonomics in Design. Advances in Intelligent Systems and Computing, pp. 677–687. Springer, Cham (2016)

Importance of Involving Children in Designing Recycling Facilities: A Case Study of Hong Kong Children Collecting and Recycling Recyclables

Kin Wai Michael Siu[1(✉)], Yi Lin Wong[1], and Mei Seung Lam[2]

[1] School of Design, The Hong Kong Polytechnic University, Hunghom, Kowloon, Hong Kong
{m.siu,yi-lin.wong}@polyu.edu.hk

[2] Department of Early Childhood Education, The Education University of Hong Kong, Tai Po, New Territories, Hong Kong
mlam@ied.edu.hk

Abstract. Given the importance of recycling to ensuring an environmentally sustainable future, it is imperative that children be taught to recycle at home. Yet in major metropolises such as Hong Kong, household recycling is still uncommon and relevant knowledge is not passed between generations. Based on a Hong Kong case study, the study reveals the importance of involving children through teaching children to design and make their own recyclable storage bags. The findings show that children who attended a lecture about environmental protection and/or workshops on making recyclable storage bags did not initially have recycling habits, but they were more eager to recycle afterwards. More involvement in designing recycling gadgets and facilities would result higher participation and motivation. The study thus provides strong evidence of the effect of children's involvement in recycling activities through designing and making recycling-related items.

Keywords: Recycling · Participatory design · Design and making · Children

1 Introduction

Environmental sustainability is one of the pressing issues that all nations in the world are considering and developing appropriate and efficient policies to maintain. Carbon dioxide emission is the hot topic that the 24th Conference of Parties (COP24) to the United Nations Framework Convention on Climate Change had worked towards agreements [1]. International agreement, government policies and educational programs have been achieved and adopted to help our planet to be more sustainable. However, most of these measures had only focused on behavior and attitudes of adults

C. S. Shin (Ed.): AHFE 2019, AISC 968, pp. 83–90, 2020.
https://doi.org/10.1007/978-3-030-20470-9_10

but not children. It can be argued that the current policies and programs should put more emphasis on children, the next owner of our planet.

Educational programs for children inevitably should start at home, children should be taught by their caretakers in daily basis in order develop environmental-friendly habits. Among all simple sustainability concepts, recycling is the easiest for children to understand and learn. However, recycling is an uncommon habit for Hong Kong households. The adults in the older generation seldom teach their children to recycle recyclables at home. Some of them may have done so but the purpose is not to be environmental-friendly but to sell the recyclables to private recycling stores to earn money. Children may learn the concept of recycling at school; however, the education is very limited, and they are not motivated to participate in recycling activities without incentives.

The paper argues that in order to facilitate and motivate children to develop habits of recycling, it is essential to involve children in not only the recycling activities but also the design process of the recycling gadgets and facilities. Experiencing the design process with hands-on activities allows children to develop a more long-lasting recycling habit [2]. Taking a case study in Hong Kong, the importance of involving children through teaching children to design make their own recyclable storage bags is studied. Including children in the design process is able to change the passive role to active participating role in recycling [3]. The findings show that although the participating children did not have a habit to recycle, they were more eager to do so after making their own bags. It was also more conceivable for them to use the public recycling facilities. More involvement in designing recycling gadgets and facilities would result higher participation and motivation. This study provides a strong basis on the effect of children's involvement in re-cycling activities through designing and making recycling-related items.

2 Method

2.1 Participants

Ninety-three children (33 boys and 58 girls) aged 9–11 were involved in the study. They are elementary school children who were selected randomly by their teachers from 26 elementary schools. They agreed to join a lecture, a hands-on workshop to make a recyclable storage bags, and use the bag at home to store recyclables. All of them did not have hands-on experience on making bags; however, they had Art lessons in their school learning some basic drawing techniques.

2.2 Instruments

A survey was conducted to obtain feedback after attending the lecture and children's behavior towards recycling. A set of questions were designed, and the children had to answer either yes-no questions or rate on a 4-Likert scale from very satisfactory/very often (4) to not satisfactory/never (1). They were also asked to prove comments about the improvement of the lecture. The two-page questionnaire was distributed to all the children attending the lecture.

2.3 Procedures

The kick-start event of the study is a lecture about environment protection. Government officers were invited to give the children a talk, and the lecture was about an hour and a half. Other research studies have also showed that adequate knowledge regarding recycling is one of the key factors for people to participate in recycling activities [4]. Questionnaires were distributed to all the children after the lecture.

After few weeks of the lecture, several workshops were conducted to teach the children making recyclable storage bags. Recycling materials such as banners were given to the children, and the children could design and make their storage bags with the facilitators in the workshop. Each of the children had to make a bag, and was asked to bring the bag home and use them to store recyclables at home. The children were also taught to bring the storage bags with the recyclables to recycling point nearest to their home. The hands-on workshops allow children to develop recycling habits more efficiently [2].

The children had recorded the number of recyclables they collected and stored everyday, and took photos to track the record. They also took photos of how they used the storage bags and the procedures of recycling. A photo diary of using the bags were compiled.

3 Findings

3.1 Survey

Seventy-seven children had attended the lecture, and 60 questionnaires were collected after the lecture. The return rate is about 77.9%. The findings of the survey showed that the children were satisfactory with the arrangement and the content of the lecture (mean = 3.32 and 3.28, SD = 0.79 and 0.75 respectively). They were also very attentive and motivated to response to the lecturers' questions. Almost all of them found that after attending the lecture they have gained more knowledge about energy conservation and waste reduction, and plastic recycling (96.7%). The findings also revealed that the children sometimes separated and recycled waste (mean = 3.05, SD = 0.76). Forty-nine of them (81.7%) also expressed that they would change the habit of handling waste after attending the lecture. Figures 1 and 2 show the findings of the survey.

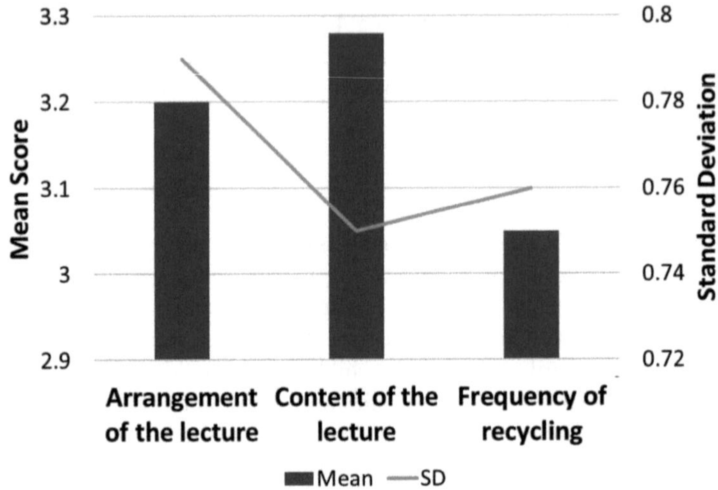

Fig. 1. Mean score and the standard deviation of the children's satisfaction level of the arrangement and the content of the lecture (4 = very satisfactory, 1 = not satisfactory), and the frequency of recycling (4 = very often, 1 = never).

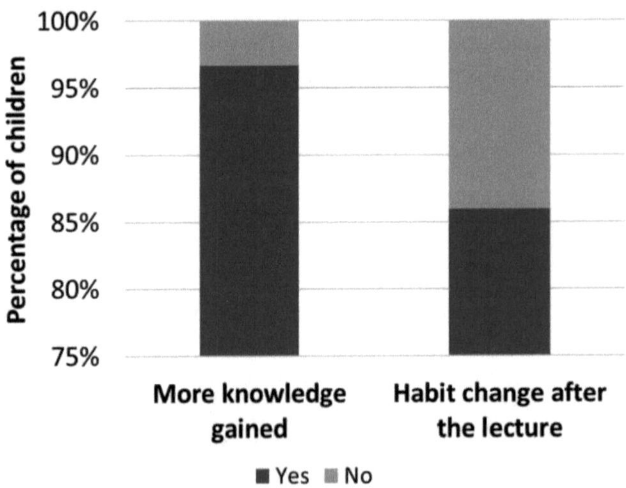

Fig. 2. Percentage of the children who had gained more knowledge and would change their recycling habits after attending the lecture.

3.2 Workshops

Seventy-nine children attended the workshops to make recyclable storage bags. Not all of them had attended the lecture. The children were divided into five groups, and the facilitator-child ratio was about 1:4. The children were briefed on the process of making the bags before they started. The facilitators helped the children to make the bags and answered any questions they raised during the process. Tools such as rubber mallets, cutting knifes, scissors, and punches were provided to the children. The children used materials such as recycled banners, double-sided tape, snap fasteners and eyelets to make their bags. Figure 3 shows the construction process in the workshops.

Fig. 3. Making process of the recyclable storage bags in the workshops.

3.3 Photo Diary

After making the storage bags in the workshops, the children were asked to take the bags home and hang them in their kitchens or next to their general waste bins. They started using the bags to collect recyclables and were asked to take photos to record the usage. Some of the children also recorded how they used the bags to bring recyclables to the recycling points near their living areas. Figure 4 shows some of the photos that the children and their family members took to record the process.

Few of the children were unable to understand how recyclable storage bags were used and the meaning of the project. They used the bags for shopping and to carry food home from the supermarket. Figure 5 shows one photo a child provided to show how the bag was used for shopping.

Fig. 4. Photos showing the recyclables the children collected and how they used the recyclable storage bags in their daily lives.

Fig. 5. Use of a recyclable storage bag for shopping.

4 Children's Involvement and Value Creation

The children were actively involved in the design process, from understanding the issues to using their designs in daily life in the final stage. Through this design process, they became more willing to use their own recyclable storage bags than other kinds of storage bags. The children became more actively involved in recycling activities than before. Their sense of belonging to the bags motivated them to engage actively with minimal support and reminders from adults [5, 6]. Due to their increased participation and motivation, the children were encouraged to use public facilities related to recycling, such as recycling bins. This was a good start for them to develop recycling habits and learn to be environmentally friendly.

Adequate understanding of the activity was important for the children to proceed to the design and construction stages and then use the bags they made. Without a thorough understanding, some children misused what they created and designed. Regardless of how good their designs were, the value of the process was lost. Attending the lecture before the workshops was important in this sense.

This design process is also a value creation process. Through active participating and the design-and-make and hands-on activity, children themselves had created values for recycling in their daily lives. They have also developed positive attitudes towards recycling, which is considered as an important factor for making sense of the recycling participation [7]. As the entire project had lasted for more than a month, it is believed that the value created would stay in their attitude towards recycling. In addition, as the children had designed and made the bags in a group with their schoolmates, some of them would participate in the recycling activities using the bag in a group as well. In this process, children's behavior and recycling habits in the same group were similar. The value may also be spread to other persons whom the children contact with in daily life or in school.

5 Conclusions

The importance of the abovementioned case study of Hong Kong is not only regarding its academic value in research but also its impact by its applied research outputs. The designs generated by the children have brought high degree of influence to their recycling practice, i.e., not only the time right after the study, but also up to the preparation of this paper. The findings of the case study of Hong Kong reveals the importance of educating children about recycling by teaching them to design and make their own recyclable storage bags. It also indicates the importance of teaching children about recycling before the bag-making process. A lecture can help children make sense of what they should do after the bags are made. Although in this case a few children were unable to understand the project's objective, that should not suggest the study failed; such children still gained experience with using recyclables in their daily lives. Positive attitudes toward recycling and realization of the value of the process were created in all participating children, and should spread to people they know in the near future.

Acknowledgements. The authors acknowledge the support provided by the Alumni Engagement Innovation Fund for the data collection and paper preparation. They also thank The Hong Kong Polytechnic University and The Education University of Hong Kong for data analysis support. The Eric C. Yim Endowed Professorship provided partial support for the data analysis and preparation of this paper.

References

1. Vox. https://www.vox.com/energy-and-environment/2018/12/14/18139402/cop24-climate-change-katowice-poland
2. Beard, C., Wilson, J.P.: Experiential Learning: A Handbook of Best Practice for Educators and Trainers. Kogan Page, London (2006)
3. Lozanovska, M., Xu, L.: Children and university architecture students working together: a pedagogical model of children's participation in architectural design. Codesign **9**, 209–229 (2013)
4. Budak, F., Oguz, B.: Household participation in recycling programs: a case study from Turkey. J. Environ. Biol. **29**, 923–927 (2008)
5. Siu, K.W.M., Kwok, J.Y.C.: Collective and democratic creativity: participatory research and design. Korean J. Think. Probl. Solving **14**, 11–27 (2004)
6. Klahr, D., Triona, L.M., Williams, C.: Hands on what? The relative effectiveness of physical versus virtual materials in an engineering design project by middle school children. J. Res. Sci. Teach. **44**, 183–203 (2007)
7. Vicente, P., Reis, E.: Factors influencing household's participation in recycling. Waste Manag. Res. **26**, 140–146 (2008)

The Role of Affective Design in Sustainability

Guillermo Juarez(✉)

Carleton University, 1125 Colonel by Dr, Ottawa, ON K1S 5B6, Canada
Guillermo.Juarez@Carleton.ca

Abstract. As the changes in global climate become more evident, *sustainable design*, which is also known as green design, environment-friendly design or eco-design has become a major topic in many of the most recent forums and conferences. *Sustainable design* is defined by Zande (2010) as "a design philosophy that seeks to maximize the quality of the built environment, while minimizing or eliminating negative impact to the natural environment" [18]. In the field of architecture, the value of sustainability has also been highlighted, as it is considered a useful tool that can help to overcome the challenges of sustainable development. Nevertheless, it is common to associate this concept with sacrifice when addressed in practice, as stated by Fiore, Phillips and Sellers (2014) "society is asked to sacrifice quality of life to achieve some sustainable goal" [8]. In order to overcome this attitude, a new approach has been generated in the last decade. This new approach named *hedonistic sustainability* by Bjarke Ingles (2011), aims to address the limited scope of sustainable design by bridging two design methodologies- hedonistic (affective design) and sustainable design.

Keywords: Human factors · Sustainability · Sustainable design · Affective design · Hedonomics · Hedonistic sustainability · Ergonomics

1 Introduction

More attention has been drawn to affective design in recent decades, however it is not a new concept. This concept was first highlighted by Kansei Engineering (KE) in the early 1970's by Professor Misuto Nagamachi [1]. Afterwards, the concepts of hedonomics and hedonistic design were introduced in the 2000's in the domains of human factors and ergonomics (HF/E). To underpin the grounds for the discussion framed further in this article, it is important to define affect (in design) as "consumers' psychological responses to the perceptual design details of the product" [3]. Affect has been subject of study in different fields, such as psychology and HCI. Research in these fields (psychology and HCI) report that evidence indicates, (affect) is a major actor in consumers' decision-making process.

This article aims to answer the question: how can affective design help to promote sustainability? The answer to this question unfolds in two parts: first, the paper summarizes sustainable and affective design. The literature review draws from databases and journals in the design field and breaks down the origins of affective (design) and sustainable design to paint a picture of their current state. Secondly, an argument is presented for the need to bridge affective design principles and sustainable design

C. S. Shin (Ed.): AHFE 2019, AISC 968, pp. 91–95, 2020.
https://doi.org/10.1007/978-3-030-20470-9_11

philosophy, to increase the acceptance of the latter. It concludes with the possible development of a model, which holistically integrates these design approaches, not only to improve users' well-being, but also positively impact the environment.

2 Background

2.1 Sustainable Design

Since the 1980's sustainability became a widely recognized concept - the importance of which has been stressed in several fields. In addition to designers, it has also drawn the attention of institutions like the United Nations (UN) and led to the emergence of World Wildlife Foundation (WWF). This organization states a clear call of action: "we can no longer ignore the impact of current unsustainable production models and wasteful lifestyles" [17]. These concerns caused the United Nations to appoint a commission. As Zande (2010) states, it was set "to propose ways to improve human well-being by studying the world's environmental problems and proposing a global agenda" [18]. Global concerns grew about the evident unsustainable path carved by almost a century of unmeasured industrial development.

Throughout the following decades, sustainable design has been approached from a variety of perspectives, resulting in a richer terminology that includes terms as: "ecological design, green design and architecture, eco-effective design, and holistic and environmentally friendly design" [15]. This tendency is understood as a reaction to over 200 years of progress in making life more comfortable, that lead to unintended consequences [18]. Furthermore, it is important for the purpose of this article, to establish the definition of what sustainable design is, as stated by Zande (2010) "sustainable design is a design philosophy that seeks to maximize the quality of the built environment, while minimizing or eliminating negative impact to the natural environment." [18]. This brief summary offers a historic background of sustainable design, with the aim of illustrating a panoramic perspective to better understand the emergence and evolution of sustainable design, up to its current state.

2.2 Affective Design

In recent decades, affective design has been acknowledged to be a powerful tool in the design field. However, affect in design is not a recent discovery, it has been studied since the 1960's and in the 1980's one of the most important models was formulated by Nagamachi, as stated by Levy (2013) "the term kansei engineering (KE) itself was used for the first time in 1986 by Yamamoto, president of Mazda Automotive" [12]. In this fashion, KE aims "at translating user's feelings into concrete product parameters." [12]. Therefore, as reported by Levy (2013) the first and possibly best-known example of applied KE is the 1995 Mazda Miata (MX-5), developed by using the KE model, the resulting product became a market success. This led to the further developing of KE within the car industry and rapidly adopted by different domains, ranging from textiles to the extent of cosmetics [12].

It is worth noting that affect has been addressed by a set of different fields, which ended up producing a rich terminology, including emotional design, product semantics, affective design, affective engineering and engineering aesthetics, name a few. Thus, it is important to provide common language, grouping the mentioned terms under the affective design umbrella. This paper uses the definition formulated by Demirbilek and Sener (2003) "the consumer's psychological response to the semiotic content of the product." [5] supporting the communication that takes place between the consumer and the product.

2.3 Semantics

Krippendorf and Butter (1984) define semantics "as the study of symbolic qualities of man-made shapes, in the cognitive and social context of their use" [11] it is then sustained by the user's respond to a product at a cognitive level, that product semantics studies the symbolic qualities that evoke these responses. The importance of this hidden language is underlined by Demirbilek and Sener (2003) "intentionally or not, all manufactured products make a statement through shape, form, colour, texture, etc." [5]. This statement offers an opportunity for a silent monologue, in which objects are given the chance to communicate through its design (shape, form, colour, texture, etc.) and function. However, this monologue may remain unheard if the message is unclear the possibility in reaching the targeted audience is low or even nil. This is not an easy task to perform since the "consumers' affective preferences are often a black box and cannot be precisely described" [13]. Although, shedding light on these processes is often difficult, the existence of this relationship has been largely acknowledge as reported by Khalid (2006) "The correlation between affect and cognition is well documented" [10] also mentioned by Ellsworth and Scherer, 2003 "emotional reactions typically involve extensive cognitive processing" [10] in the context where "cognition is used to interpret, make sense of, and understand user experience" [10] this serves to highlight the role of semantics in the decision-making process of the user.

2.4 Hedonistic Sustainability

Hedonistic sustainability, a term that has been produced in the field of architecture, has brought a positive attitude toward sustainability, as stated by Mohtadi (2016) "global innovations have reimagined sustainability not only in terms of the environment, but also in terms of the quality of life and prosperity of its inhabitants" [16]. This concept builds on the concept of Hedonomics, which was developed in recent decades in the fields of human factors and ergonomics (HF/E), defined by Hancock (2003) as "that branch of science and design devoted to the promotion of pleasurable human-technology interaction" [9] and "derived from two Greek roots: hedon(e), meaning joy or pleasure, and nomos, meaning law-like or collective." [9]. However, we must denote that pleasure is not referred in this article as an interchangeable term to affect. This recent approach has been interchanged in different domains in recent decades, noting the importance that positive experiences have in the many branches of design. Therefore, hedonistic sustainability "a term popularized by Danish architect Bjarke

Ingels" [16] emphasizes the integration of sustainable development with positive outcomes.

These interactions have been framed by Mohtadi, (2016) and Fiore, Phillips and Sellers, (2014), where they examine the example of the Amager Bakke waste incinerator situated in the capital city of Copenhagen, designed by Bjarke Ingles Group (BIG). This case is reported to have changed the general perception of waste management, that very often can bring a positive experience to the community, by designing a "waste-to-energy plant that also serves as a human-made ski slope, a feedback mechanism/public art installation, and an integral part of a plan for Copenhagen to become carbon neutral by 2025" [7]. On the other hand, product design has yet the task to frame cases where hedonistic sustainability can be observed.

3 Discussion

This article has presented a frame for the pertinent integration of affective design in the sustainable design model. Based on the current climate reports, sustainable design is in the need of immediate action. The introduction and further dissemination of this practice in order to sustainable design becomes just design, calls for a set of tools that helps it propagate in a successful manner. This article argues for further research in order to integrate affective design models within sustainable design, having proved the role of affect as a major actor in the decision-making processes of the users. Nevertheless, this article does not diminish the importance of a user-centred approach. Instead, it argues for an integral approach that relies on the advantages that affective design can offer. Additionally, as stated by Fiore, Phillips and Sellers (2014) the need for "involving community participants, not as end users but as team contributors" [8], calling architecture practitioners for a co-design model, which already exists in design practice. We must take advantage of the already-proven tools, in order to be able to overcome future challenges. As other domains have already started noticing the resulting benefits of integral approaches, the outcomes have just started to appear and being studied.

4 Conclusion

Given these points, this paper presented a synthesized background of affective and sustainable design that underpins the existing relationship, which has been recently highlighted and discussed in the architecture field. This paper aims to answer the question: what is the role of affective design in sustainability? Arguing for the need to incorporate affective design methods into sustainable design, that has been recognized to turn the existing attitude that persists sacrifice must be made to obtain sustainability, into the positive concept of hedonistic sustainability. Drawing from the literature review, we suggest further research is needed to be able to progress toward a sustainable future development, which evokes affect in the users as a manner to propagate and accelerate the assimilation of sustainable design in consumers, underlining the need for immediate action.

References

1. Axelsson, J.R.C., Nagamachi, M.: Concepts, Methods and Tools in Kansei Engineering, May 2004. https://doi.org/10.1080/1463922021000049980
2. Basulto, D.: Designing a hedonistic and sustainable future. Big Think, 10 May 2011 (2011). http://bigthink.com/endless-innovation/designing-a-hedonistic-and-sustainable-future
3. Chang, M.-L., Lee, J.-H.: Symbiosis: creativity with affective response. Lecture Notes in Computer Science (Including Subseries Lecture Notes in Artificial Intelligence and Lecture Notes in Bioinformatics). LNAI, vol. 4562, pp. 32–41 (2007). http://www.scopus.com/inward/record.url?eid=2-s2.0-37249032473&partnerID=40&md5=8394aa924526825deff9eaac69c661b7
4. Chen, C.-W., Ugarte, R.C., Wu, C., Aghajan, H.: Discovering social interactions in real work environments. In Face and Gesture 2011, pp. 933–938 (2011). IEEE. https://doi.org/10.1109/FG.2011.5771376
5. Demirbilek, O., Sener, B.: Product design, semantics and emotional response. Ergonomics **46**(13–14), 1346–1360 (2003). https://doi.org/10.1080/00140130310001610874
6. Desmet, P.: A multilayered model of product emotions. Hell. J. Cardiol. **54**(4), 337–338 (2013). https://doi.org/10.2752/146069203789355480
7. Grose, T.: Urban ski slope to raise profile of Europe's waste- to-energy drive. National Geographic, August 2013. http://www.nationalgeographic.com
8. Fiore, S.M., Phillips, E., Sellers, B.C.: A transdisciplinary perspective on hedonomic sustainability design. Ergon. Des. **22**(2), 22–29 (2014). https://doi.org/10.1177/1064804613516762
9. Hancock, P.A., Pepe, A.A., Murphy, L.L.: Hedonomics: the power of positive and pleasurable ergonomics. Ergon. Des.: Q. Hum. Factors Appl. **13**, 8–14 (2005). https://doi.org/10.1177/106480460501300104
10. Khalid, H.M.: Embracing diversity in user needs for affective design. Appl. Ergon. **37** (4 SPEC. ISS.), 409–418 (2006). https://doi.org/10.1016/j.apergo.2006.04.005
11. Krippendorff, K., Butter, R.: Product semantics: exploring the symbolic qualities of form (March). Industrial Design 139–140, 10–13 (1984)
12. Lévy, P.: Beyond Kansei engineering: the emancipation of Kansei design. Int. J. Des. **7**(2), 83–94 (2013). https://doi.org/10.1029/2009JB006743
13. Lin, Y., Wei, C.: A hybrid consumer-oriented model for product affective design: an aspect of visual ergonomics, 17–29, April 2015 (2017). https://doi.org/10.1002/hfm.20403
14. Lokman, A.M., Kamaruddin, K.A.: Kansei affinity cluster for affective product design. In: Proceedings - 2010 International Conference on User Science and Engineering, i-USEr 2010, pp. 38–43 (2010). https://doi.org/10.1109/IUSER.2010.5716719
15. McDonough, W., Braungart, M.: Cradle to Cradle. North Point Press, New York
16. Mohtadi, T.: The complementarity of improving quality of life and reducing environmental footprints in urban spaces: the argument for "hedonistic sustainability." Cons.: J. Sustain. Dev. **16**(1), 14–28 (2016). https://doi.org/10.1073/pnas.0707749105
17. Grooten, M., Almond, R.E.A. (eds.): WWF. Living Planet Report - 2018: Aiming Higher. WWF, Gland, Switzerland (2018)
18. Zande, R.V.: Creating the urban village: teaching pre-service teachers about sustainable design in architecture and community planning. Int. J. Art Des. Educ. **29**(3), 321–329 (2010). https://doi.org/10.1111/j.1476-8070.2010.01659.x

Using Gamified Solutions in Pediatric Diabetes Self-management: A Literature Review

Maryam Attef(✉) and Eileen Harris

School of Industrial Design, Carleton University, Ottawa, Canada
Maryamattef@cmail.carleton.ca

Abstract. Gamification, the use of game elements and techniques, is increasingly being used in healthcare field. This paper reviews the literature on the application of design and technology in improving the quality of healthcare services using gamification, specifically on diabetes self-management. It proposes the use of game elements to help diabetic children better manage their lifelong condition. The paper concludes by presenting a design proposal for further research on the potential use of gamification to promote adherence to a recommended diabetic protocol.

Keywords: Chronic illness · Gamification · Gamification for healthcare · Interdisciplinary design · Pediatric diabetes · Self-management

1 Introduction

Diabetes is a major chronic disease that is growing globally at an epidemic rate [3]. Patients are either unable to produce insulin (type 1 diabetes or insulin-dependent diabetes) or produce some insulin, but their ability decreases over time (type 2 diabetes). If a diabetic patient (type 1 or type 2) fails to manage their blood glucose consistently, it can lead to severe consequences: blindness, kidney diseases, stroke, heart disease, and amputations.

A growing problem in healthcare is the lack of adherence to treatment [8]. One approach to addressing this issue is the use of gamification. The application of gamification in health and wellness contexts is used as a motivational tool to change behaviors related to self-management and monitoring, in chronic diseases such as diabetes.

Diabetes requires a high degree of adherence to medical treatment, which is often challenging for diabetics, especially children [10]. The use of gamification approach to promote better self-management for children with diabetes is a new approach in game designing. It requires users to incorporate their needs in the design system, taking into consideration each user's qualities, abilities and interest to help these children better manage with their lifelong condition. It is critical to find ways to help children engage in their self-management plans by helping them to see 'the big picture' of their chronic illness [10].

C. S. Shin (Ed.): AHFE 2019, AISC 968, pp. 96–102, 2020.
https://doi.org/10.1007/978-3-030-20470-9_12

2 Background

2.1 Gamification

The first documented use of the term "Gamification" can be dated back to 2008 [11]. Kim and Werbach (2016) define gamification as the "use of elements and techniques from game design in non-game contexts" [4]. The term gamification means work as play [11].

The gamification approach allows users to incorporate their requirements in the design system, taking into consideration each user's qualities, abilities and interest [2]. Gamification is interdisciplinary, providing a common language to enable specialists in several disciplines as stakeholders and end users to work together. It emphasizes the importance of involving all groups of stakeholders in the design process [9]. Gamification is a creative approach to problem-solving that starts with understanding the need of the people and ends with individual solutions to meet these needs [5].

2.2 Gamification in Healthcare

The increase in user engagement in the design process makes gamification useful in different domains such as healthcare practice [10]. While gamification is still an evolving term for healthcare practice, the evidence for its benefits is mounting [11]. Gamification in the healthcare transforms ordinary therapeutic experiences into more gratifying ones that help patients adhere better to the treatment process [9]. The main purpose of the use of gamification is to engage learners/players, enhance their motivation and influence a positive behavioral change within a well-designed environment [9].

Gamification in healthcare serves as an interactive communication process between patients and healthcare providers using problem-solving and shared decision-making strategies [3]. Gamification uses "design thinking, human-centered approach, in developing new solutions … focuses on rapid prototyping, which means turning ideas into actual products that are tested, iterated and refined based on user feedback" [6, p. 2]. A well-designed game can improve learning, improve skill development, self-concepts, emotions, social relationships, social support, and enhance motivation to promote healthy behaviours and outcomes [5].

Gamification design depends on multiple stakeholders and different levels of motivation to achieve the therapeutic objectives [9]. Examples of these objectives include the motivation to achieve long-term behavior change, the motivation to engage with the treatment, or accomplish of short-term therapy objectives along with the motivation to use gamification within the therapeutic procedure [9]. There are, however, challenges associated with the application of gamification for healthcare practice. It is difficult to find an appropriate to balance between the health goal and game appeal when developing gamified intervention [9], which could be a result of ineffectively transferring the knowledge to design games for healthcare problems [10].

There are various methods from the field of design (game design and user-centered design) that could be useful in helping to address the challenges encountered at each stage of the gamification design process. While gamification is being used as method to help improve healthcare outcomes, Wen et al. (2015) note three main issues in the use of gamification in healthcare. First, the quality of existing work has a lot of unsupported arguments from academics and practitioners. Second, the involved process of gamification design for healthcare is still unclear in terms of knowing which process of the design is the most popular in the field. Last, lack of information on the most widely implemented elements of gamification design for healthcare.

Furthermore, the reviewed literature revealed that it is difficult to design for healthcare practices and to find a balance between the health goal and game appeal.

If the implementation of the approach is poor, it will result in failing the gamified application to meet its goals due to poor design.

3 Gamified Solutions in Diabetes Self-management

Living with diabetes demands an understanding of the condition, which requires long term treatment by the diabetic patient including medication taking, lifestyle changes and self-monitoring [3]. Self-management starts with monitoring blood glucose levels, food intake, taking medication on time and exercise [1]. The approach of using information and technology with association of glucose self-monitoring systems reveals its efficiency in enabling individualized evaluation of sugar profiles in connection with metabolic intervention [7]. There are four elements of gamification design that are important for improving levels of hemoglobin A1c: communication, patient-generated health data, education, and feedback [3].

The development of new technologies and design models enhances diabetes self-management knowledge and skills and may lead to reduced diabetes-related urgent care and emergency clinical utilization. This helps lower the cost of care and improves the patient's quality of life. The use of modern technology and the developing the healthcare system will change the relationship between patients and healthcare providers and become more of partnership [7].

While there are existing diabetes management technologies, they do not follow medical guidelines or the incorporation of clinical practices recommended by diabetes professionals [11]. Some of them have usability issues, or focus narrowly on insulin dosage suggestions, recording medications, diet, and weight management. Of particular concern is that the tools that do exist are not available to pediatric diabetes.

Patients are drained by the number of repetitive daily tasks they have to perform such as taking medication, taking notes of their food intake, their exercise [1]. The use of gamified tools can provide assistance, allowing patients to spend less money and save more time on their self-management processes [1]. Gamification applies a positive reinforcement which can be developed through the use of rewards [1]. Rewarding patients for taking their medication is more effective than punishing them if they do not take their medication regularly [1].

Gamification elements stimulate internal and external motives toward boosting self-esteem when designing for the right context [1, 11]. There are guidelines for using these elements to engage and motivate users in the best way. The following elements are in order of their relevance by gamification studies [11, p. 120]:

1. Points: It is the most rewarding system in the most implemented elements of the gamification in the healthcare practice. It is in the form of scores or points on the performance or completion of a certain task.
2. Badges: A reward that uses to present certain achievements.
3. Levels: A point-based reward that increases as the participant obtains a certain amount of points.
4. Voting: Each participant can evaluate behaviors of other players.
5. Quests: Using attractive story to simulate game environment.
6. Ranking: Board of achievements promotes competition between all participants through points or levels.
7. Betting: The most basic and popular form of gaming in human history. A wager on a certain task to motivate positive competitions amongst all participants.

4 Proposal for a Game to Promote Adherence to Treatment for Children with Diabetes

Below is a design proposal for the development of a game for children to increase adherence to treatment. It follows the four stages of a design process as proposed by Siriarya et al. (2018).

1. The Transfer Effect Identification Stage:
The first step in a design process is to determine the desired effect or outcomes that gamification design should achieve. In this case, the potential outcome is to improve patient's quality of life by improving their diabetes-related knowledge, increasing self-efficacy, increasing communication with family and friends, and increasing self-care behaviors. As well as, using discussion sessions with healthcare staff to determine the transfer effect, and the level of activity needed to achieve this impact within gamification.

2. The Real-World Context Exploration Stage:
The second step is to examine the real-life context by considering the following aspects:

- Characteristics of the targeted user – children with diabetes.
- The implementation of gamification in the therapeutic context – self-management knowledge and skills.
- Goals setting for the user context – to provide these children with basic information about diabetes self-management through the following approaches:

  a. Small steps – children will view the goal at the start of each level, starting from easy achievable steps to more challenging levels. This approach will help these children perform better and enjoy achieving their goals and will provide them enough challenge to match their current skills.
  b. Feeling of accomplishments – after a child completes the previous goal; he/she will feel encouraged to pursue new goals through rewarding systems (i.e., points). Children can use their reward points to unlock new game items and game levels.

3. The Gamification Concept Design Stage:
The third step is to identify the appropriate concept of gamification to implement the factors in the previous stages and to construct a core game loop. This could be done in different approaches involving all the groups of stakeholders in the design process. An appropriate method is brainstorming sessions in collaboration with game designers to generate ideas using gamification elements to motivate users. Another approach is to evaluate the designed product through questionnaire study, and comparing the users' and healthcare team's perspective of the gamification concepts.

4. The Iterative Development and User Testing Stage:
The fourth step is iterative developments of design through user testing and repeating processes to drive the core interaction between users and the game. The designed game is first used in a user testing session to investigate how users perceive the implementation of the game mechanics and investigate the issues related to general accessibility, which leads to structural improvements to the content structure and follows the design plan.

The game will involve children problem-solving and decision-making in a simulation of diabetes self-management. The aim is to balance food intake and insulin to keep the gameplay character's blood glucose within the normal range. It will require players to rehearse skills repeatedly until they win the game, with the help of a storytelling approach. For example, if the gameplay character skipped sugar, then the storytelling voice will remind the player that good actions can have a great impact on the child. This approach provides practice and shows cause and effect. The gameplay character will provide the players with basic information about diabetes self-management, and it will have demographical data which can relate to the level of the challenge.

The design will incorporate game elements to make healthcare more fun and attractive. Players will learn to make good food choices through a decision-making approach; where they have to keep the gameplay character's blood glucose in the normal range by eating healthy foods and using insulin. The game will engage players in detecting diabetes through fast-based exercise, through the developed Nintendo Wii game. The player will run with the gameplay character to and at the same time must take insulin and choose the right foods to remain healthy and strong. This approach will help children learn how insulin and food affect blood glucose levels as they observe the effects of their choices. Through online playing, the game will also bring other children with diabetes together to reduce isolation that diabetics often experience. It will also enable parents and clinicians to monitor the patient's condition over time. Since better outcomes come when communication occurs between patients and care providers, participants may need training in how to better construct appropriate communication (i.e. messages) to their healthcare team, engaging effective 2-way communication [3].

5 Discussion

The concept of gamification uses the need for recognition and instant feedback to promote change in behavior [1]. The design has to provide feedback which provides evidence of progress even in the face of failed attempts [10]. Feedback is one of the most important aspects of the success of gamification through rewards, leveling up and gaining points.

Using gamification design, it is possible to shift the "real world" experience towards a "game-like" experience [9]; by adding game elements by adding game elements to real world tasks [9] Players discover and learn through exploration and experimentation [10]. It rewards accomplishment of action toward assigned goal or task [11]. Gamification uses the desire for recognition and instant feedback to promote change in human behavior [1].

In gamification, developers can tune design elements to reach a preferred outcome [11]. For example, if the gameplay character skips a meal, it will become hypoglycemic which will result in trembling and weakness. If, however, the gameplay character eats regularly, it will maintain glycemic control and will result in feeling more energetic and increase the child's knowledge and skills gained in the game world and applied in the real world [10].

It is important to manage the user's expectations and control the influence of the real-world aspects in the game; through personalization, adaption and tailoring techniques on the gamification design [9].

6 Conclusion

Designing games for healthcare is an active segment of gamification with games that address different topics starting from lifestyle improvement to prevention, self-care and self-management [5]. The use of games can help change health behavior and improve player's diabetes self-management. Gamification can help children make meaningful connections between events through observing –which is the most common way in learning a new behaviour– the gameplay character engaging in a sequence of actions [10].

Gamification design may influence the behavior of diabetics and motivate them to better self-manage their condition [1]. Creating a game-like environment will promote the need for satisfaction, increase the likelihood of behaviors to become more internalized and integrated into one's sense of self, which will be maintained overtime: "I am a person with diabetes who self-manages" [10].

References

1. AlMarshedi, A., Wills, G.B., Ranchhod, A.: The wheel of Skur: a framework for gamifying diabetes self-management in Saudi Arabia. In: The 5th International Conference on Current and Future Trends of Information and Communication Technologies in Healthcare ICTH 2015, vol. 63, pp. 475–480 (2015)
2. Gardsten, C., Mortberg, C., Blomqvist, K.: Designing an ICT self-management service: suggestions from persons with type 2 diabetes. Health Technol. **7**, 197–206 (2017)
3. Greenwood, D.A., Gee, P.M., Fatkin, K.J., Peeples, M.: A systematic review of reviews evaluating technology-enabled diabetes self-management education and support. J. Diabetes Sci. Technol. **11**(5), 1015–1027 (2017)
4. Kim, T.W., Werbach, K.: More than just a game: ethical issues in gamification. Ethics Inf. Technol. **18**(2), 157–173 (2016)

5. Lieberman, D.A.: Video games for diabetes self-management: examples and design strategies. J. Diabetes Sci. Technol. **6**(4), 803–806 (2012)
6. Petersen, M., Hempler, N.F.: Development and testing of a mobile application to support diabetes self-management for people with newly diagnosed type 2 diabetes: a design thinking case study. BMC Med. Inform. Decis. Mak. **17**(91), 1–10 (2017)
7. Salzsieder, E., Augstein, P., Vogt, L., Kohnert, K.D., Heinke, P., Freyse, E.J., Ahmed, A., Metwali, Z., Salman, I., Attef, O.: Telemedicine-based KADIS® combined with CGMSTM has high potential for improving outpatient diabetes care. J. Diabetes Sci. Technol. **1**(4), 511–552 (2007)
8. Sardi, L., Idri, A., Fernández-Alemán, J.L.: A systematic review of gamification in e-Health. J. Biomed. Inform. **71**, 31–48 (2017)
9. Siriaraya, P., Visch, V., Dooren, M.M., Spijkerman, R.: Learnings and challenges in designing gamifications for mental healthcare: the case study of the ReadySetGoals application. In: 2018 10th International Conference on Virtual Worlds and Games for Serious Applications (VS-Games) (2018)
10. Thompson, D., Baranowski, T., Buday, R.: Conceptual model for the design of a serious video game promoting self-management among youth with type 1 diabetes. J. Diabetes Sci. Technol. **4**(3), 744–749 (2010)
11. Wen, C., Hsien, H., Huang, H.: A survey of gamification for healthcare system. In: 2015 International Conference on Orange Technologies (ICOT) (2015)

Application of 3D Scanning to Product Design

The Application of the Performance Hand Wear and Tools Innovation Approach: Road Cycling Gloves

Susan L. Sokolowski(✉) and Christel Thue Hoegsted

Sports Product Design, University of Oregon, 70 NW Couch Street, Portland, OR 97209, USA
ssokolow@uoregon.edu

Abstract. Over the last twenty years, the apparel industry has modernized the design process, by integrating digital design tools and technologies to collect and analyze body measurements, draft patterns, conduct fittings and manufacture. However, for performance hand wear and tool design, many of the methods used today are outdated and limit innovation opportunities. By integrating digital design tools from the apparel industry, there is an opportunity to develop products, to improve fit, performance and safety. This is especially important for users that rely on products for hand protection and accuracy, including, athletes, military personal, surgeons, construction and fire service workers. The following paper demonstrates how a performance hand wear and tools innovation approach developed by the author, was used by a graduate student to create a new glove for road cycling. Future work using the approach will aid in designing products for other users and help manufacturers improve safety and overall performance.

Keywords: Innovation approach · Hand wear · Tools · Design · Road cycling gloves

1 Introduction

Athletes wear and use equipment with their hands to acquire thermal comfort, dexterity, impact and abrasion protection, support and assist with grip [1]. There are a variety of hand wear and tool genres used for sport, including: mitts, mittens, gloves, wraps, braces sticks, paddles, racquets, etc. Compared to the sports apparel and footwear industries, the innovation approach used to create these types of products has not been updated to fully integrate digital tools.

The skills needed today to design innovative sports equipment are multi-faceted, including the knowledge of soft and hard goods design, research/design process methodology, materials science, manufacturing, engineering, human physiology, biomechanics, business and marketing [1]. Because academic programs in the U.S. are typically soft goods (apparel) or hard goods (industrial design/engineering) based, designers often do not learn academically all of the available technologies to create performance products for the hand, because the skills are pedagogical split between two different disciplines [1]. This paper will demonstrate how a performance hand wear

C. S. Shin (Ed.): AHFE 2019, AISC 968, pp. 105–111, 2020.
https://doi.org/10.1007/978-3-030-20470-9_13

and tools innovation approach developed by the author, was used by a graduate student to create a new glove for road cycling. The intention of the work is to inspire designers from different pedagogical backgrounds the possibilities available to innovate new hand wear and tool products. Future work using the approach will aid in designing hand wear and tools for other user groups and help manufacturers to improve overall product safety and overall performance.

2 Performance Hand Wear and Tools Innovation Approach

The Performance Hand Wear and Tools Innovation Approach was developed to assist designers with opportunities during the development process to create better products for users that need thermal comfort, dexterity, impact and abrasion protection, support and grip [2] (Fig. 1).

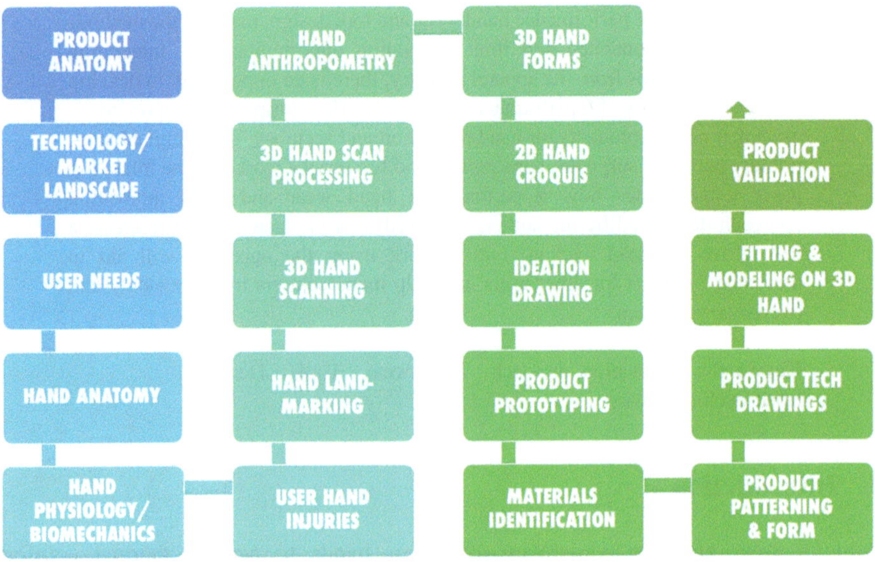

Fig. 1. Performance hand wear and tools innovation approach [2].

The approach includes three phases, with considerations for designers to utilize throughout the entire product creation process [2]. The considerations are recommendations, however the more of them that are used, the more robust the design work. The phases include: understanding the hand wear and tool project background, defining the user's 3D and 2D hand, and hand wear and tool product innovation [2].

2.1 Understanding the Hand Wear and Tool Project Background

The first phase of the approach includes understanding the background of the hand wear or tool design project [2]. The designer will conduct research about existing products the user may currently interface with, and how those products are built, the components involved, materials used, and method of make. The designer can identify the user's specific hand wear or tool needs, along with basic hand anatomy, physiology and biomechanics related to the product to be designed. In the traditional design process, this work would be conducted during problem identification [3].

2.2 Defining the User's 3D and 2D Hand

With the development of more accessible 3D scanning technologies, this phase of the approach allows designers to work more like apparel designers, by capturing, processing and defining the user's 3D hand to understand relevant anthropometric measurements, hand shapes, forms and 2D croquis to base designs upon [2].

2.3 Hand Wear and Tool Product Innovation

By defining the user's 3D and 2D hand, product creation can be based upon more accurate and relevant information [2]. With the innovation approach, designers are able to ideate, prototype, identify materials, pattern/model, develop tech drawings, fit and validate new ideas in the lab and on user, all while referencing the actual 3D and 2D user's hand [2].

3 Case Study: Road Cycling Glove Innovation

Students in a graduate level sports equipment design course innovated new sport gloves, to experiment with the Performance Hand Wear and Tools Innovation Approach. This paper will outline the results of a project, where a student designed a new road cycling glove.

3.1 Understanding the Hand Wear and Tool Project Background

For the course project, the student was assigned a base line road cycling glove from Pearl Izumi, to innovate upon. The glove was analyzed to understand how it was fabricated, along with the components and technologies used [4] (Fig. 2).

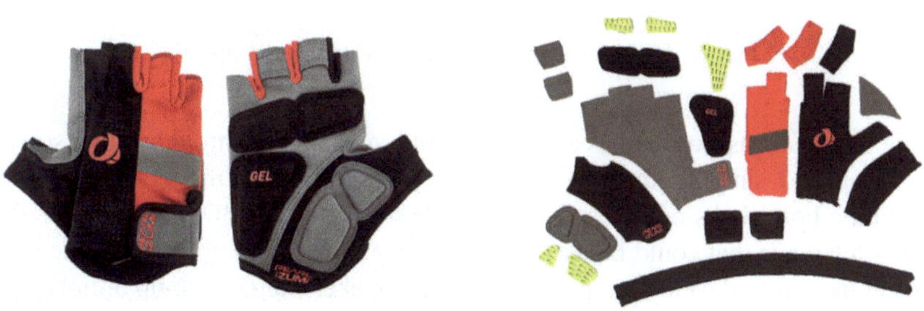

Fig. 2. Baseline product anatomy, components and technology identification.

The student also conducted research to understand the user's needs while cycling, by studying the three handlebar holds and grip patterns. Through painting the handle bars white, the grip contact surface was able to be transferred onto black disposable rubber gloves, where it identified where palm protection should be placed for the new product design [4] (Fig. 3).

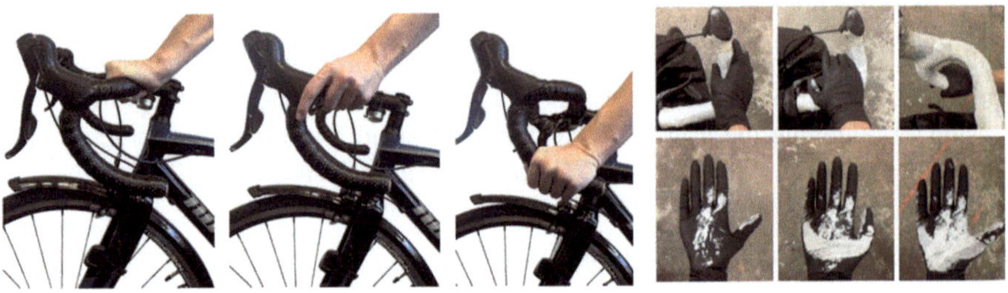

Fig. 3. Road cyclist hand needs.

3.2 Defining the User's 3D and 2D Hand

In order to accurately collect anthropometric data of the hand, landmarks were applied to the skin surface with a sticker and marking pen, in contrast color [5]. Once landmarked, the hand was 3D scanned. For the course, students used an Occipital Structure Sensor (mounted to an iPad) and Skanect software to capture the 3D hand scan (Fig. 4).

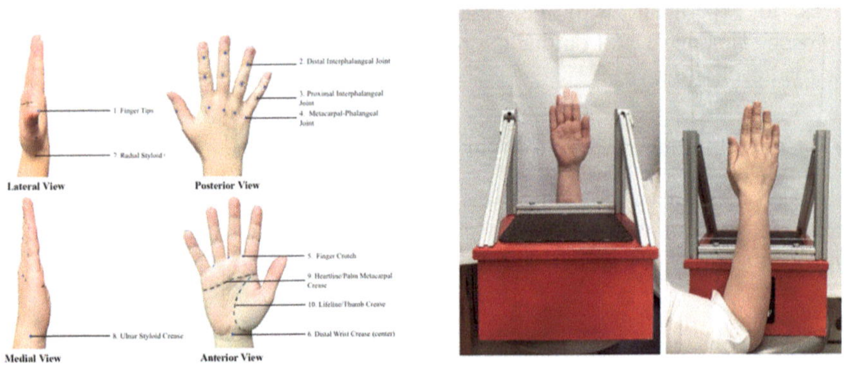

Fig. 4. Hand landmarking and 3D scanning set-up [5].

The 3D scan was exported as an OBJ file and manipulated through Rhino software, to clean-up and simplify the mesh [6]. Completed 3D scan model files were then printed on a Form Labs printer, using a flexible resin [4]. The 3D printouts provided a physical reference that could be used through the entirety of the project by the student to collect measures, develop croquis proportions for sketching, mock-up prototypes and conduct fittings (Figs. 5 and 6).

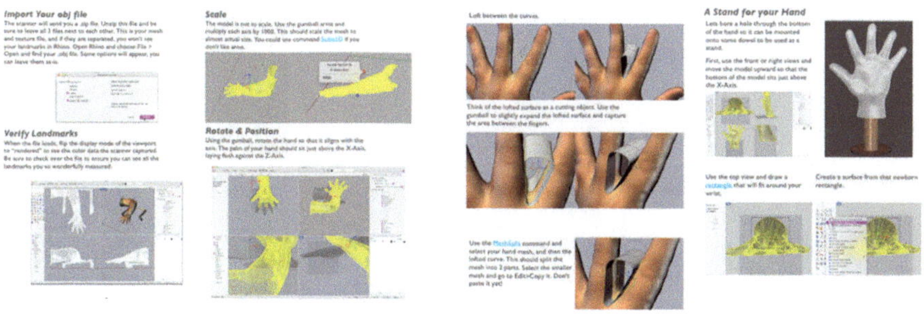

Fig. 5. 3D hand scan processing [6].

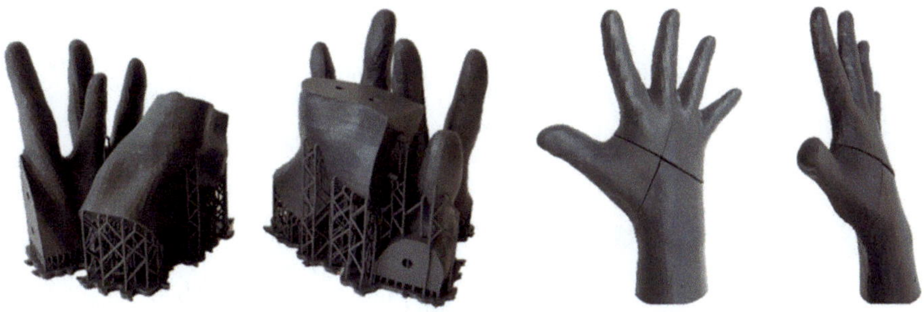

Fig. 6. 3D hand print and form.

3.3 Hand Wear and Tool Product Innovation

Once the user's 3D and 2D hand was defined, the student was able to commence the ideation process [4]. Pencil sketches were created, along with the development of rapid prototypes to address fit, mobility, thermal comfort and palm cushioning while gripping the handlebars of the bicycle [4] (Figs. 7, 8 and 9).

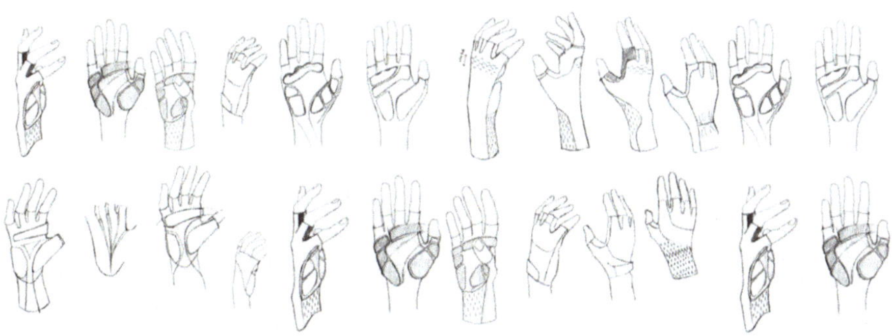

Fig. 7. Cycling glove sketch ideation.

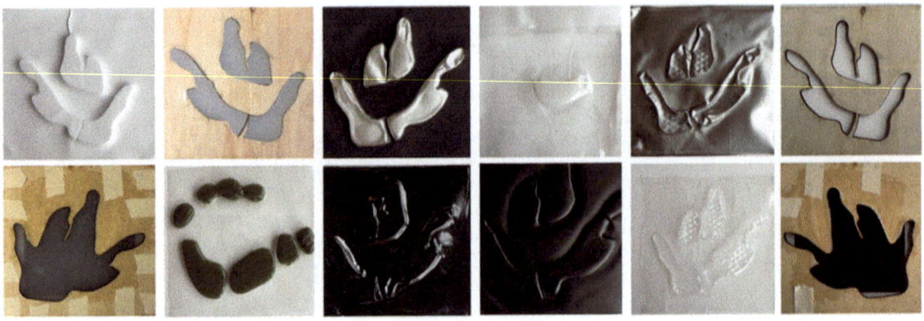

Fig. 8. Palm cushioning prototypes.

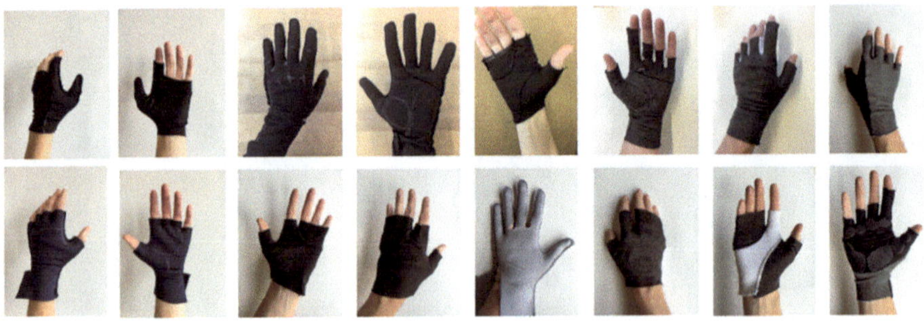

Fig. 9. Cycling glove prototypes.

Upon understanding the design direction, materials were identified and final patterns were drafted to fabricate the final glove concept [4] (Fig. 10).

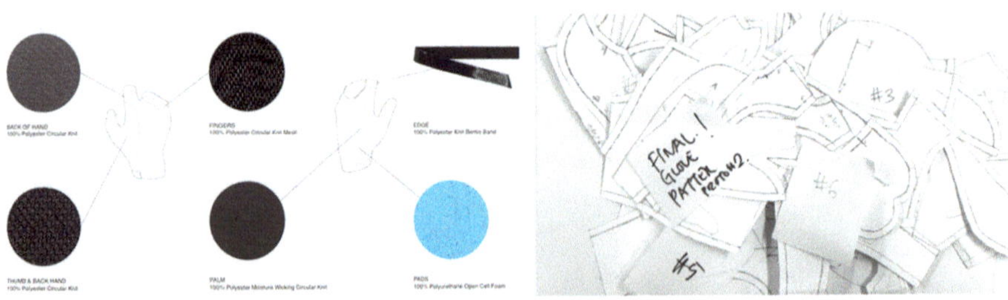

Fig. 10. Glove materials identification and patterning.

The final concept was fitted on the user's hand and evaluated in the three road cycling handlebar positions for grip comfort, fit, mobility and thermal comfort [4] (Fig. 11).

Fig. 11. Glove fitting and validation.

4 Summary and Conclusion

By using the Performance Hand Wear and Tools Innovation Approach, the graduate student was able to create a new, innovative road cycling glove. Each phase of the approach, provided the student with an abundance of information to understand the user's needs and their 3D and 2D hand, in order to ideate and prototype a new glove design. Through the success of the project, other students interested in hand wear and tool design could reference this process, to guide their work. Future work using the approach may also aid in designing hand wear and tools for other athletes, military personal, surgeons, construction and fire service workers. It is also intended to help manufacturers with their internal design processes to bolster innovative solutions to improve user safety and performance.

References

1. Sokolowski, S.L.: Sports industry meets academia: the pedagogical development of a MS degree program in sports product design. Special issue of Technology and Innovation, Journal of the National Academy of Inventors (accepted, to publish in 2019)
2. Sokolowski, S.L.: The development of a performance hand wear and tools innovation approach. New Approaches to Apparel Making in the 21st Century. Fashion and Textiles Journal (under review)
3. LaBat, K.L., Sokolowski, S.L.: A three-stage design process applied to an industry-university textile product design project. Cloth. Text. Res. J. **17**(1), 11–20 (1999)
4. Sokolowski, S.L., Hoegsted, C.T.: The Future of Performance Glove Design: Tools and Technologies for Designers. Outdoor Retailer Snow Show: Design and Trend Presentation (2019)
5. Sokolowski, S.L., Griffin, L., Carufel, R., Kim, N.: Drawing hands for glove design: does the data match-up? In: Chung, W., Shin, C. (eds.) Advances in Interdisciplinary Practice in Industrial Design. AHFE 2018. Advances in Interdisciplinary Practice in Industrial Design, vol. 790, pp 69–77. Springer, Cham (2018)
6. Sokolowski, S.L., Corr, G.: Pedagogical considerations for teaching modern performance glove design. In: Proceedings of the International Textiles and Apparel Association, Cleveland, Ohio, USA, 75 (2018)

3D Hand Scanning to Digital Draping for Glove Design

Linsey Griffin[1(✉)], Emily Seifert[1], Chris Curry[1], and Susan Sokolowski[2]

[1] College of Design, University of Minnesota, 1985 Buford Ave., 240 McNeal Hall, Saint Paul, MN 55108, USA
lgriffin@umn.edu

[2] Sports Product Design, University of Oregon, 70 NW Couch Street, Portland, OR 97209, USA

Abstract. New technologies such as 3D simulation and digital prototyping offer endless design opportunities for gloves. One potential application is digitally draping, which can be used to improve glove fit. A pilot study with one subject was conducted to examine the feasibility of this approach applied to glove design and fit. The subject completed a fit survey and then had their hand 3D scanned. A glove was digitally draped on the subject's hand scan in EFI Optitex®. The pressure map and mesh tool in Optitex were utilized to aid expert fit evaluations.

The findings of this pilot study show that the 3D glove simulation can provide a robust understanding of fit. Designers can use this visual and quantifiable information as a way to improve glove fit during early design iterations, thereby potentially sparing resources. Additionally, better fitting gloves could potentially be an effective way at combating common hand injuries.

Keywords: Glove design · Digital draping · 3D hand scans · Design process

1 Introduction

Workers in various industrial domains rely on gloves to protect their hands from physical harm, as well as act as a barrier to the transmission of microorganisms [1]. While gloves may prevent the worker from experiencing either of these harmful conditions, poor fitting gloves can imperil the worker's well-being and ability to perform their job adequately. When a worker wears poor fitting gloves, they may take compensatory strategies, such as overusing their shoulder, for example [2]. Such repetitive movements could lead to an injury [3]. It was reported that in 2014, nearly 200,000 workers in the United States had wrist and hand injuries that resulted in them missing multiple days of work [4]. Of those 200,000 workers, 30% of them reported wearing gloves.

In addition to injury concerns, poor fitting gloves reduce the workers' dexterity, and grip strength, to name a couple concerns [5, 6]. This reduction in dexterity could reduce the worker's ability to use their communication gear, thereby reducing their ability to

C. S. Shin (Ed.): AHFE 2019, AISC 968, pp. 112–123, 2020.
https://doi.org/10.1007/978-3-030-20470-9_14

communicate with team members in life and death situations; such is the case with firefighters [7].

Poor fitting gloves does not equally impact males and females. For instance, a study conducted with firefighters found that 30% of males and 62% of females mentioned their gloves were poor fitting [8]. Gender differences such as these can result in discriminatory repercussions in the workplace, as female workers may be dissuaded to work in such environments. Therefore, it is imperative that workers are provided with properly fitting gloves.

Traditional glove fitting techniques can include methods of assessing the glove's mobility and dexterity, and Likert scale fit surveys. Mobility and dexterity testing methods can consist of using a Cylinder task, a Modified Kapanji Index and Knot tying tasks [9]. Subject fit assessments can include user feedback in the form of a survey, where the user/wearer provides feedback on a Likert scale of crucial glove fit areas [10]. While these methods are helpful, they occur after glove development, thereby doing little to help designers understand the hand-product relationship. Traditional glove fitting techniques also do not offer concrete information about how much and where to alter the pattern to improve fit, making it challenging to translate fit test data into a better fitting and designed gloves.

The design of gloves are improved by using anthropometric measurements of representative users. Typically, such measurements have been carried out using calipers and tape measures, while the hand is in either a flat or splayed position [11]. However, such measurements may not be indicative of work postures, as skin deformation and measurement change can occur depending on the position of the hand [12]. In previous years, it may not have been possible to measure the hand during these representative postures. Nonetheless, this limitation can be overcome with the aid of 3D scans.

In addition to collecting measurements from 3D scans, new technologies such as rapid prototyping, 3D simulation, and digital prototyping offer endless design opportunities for gloves [13, 14]. The advantages of virtually draping include a reduction in cost and time, as well as the ability to change the materials before prototypes are manufactured [14]. Designers and engineers can use the 3D scans to digitally drape glove designs to test glove fit and a sizing system before prototypes are made, as well as examine design features in relation to anatomical hand features.

The ability to test the fit of gloves digitally on 3D hand scans has been explored by Ashdown and Stull [16] where they scanned the hand with and without a firefighter glove and superimposed the scans to understand the hand-glove interface [15]. While this was effective, the process was time consuming and difficult to align the scanned hand and the glove scan properly. Digitally draping gloves on real 3D scans (as opposed to avatars or models) can enhance our understanding of glove fit across different populations, and can be an effective design tool to test functional design features that require precise anatomical placement such as seams in the palmar region, padding for protection, or design contour to improve movement at finger joints.

The purpose of this research was to compare a traditional glove fit test to digital draping methods using 3D hand scans. This research explores a method of digital draping gloves and methods of analyzing digital fit to improve the glove design and hand-product interface.

2 Method

This research compares the traditional and digital methods of fitting gloves. First, one traditional work glove was selected in size large, containing leather as the base material (Fig. 1). Using hand circumference and hand length measurements, one male Caucasian subject, aged 32 was identified that would be classified as wearing a size large in the selected gloves. This classification was based on the glove manufacturers' sizing system recommendation. The subject was asked to do a variety of tasks while wearing the glove, which included picking up a cylinder object and using scissors. These tasks were chosen as it allowed the subject to gain a general understanding of the mobility that was permitted with these gloves.

After wearing these gloves and performing basic tasks, the participant then completed a glove fit survey that was developed by Tremblay-Lutter et al. [10]. This survey contained questions that assessed fit for each of the digits on a 5-point Likert scale. The questions investigated on the survey, as well as the participant's responses, are presented in Table 1. The last two questions in the Tremblay-Lutter et al. survey were not utilized, as the subject was not a representative user and it would have been difficult to ascertain how performance may be impacted.

Table 1. Tremblay-Lutter et al. [10] fit survey.

(1) Overall fit				
1-Very good	2-Good	3-Neither good nor poor	4-Poor	5-Very poor
(2–6) Digit 1–5 length				
1-Too long	2-Too long but acceptable	3-Just right	4-Too short but acceptable	5-Too short
(7–11) Digit 1–5 circumference				
1-Too loose	2-Too loose but acceptable	3-Just right	4-Too tight but acceptable	5-Too tight
(12) Palm circumference				
1- Too Loose	2-Too loose but acceptable	3-Just right	4-Too tight but acceptable	5-Too tight
(13) Wrist circumference				
1-Too loose	2-Too Loose But Acceptable	3-Just right	4-Too tight but acceptable	5-Too tight
(14–17) Crotch of thumb, 2, 3, 4				
1-Too high	2-Too high but acceptable	3-Just right	4-Too low but acceptable	5-Too low
(18) Rate the extent that you feel your performance was affected by this glove (compared to bare hand performance)				
1-Extremely degraded my performance	2-Degraded my performance a lot	3-Degraded my performance a little	4-Had no effect on my performance	5-Enhanced my performance
(19) Rate the comfort of this glove:				
1-Uncomfortable	2	3	4	5

Next, the subject's hand was 3D scanned in a splayed position. The hand position was chosen based on ISO 7250-1:2017 recommendation to collect anthropometric measurements [16]. Once the subject was positioned, a 3D scan was captured using an Occipital Structure Sensor attached to an Apple iPad. The scan was cleaned and oriented in MeshLab.

The 3D modeling application selected for this study was EFI Optitex®, a software package predominantly used in the apparel industry. Optitex was chosen because it is better at modeling soft textiles than more commonly used product design modeling software, such as Rhino. Optitex also allows the user to stitch papers and digitally drape, a function not available in Rhino.

Next, the glove was disassembled (Fig. 1c), patterns were digitized, and imported into Optitex. The glove patterns were refined and dimensions compared to the original glove to ensure accuracy. The 3D hand scan was imported into Optitex 3D as an.obj file. Textile and material specifications from the glove were imported into Optitex. Once in Optitex 3D, a glove was digitally draped on the hand scan.

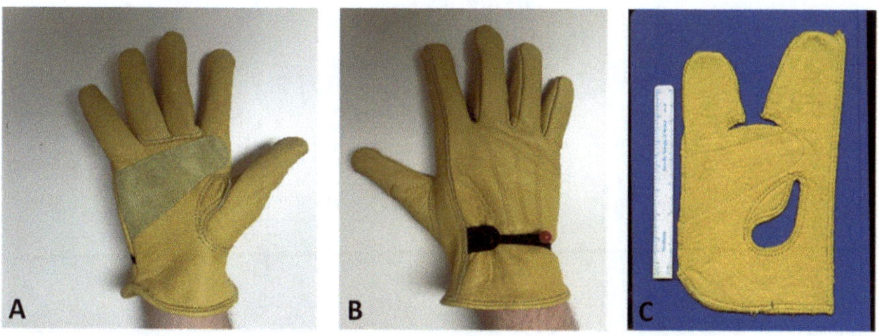

Fig. 1. (a–b) Original glove (c) glove taken apart and digitized.

In order to properly drape the pattern onto the 3D hand scan, the original pattern (Fig. 2) was split into different sections to create a final pattern (Fig. 3) that can drape around the complex shape of the hand. These sections are created by adding seams in the pattern. Seams added to the pattern do not include contour and are only there to assist the pattern as it moves around the 3D model. Unlike traditional glove seams, the digital seams do not add bulk or influence the fit of the product.

Each piece of the final pattern was then carefully placed near and rotated around the 3D model to ensure that the pieces drape smoothly around the 3D hand scan (Fig. 4).

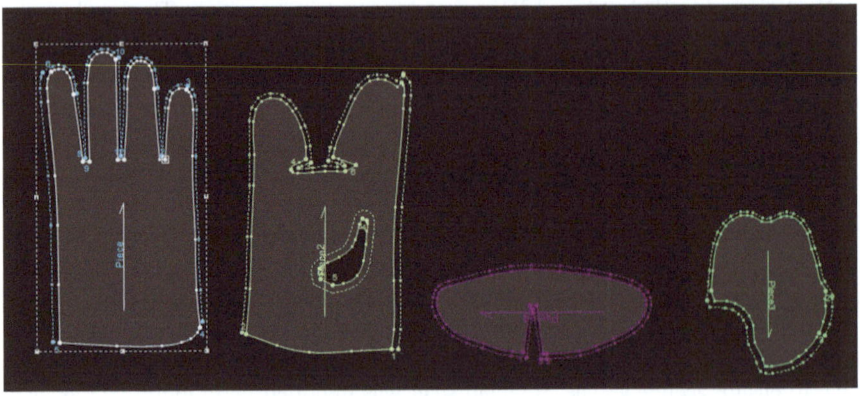

Fig. 2. Original digitized glove pattern

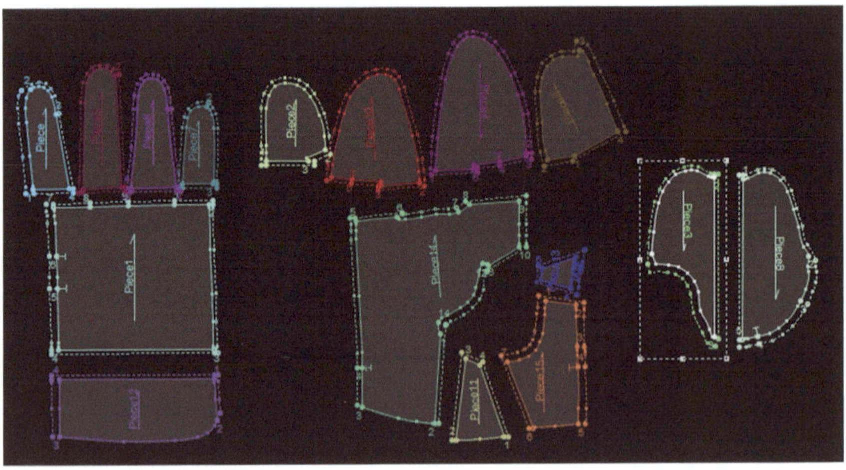

Fig. 3. Final pattern used to digitally drape the glove on the 3D hand scan.

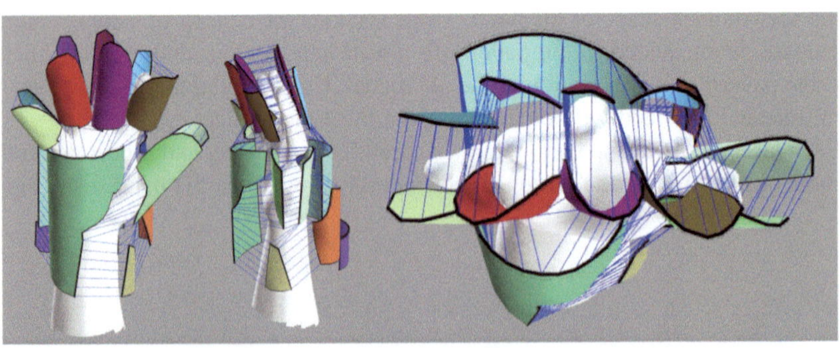

Fig. 4. Patterns in final placement for 3D simulation from various angles.

Researchers analyzed the fit of the glove by using the Pressure Map function and mesh tool. Two fit experts completed the Tremblay-Lutter et al. [10] glove fit survey while analyzing the digital drape.

3 Results

Overall, both the participant and the fit experts perceived the glove as being overall poor fitting (Table 2). Each of the finger crotch areas were rated as "Too High." Additional concerns were expressed in each of the digit's circumference and length questions, but with some variability. The results of the survey are found in Table 2.

Researchers further analyzed the fit of the glove by using the Pressure Map function in Optitex to see areas of strain (red) and loose fit (blue) (Fig. 5). Additionally, the digitally draped glove was examined for areas that were draped/sewn improperly due to poor fit. Finally, the digital drape was analyzed using the Mesh Tool to see how different areas of the glove fit in relation to the anatomical location of the hand and to examine the body-product relationship in depth (Fig. 6). Table 2 discusses the areas of most concern that do not appear in the survey but are visible in the 3D glove simulation using the Pressure Map and Mesh Tool functions. The table also provides recommended pattern adjustments to improve the glove fit.

Table 2. Results of the subject fit test survey, fit expert survey of 3D simulation, summary of visual analysis of 3D simulation, and recommended changes to glove for improved fit.

Area of hand	Subject survey results	Fit Expert survey results of 3D simulation	Visual analysis of 3D simulation	Recommended changes to glove for improved fit
Overall fit	Overall fit was assigned a score of poor	Overall fit was assigned a score between poor and very poor	Areas of strain are occurring at the full circumference of the palm and near the crotch of the digits (see Fig. 7a). There is an inconsistency in the lengths and circumferences of the digits (see Figs. 7a and c, 8a, and c). Note that extended length in the pattern on digit 1 (see Fig. 8b) could be caused by distortion in the hand model	The overall palm circumference should be adjusted to account for the areas of strain, especially at the Ulnar side and at the pad of the palm near the 1st digit (see Fig. 7a). Lengths and circumferences of the digits need to be adjusted based on feedback

(*continued*)

Table 2. (*continued*)

Area of hand	Subject survey results	Fit Expert survey results of 3D simulation	Visual analysis of 3D simulation	Recommended changes to glove for improved fit
Digit length	Digit 1: too long but acceptable Digit 2: just right Digit 3: too long but acceptable Digit 4: just right Digit 5: too long but acceptable	Digit 1: too long Digit 2: too long but acceptable Digit 3: too long but acceptable Digit 4: too short Digit 5: between too short but acceptable and too short	There is inconsistency in the lengths of the digits (see Figs. 7a and c, 8a, and c). Extended length in the pattern on digit 1 (see Fig. 8b) and contrasting feedback on digit 5 could be caused by distortion in the hand model	The lengths of the digits need to be adjusted
Digit circumference	Digit 1, 3–5: too loose but acceptable Digit 2: too tight but acceptable	Digit 1: too tight but acceptable Digit 2: just right Digit 3: too loose Digit 4: too loose but acceptable Digit 5: too tight	Strain can be seen in the digital drape around the lower circumferences for digit 1 and 5 (see Fig. 7b and d). In digit 2 and 3, less strain is seen in the circumferences (Fig. 7a and c). Digit 4 show strain at the tip of the digit (Fig. 7c), which appears to be more related to an issue in the length than in the circumference	The circumferences of the digits need to be adjusted
Palm circumference	Too tight but acceptable	Too tight	The digital drape shows that the glove is strained over the palm and 1st digit region (see Fig. 7a). While draping, the glove was so tight that the seams pulled to fit over the palm and around the thumb (see Fig. 7a and d)	Improvement could be made in the overall palm circumference to account for the areas of strain, especially at the Ulnar side and at the pad of the palm near the 1st digit (see Fig. 7a)
Palm length	Not listed on survey	Not listed on survey	The digital drape of the glove shows that the palm length is too short overall and may be too tight (see Fig. 7a)	Palm length needs to be lengthened and ease should be added for areas of strain around the thumb.

(*continued*)

Table 2. (*continued*)

Area of hand	Subject survey results	Fit Expert survey results of 3D simulation	Visual analysis of 3D simulation	Recommended changes to glove for improved fit
Wrist circumference	Too loose but acceptable	Too loose but acceptable	The digital drape show that the wrist circumference is loose (Fig. 7). The Mesh Tool shows a slight amount of excess fabric on the radial side (Fig. 8a and c)	Wrist circumference could be reduced or more adjustability added to account for the extra material
Finger crotches of the digits	Thumb: too high Crotch 2–4: Too high	Thumb: too high but acceptable Crotch 2–3: too high Crotch 4: too high but acceptable	The finger crotches of all of the digits appear to be webbing both in the front (see Fig. 8a) and the back (see Fig. 8c) of the hand at the finger crotches of the digits. Note that the crotch 3 appears to have a distortion in the hand model	Lowering of the finger crotches to ensure webbing does not occur

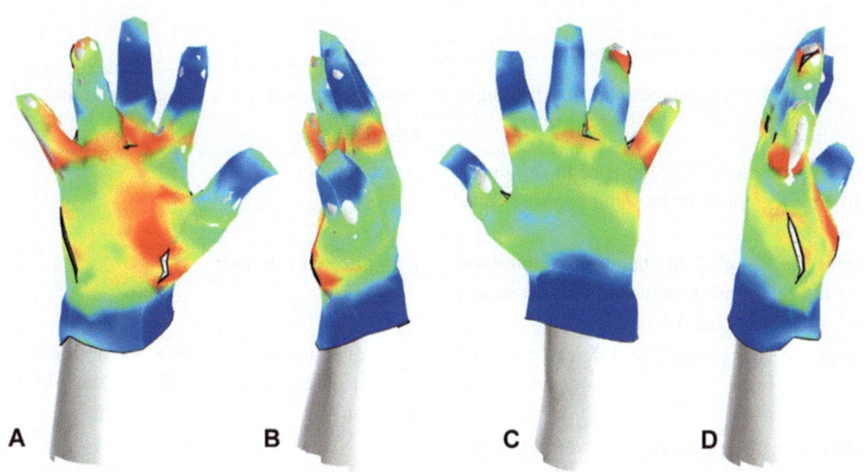

Fig. 5. Pressure map of digital glove shown from various views (a) Palm (b) Radial (c) Dorsal (d) Ulnar

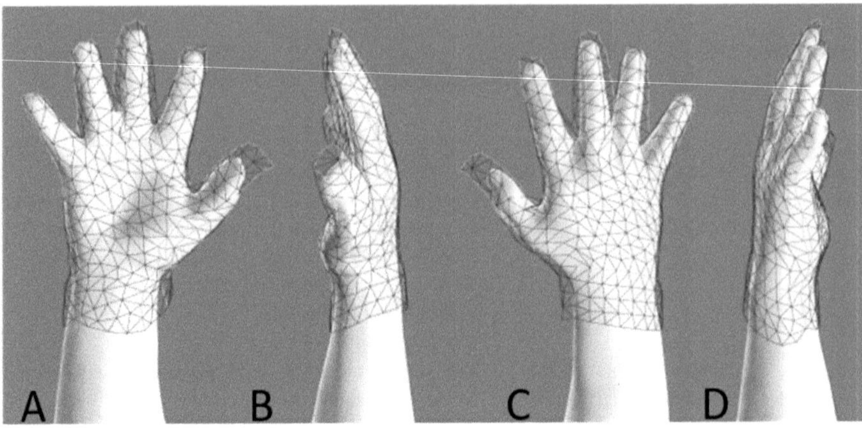

Fig. 6. Mesh Tool shown from various views (a) Palm (b) Radial (c) Dorsal (d) Ulnar

4 Discussion

4.1 Optitex

In this pilot study, the feasibility of digitally draping a glove pattern onto a subject's 3D splayed hand scan was demonstrated. The advantages and disadvantages of using this method in Optitex are compared in Table 3.

Table 3. Advantages and disadvantages of digitally draping glove designs in optitex.

Advantages	Disadvantages
Robust and clear indications of fit through utilizing the mesh feature and pressure map functions	Not beginner friendly
Ability to visualize the hand-glove interface and the relationship between design features and anatomical hand location	Complicated interface
Ability to change and test different materials	Lack of intuitiveness of how the designer needs to drape a glove until they start fitting the scan
Reduction in cost based on fewer physical prototypes needed to be produced	

The advantages of digital draping the glove far outweigh the disadvantages. However, Optitex 3D has a complicated interface and there is a lack of intuitiveness in terms of 'how' to digitally drape a glove on a hand scan. To address the limitations, it may be beneficial to conduct research with the goal of streamlining the digital draping process and making it more user-friendly for glove designers.

Now that the feasibility of this approach has been demonstrated, in future studies, 3D scans from representative users, in multiple hand positions could be investigated. The decision to include multiple positions will provide a better understanding of the hand and glove fit in different positions, as skin deformation and measurement changes occur depending on the position Nasir et al. [12]. Additionally, in future studies, landmarking and use of color scans could be explored. Landmarking is a process of improving the quality of 3D scan by placing markers on various body areas Griffin et al. [14]. This improved 3D quality allows the researcher to identify areas of the hand such as finger crotches and crease lines more efficiently but also allows more accurate measurements to be taken. Lastly, in future studies, a structure to stabilize the hand will be utilized. A 3D hand scan can be obtained using the Occipital Structure Sensor and takes 1 to 2 min per scan. Without the use of a stability structure, it is difficult to ensure that the hand scan is aligned correctly in the splayed position. The ease of capturing hand scans can help gather in depth anthropometric measurements of the hand, as well as enhance the glove design and fit process through digital draping. Since 3D scans are edited as little as possible, any distortion to the hand position in terms of movement or proper alignment can introduce variables to the digital draping process that hinder the fidelity of both the digital drape and the evaluation of glove fit.

4.2 Tremblay-Lutter et al. [10] Fit Survey

The results of the survey demonstrated that both the participant and the fit experts perceived the glove as being overall poor fitting. Each of the crotch areas were rated as "Too High." Concerns were expressed in each of the digit's circumference and length questions, but with some variability. After verbal discussions with the participant and visual analysis of the scans, it is evident that there is a need to refine the survey to capture a more in depth understanding of fit. In future studies, the survey will need to be expanded and refined, thereby allowing researchers to capture more comprehensive fit feedback. While some elements of the product-hand interface are difficult to capture in terms of participant feedback, a more detailed fit survey can improve the evaluation of the digitally draped glove, as well as collect more data from subject fit-tests. A comparison between the survey and the visual analysis conducted by two experts based on the Optitex glove simulation are found in Table 2.

Based on Table 2, there is value in using digital draping techniques on real hand scans for improved glove design. Digital draping captures a significant amount of fit data that would be missed if designers and researchers focused solely on survey methods of fit analysis. However, an improved survey could aid in both the examination of the digital glove drape and provide more areas of fit feedback for subject testing. A future survey should include questioning on digit circumference at various points, palm length, length at the side of the glove, finger crotch at both palmar and dorsal side, and location placement of glove features. The inclusion of such questions could result in a better understanding of the fit of the glove when used in combination of 3D simulations and visual analysis techniques.

5 Conclusion

This pilot study demonstrated the feasibility of digitally draping a 3D hand scan to analyze the fit of a glove. The ability to digitally drape a 3D scan has the potential to develop better-fitting gloves that could aid in the reduction of workplace injuries. Future studies will look at conducting a large scale study with representative users by scanning hands of a selected occupational group in different positions, digitally draping gloves on the 3D scans, and analyzing fit. While digital draping may not replace traditional fit tests, it does provide designers the ability to make adjustments to patterns and develop a more accurate fitting glove early in the design process. Such an approach will save time and resources, as the findings of a subject fit test late in the design process may result in only minor adjustments to fit, rather than significant modifications that the 3D simulations enable.

Overall, the 3D glove simulations using subject's 3D scans were able to show more concrete evidence of the exact location of the fit issue for each design and provided quantifiable information as to how to improve the design. This immediate design feedback has the potential to profoundly impact the way in which gloves are designed in the future and how the glove fit is analyzed across a wide population of wearers.

Acknowledgement. This material is based upon work supported by the Minnesota Agricultural Experiment Station, and the National Institute of Food and Agriculture, U.S. Department of Agriculture, Hatch under MIN-53-088.

References

1. Olsen, R.J., Lynch, P., Coyle, M.B., Cummings, J., Bokete, T., Stamm, W.E.: Examination gloves as barriers to hand contamination in clinical practice. JAMA **270**(3), 350–353 (1993)
2. Yoo, I.G., Lee, J., Jung, M.Y., Lee, J.H.: Effects of wearing the wrong glove size on shoulder and forearm muscle activities during simulated assembly work. Ind. Health **49**(5), 575–581 (2011)
3. da Costa, B.R., Vieira, E.R.: Risk factors for work-related musculoskeletal disorders: a systematic review of recent longitudinal studies. Am. J. Ind. Med. **53**(3), 285–323 (2010)
4. U.S. Bureau of Labor Statistics, 2016 Survey Of Occupational Injuries & Illnesses. U.S Department of Labor, November 2017
5. Drabek, T., Boucek, C.D., Buffington, C.W.: Wearing the wrong size latex surgical gloves impairs manual dexterity. J. Occup. Environ. Hyg. **7**(3), 152–155 (2010)
6. Kovacs, K., Splittstoesser, R., Maronitis, A., Marras, W.S.: Grip force and muscleactivity differences due to glove type. AIHA J. **63**(3), 269–274 (2002)
7. Park, H., Park, J., Lin, S.H., Boorady, L.M.: Assessment of Firefighters' needs for personal protective equipment. Fash. Text. **1**(1), 8 (2014)
8. Hsiao, H., Whitestone, J., Kau, T.Y., Whisler, R., Routley, J.G., Wilbur, M.: Sizing firefighters: method and implications. Hum. Factors **56**(5), 873–910 (2014)
9. Korona, F., McFarland, S., Walsh, S.: NASA EVA glove characterization protocol development. In: 48th International Conference on Environmental Systems (2018)

10. Tremblay-Lutter, J.F., Weihrer, S.J.: Functional fit evaluation to determine optimal ease requirements in canadian forces chemical protective gloves (No. DRES-SR-635). Defence Research Establishment Suffieldralston (ALBERTA) (1995)
11. Gordon, C.C., Blackwell, C.L., Bradtmiller, B., Parham, J.L., Barrientos, P., Paquette, S.P., Mucher, M.: 2012 anthropometric survey of US army personnel: methods and summary statistics (No. NATICK/TR-15/007). Army Natick Soldier Research Development and Engineering Center, MA (2014)
12. Nasir, S.H., Troynikov, O., Watson, C.: Skin deformation behavior during hand movements and their impact on functional sports glove design. Proc. Eng. **112**, 92–97 (2015)
13. Smith, T.F., Waterman, M.S.: Identification of common molecular subsequences. J. Mol. Biol. **147**, 195–197 (1981)
14. Griffin, L., Sokolowski, S., Seifert, E.: Process considerations in 3D hand anthropometric data collection. In: Proceedings of 3DBODY.TECH 2018 - 9th International Conference and Exhibition on 3D Body Scanning and Processing Technologies, Lugano, Switzerland, 16–17 October 2018, pp. 121–130 (2018). http://dx.doi.org/10.15221/18.121
15. Mesuda, Y., Inui, S., Horiba, Y.: Virtual draping by mapping. Comput. Ind. **95**, 93–101 (2018)
16. Ashdown, S.P., Stull, J.M.: Analysis of fit and sizing of firefighter gloves using 3D data. In: International Conference on Applied Human Factors and Ergonomics, Springer, Cham (2016)
17. ISO 7250–1: basic human body measurements for technological design—part 1: body measurement definitions and landmarks. International Organization for Standardization, Geneva, Switzerland (2017)

The Variability of U.S. Women's Plus Size Product Sizing and Self-Identified Size 18 Bodies

Susan L. Sokolowski[1](✉), Linsey Griffin[2], and Jessie Silbert[1]

[1] Sports Product Design, University of Oregon, 70 NW Couch Street, Portland, OR 97209, USA
ssokolow@uoregon.edu
[2] Design Housing and Apparel, University of Minnesota, 350 McNeal Hall, Saint Paul, MN 55108, USA

Abstract. In 2016, the apparel market value for plus sizes in the U.S., was estimated at 20.4 billion dollars. As there is a lack of accessible measurement and sizing standardization in the U.S. for this body type, retailers have developed their own unique systems. This pilot study will explore how a sample of 65 plus size women, through 3D body scans, fit into the top U.S. retailer's measures and sizes. The retailers investigated, included: Walmart, Kohl's, JC Penney, Target, Macy's and Lane Bryant. The findings established that none of the retailers are meeting the needs of the plus size body. Future research must consider methods to improve access to measurements, apparel sizing and product development for this growing demographic.

Keywords: Plus size U.S. women · Sizing systems · Fit · Apparel

1 Introduction

Between 2015–2016, the Centers for Disease Control and Prevention (CDC), reported the prevalence of U.S. adult obesity at 39.8% of the population, which affects about 93.3 million [1]. Of those adults, 41% were estimated to be women [1]. In 2016, the apparel market value for this plus size demographic was estimated at 20.4 billion dollars [2]. Plus sizes typically range from 16–30 in the U.S., although some retailers may define their spectrums, by starting with a size 14. Some may also bookend their plus size range at a 24, 26, 28, 32 and 34.

In 2012, the Mintel Group reported that women who shopped for plus sizes in the U.S. preferred the following retailers: Walmart (47%), Kohl's (20%), JC Penney (19%), Target (15%), Macy's (11%) and Lane Bryant (9%) [3]. Mintel's report stated that these preferences were based upon plus size women being able to shop locally, conveniently, with anonymity. Figure 1 shows these preferences.

C. S. Shin (Ed.): AHFE 2019, AISC 968, pp. 124–133, 2020.
https://doi.org/10.1007/978-3-030-20470-9_15

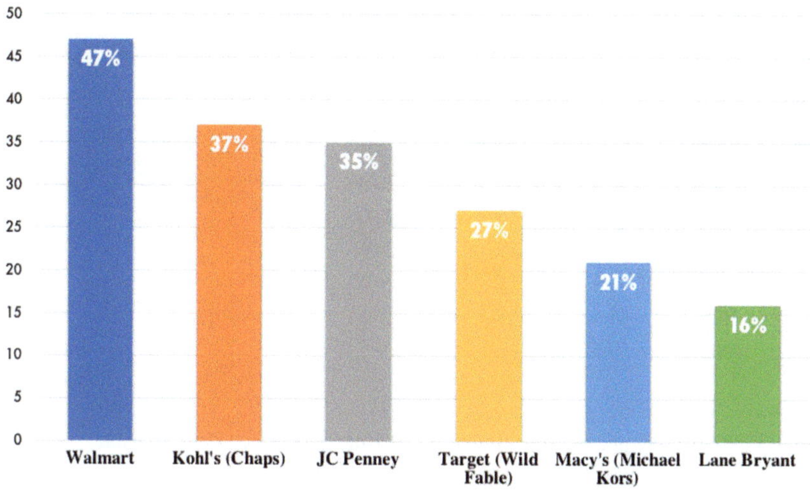

Fig. 1. Plus size retailer preferences, for women in the U.S.

As there is a lack of accessible measurements and size standardization in the U.S. for this body type, retailers can develop their own size ranges and measurements. They may even have a variety of sizing charts, one for each brand they sell. In some cases, the retailer may split its' size charts into ones for tops-only, bottoms-only and dresses/coats-only, for each brand. Size charts typically contain information regarding measurements for the bust, waist and hip. Commonly these charts will not have detailed measurement definitions; they will just list the size, measurement in inches, along with a basic diagram of where the measures lie on the body.

For the preferred U.S. retailers presented in Fig. 1, Walmart, JC Penney and Lane Bryant have one plus size chart for their entire store, meaning that anyone shopping at these retailers would size themselves to one common chart (even if the retailer sells multiple brands). Kohl's and Macy's are however, a bit more complicated, in that each brand they sell has a separate size chart. For Target, they split their sizing charts into separate tops and bottoms charts, by brand. The more charts a retailer has, the more variability and confusion it creates for the consumer.

Tables 1, 2, 3, 4, 5 and 6 present the size chart measures, for each of the preferred plus size women's retailers in the U.S. All charts include bust, waist and hip measures. The charts were acquired on-line. For Kohl's and Macy's, size charts for their biggest brand assortments (Chaps and Michael Kors, respectively) are presented. Since size charts for Target are subdivided by tops and bottoms, by brand, one combined chart was made for their biggest women's plus size brand assortment (Wild Fable).

Table 1. Walmart plus size measures.

Size	Bust	Waist	Hip
16W	42.5"	34.5"	44.5"
18W	44.5"	36.5"	46.5"
20W	46.5"	38.5"	48.5"
22W	48.5"	40.5"	50.5"
24W	50.5"	42.5"	52.5"
26W	52.5"	44.5"	54.5"
28W	54.5"	46.5"	56.5"
30W	56.5"	48.5"	58.5"
32W	58.5"	50.5"	60.5"

Table 2. Kohl's (Chaps) plus size measures.

Size	Bust	Waist	Hip
14W	42"	37"	45"
16W	44"	39"	47"
18W	46"	41"	49"
20W	48"	43"	51"
22W	50"	45"	53"
24W	52"	47"	55"
26W	54"	49"	57"

Table 3. JC Penney plus size measures.

Size	Bust	Waist	Hip
14W	42"	36"	45"
16W	44"	38"	47"
18W	46"	40"	49"
20W	48"	42"	51"
22W	50"	44"	53"
24W	52"	46"	55"
26W	54"	48"	57"
28W	56"	50"	59"
30W	58"	52"	61"
32W	60"	54"	63"
34W	62"	56"	65"

Table 4. Target (Wild Fable) plus size measures.

Size	Bust	Waist	Hip
14W	41"	36"	43"
16W	43"	38"	45"
18W	45"	40"	47"
20W	47"	42.25"	49"
22W	49"	44.5"	51"
24W	51"	46.75"	53"
26W	53"	49"	55"
28W	55"	51.25"	57"
30W	57"	53.5"	59"

Table 5. Macy's (Michael Kors) plus size measures.

Size	Bust	Waist	Hip
14W	43.5"	35"	44"
16W	45.5"	37"	46"
18W	47.5"	39"	48"
20W	49.5"	41"	50"
22W	51.5"	43"	52"
24W	53.5"	45"	54"

Table 6. Lane Bryant plus size measures.

Size	Bust	Waist	Hip
14	42"	36"	44.5"
16	44"	38"	46.5"
18	46"	40"	48.5"
20	48"	42"	50.5"
22	50"	44"	52.5"
24	52"	46"	54.5"
26	54"	48"	56.5"
28	56"	50"	58.5"
30	58"	52"	60.5"
32	60"	54"	62.5"

From the size charts presented, is it clear how variable the sizes and measures are for plus size women in the U.S. With such variability, how is it possible for women to successfully find and fit into apparel? This study will explore how a sample of plus size women, through 3D body scans fit into these retailer's sizes and measurements. The findings will help identify which retailers are fitting plus sizes better and make recommendations for future apparel sizing and product development research, for this growing demographic.

2 Literature Review

Since the 1940's, several attempts have been made to standardize apparel sizing for women, in the U.S. The pioneers of standardization, were O'Brien and Shelton – statisticians who received a grant from Works Progress Administration to enable the growth of the U.S. ready-to-wear apparel market [4]. Their research involved collecting 58 body measurements from 14,698 women across the U.S. [4]. The results of their research, published in 1941 identified several different body shapes, along with height and girth indicators, which netted to nine different sizes for each numerical bust measurement [4]. Unfortunately, the work was never commercialized [5]. The large number of sizes was daunting to manufacturers, and the women who volunteered as subjects, were not diverse enough socioeconomically or ethnically to accurately represent the population [6].

At the request of the Mail Order Association of America, Commercial Standard CS 215-58: Body Measurements for the Sizing of Women's Patterns and Apparel, was developed between the years of 1949–1952, and published in 1958 [5–8]. The standard included sizes 8 to 38 with height identifiers of tall (T), regular (R) and short (S). It also included plus (+) and minus (−) signifiers when referring to girth [6]. Although the standard was commercialized, it was not well accepted, as around the same time corseted body shapes started evolving to natural, unbound, fuller shapes influenced by processed and fast foods [7].

In the late 1960's, CS 215-58 was revised to the Voluntary Product Standard PS 42-70, in an attempt to better describe the U.S. female population, with newer anthropometric data [9]. Since the standard was voluntary, the majority of the apparel industry disregarded the update as published in 1970 [9]. The standard was completely withdrawn in 1983 [5]. The American Standards for Testing and Materials (ASTM) International, in 1995 published its' own voluntary sizing standard, which like the others before, was not widely adopted [5, 8]. ASTM also produced a Plus Size standard, named Standard Tables for Body Measurements for Plus Women's Figure Type, Size Range 14W–40W (2016), but unless the retailer belongs to ASTM and pays for the report, it is not publicly accessible, creating the current environment where retailers devise their own systems [8]. Some may also argue that that data for which this standard was create with is outdated.

Because retailers are not held accountable to adhering to any size standards in the U.S., products are inconsistently and unreliably sized. This challenge not only affects the consumer, but it also affects retailers and the environment. As an example, The Wall Street Journal reported that on-line returns average around 30%, because of fit [10].

The Retail Equation and National Retail Federation additionally reported that 260.5 billion dollars in returns are made globally [11]. They quoted, "if merchandise returns were a corporation, it would rank #3 on the Fortune 500 list" [11]. Returns also affect the environment. Retailers discard more than 25% of their returns, which can weigh over 5 billion pounds in landfills a year [12].

The intent of this paper is to begin looking at the complexities of the U.S. plus size female consumer's body, to provide better sizing system direction for retailers. It is anticipated that the results of this study will help identify which retailers are fitting plus sizes better and make recommendations for future apparel sizing and product development for this growing demographic.

3 Methodology

The goal of the paper was to understand how actual plus size women fit into the most preferred U.S. plus size retailer's size charts. For purposes of the study, it was determined to focus on one plus size only, to fully understand the complexities women face within a particular size.

Women who self-identified as plus size 18 participated in the study. Subjects were 3D body scanned, and bust, waist and hip measurements were collected, using AnthroScan software. The bust measurement was defined as the largest circumferential measurement of the bust taken parallel to the floor. The waist circumference measurement was taken at the natural waist crease on the torso, also parallel to the floor. The hip circumference was taken at the widest portion of the bottom (parallel). All measurements were taken in inches to match the unit of measure from the retailer's size charts.

Each subject's collected bust, waist and hip measures, were then matched to each of the six retailer's size charts, to see if and where they fit. In order to fit perfectly in the size, the subject had to completely fit within, each measure of the size range presented by the retailer. From the data matches, percentages were calculated for each retailer.

4 Results

3D scans from 65 plus size female subjects, with a mean age of 38.5 years were analyzed for the study. Subjects identified as one of the following races: African American/Black (46.2%), Asian/Pacific Islander (1.5%), Caucasian (41.5%), Latino/Hispanic (4.6%), Native American (1.5%), and Other (4.6%). Scans were collected from women who self-identified as a size 18, from two U.S. markets: New York City and Minneapolis. The New York scans were collected with a Human Solutions VITUS Smart XXL 3D LASER Body Scanner, and the Minneapolis scans with a 3DMD scanner. Basic statistics (mean, median and range) of the subject's measurements are presented in Table 7.

Table 7. Subject 3D scan measures.

Self-identified size 18 (n = 65)	Bust	Waist	Hip
Mean	45.83"	42.58"	49.99"
Median	45.92"	43.07"	49.51"
Range	33.46" to 63.46"	30.12" to 55.09"	41.97"to 63.23"

For each retailer's size chart presented in Tables 1, 2, 3, 4, 5 and 6, bust, waist and hip measurements ranges were calculated, for each size (Tables 8, 9, 10, 11, 12 and 13).

Table 8. Walmart plus measurement ranges.

Size	Bust	Waist	Hip
16W	>41.5" to <43.5"	>33.5" to <35.5"	>43.5" to <45.5"
18W	>43.5" to <45.5"	>35.5" to <37.5"	>45.5" to <47.5"
20W	>45.5" to <47.5"	>37.5" to <39.5"	>47.5" to <49.5"
22W	>47.5" to <49.5"	>39.5" to <41.5"	>49.5" to <51.5"
24W	>49.5" to <51.5"	>41.5" to <43.5"	>51.5" to <53.5"
26W	>51.5" to <53.5"	>43.5" to <45.5"	>53.5" to <55.5"
28W	>53.5" to <55.5"	>45.5" to <47.5"	>55.5" to <57.5"
30W	>55.5" to <57.5"	>47.5" to <49.5"	>57.5" to <59.5"
32W	>57.5" to <59.5"	>49.5" to <51.5"	>59.5" to <61.5"

Table 9. Kohl's (Chaps) plus measurement ranges.

Size	Bust	Waist	Hip
14W	>41" to <43"	>36" to <38"	>44" to <46"
16W	>43" to <45"	>38" to <40"	>46" to <48"
18W	>45" to <47"	>40" to <42"	>48" to <50"
20W	>47" to <49"	>42" to <44"	>50" to <52"
22W	>49" to <51"	>44" to <46"	>52" to <54"
24W	>51" to <53"	>46" to <48"	>54" to <56"
26W	>53" to <55"	>48" to <50"	>56" to <58"

Table 10. JC Penney plus measurement ranges.

Size	Bust	Waist	Hip
14W	>41” to <43”	>35” to <37”	>44” to <46”
16W	>43” to <45”	>37” to <39”	>46” to <48”
18W	>45” to <47”	>39” to <41”	>48” to <50”
20W	>47” to <49”	>41” to <43”	>50” to <52”
22W	>49” to <51”	>43” to <45”	>52” to <54”
24W	>51” to <53”	>45” to <47”	>54” to <56”
26W	>53” to <55”	>47” to <49”	>56” to <58”
28W	>55” to <57”	>49” to <51”	>58” to <60”
30W	>57” to <59”	>51” to <53”	>60” to <62”
32W	>59” to <61”	>53” to <55”	>62” to <64”
34W	>61” to <63”	>55” to <57”	>64” to <66”

Table 11. Target (Wild Fable) plus measurement ranges.

Size	Bust	Waist	Hip
14W	>41” to <43”	>35” to <37”	>42” to <44”
16W	>43” to <45”	>37” to <39”	>44” to <46”
18W	>45” to <47”	>39” to <41”	>46” to <48”
20W	>47” to <49”	>41” to <43.38”	>48” to <50”
22W	>49” to <51”	>43.38” to <45.63”	>50” to <52”
24W	>51” to <53”	>45.63” to <47.88”	>52” to <54”
26W	>53” to <55”	>47.88” to <50.13”	>54” to <56”
28W	>55” to <57”	>50.13” to <52.38”	>56” to <58”
30	>57” to <59”	>52.38” to <54.63”	>58” to <60”

Table 12. Macy’s (Michael Kors) plus measurement ranges.

Size	Bust	Waist	Hip
14W	>42.5” to <44.5”	>34” to <36”	>43” to <45”
16W	>44.5” to <46.5”	>36” to <38”	>45” to <47”
18W	>46.5” to <48.5”	>38” to <40”	>47” to <49”
20W	>48.5” to <50.5”	>40” to <42”	>49” to <51”
22W	>50.5” to <52.5”	>42” to <44”	>51” to <53”
24W	>52.5” to <54.5”	>44” to <46”	>53” to <55”

Table 13. Lane Bryant plus measurement ranges.

Size	Bust	Waist	Hip
14	>41” to <43”	>35” to <37”	>43.5” to <45.5”
16	>43” to <45”	>37” to <39”	>45.5” to <47.5”
18	>45” to <47”	>39” to <41”	>47.5” to <49.5”
20	>47” to <49”	>41” to <43”	>49.5” to <51.5”
22	>49” to <51”	>43” to <45”	>51.5” to <53.5”
24	>51” to <53”	>45” to <47”	>53.5” to <55.5”
26	>53” to <55”	>47” to <49”	>55.5” to <57.5”
28	>55” to <57”	>49” to <51”	>57.5” to <59.5”
30	>57” to <59”	>51” to <53”	>59.5” to <61.5”
32	>59” to <61”	>53” to <55”	>61.5” to <63.5”

Table 14. Size 18 subject fit, by retailer.

Brand	# of Subjects	Percentage	Where subjects did fit
Walmart	1	1.54%	Bust only 10.77% Waist only 9.23% Hip only 9.23% Bust & waist 1.54% Waist & hip 1.54% Bust & hip 1.54% None 64.62%
Kohl's (Chaps)	0	0%	Bust only 7.69% Waist only 3.08% Hip only 15.38% Bust & waist 4.62% Waist & hip 0% Bust & hip 4.62% None 64.62%
JC Penny	0	0%	Bust only 9.23% Waist only 3.08% Hip only 13.85% Bust & waist 3.08% Waist & hip 1.54% Bust & hip 4.62% None 64.62%
Target (Wild Fable)	0	0%	Bust only 15.38% Waist only 4.62% Hip only 7.69% Bust & waist 1.54% Waist & hip 1.54% Bust & hip 1.54% None 67.69%
Macy's (Michael Kors)	0	0%	Bust only 16.92% Waist only 1.54% Hip only 4.62% Bust & waist 0% Waist & hip 4.62% Bust & hip 3.08% None 69.23%
Lane Bryant	0	0%	Bust only 10.77% Waist only 3.08% Hip only 10.77% Bust & waist 3.08% Waist & hip 1.54% Bust & hip 3.08% None 67.69%

The ranges presented were then used to understand how well each self-identified size 18 subject fit into each retailer's sizing system. Fitting into a retailer's sizing required that each subject had to perfectly into each range posted for bust, waist and hip (Table. 14).

Sixty-four women (98.46%) did not fully fit into any of the leading plus size brand's size 18. Only one subject fit completely into Walmart's size 18 chart. To understand the results further, percentages were calculated for each measurement (bust, waist, hip), to determine where subjects actually matched. In all cases between the retailers, subjects were less able to meet the waist measurement range. When two measurements were in consideration, subjects were also less likely to fit into the retailer's size measurement range.

Since subjects self-identified their size, it was possible they vanity sized themselves. With that hunch, measurement ranges (bust, waist, hip) were calculated for all sizes, for all retailers to determine if the subjects fit into other sizes better. Surprisingly, only two more subjects were matched fully into a size – one fitting into a Target size 26 and the other fitting into a Walmart size 28.

5 Summary and Conclusion

For the preferred retailers in the U.S. that sell plus size apparel, there is an opportunity to improve how products are sized. Recommendations for future research would include conducting a study with an increased sample size, looking at scans from other regions of the U.S. and establishing better representation from Hispanic, Asian and Native American participants. Given how the plus size demographic is growing, work on this topic could help improve sales, help identify key styles and decrease the number of returns from consumers.

Acknowledgments. The authors would like to thank Body Labs (now Amazon) for their support in this study.

References

1. Hales, C.M., Carroll, M.D., Fryar, C.D., Ogden, C.L.: Prevalence of obesity among adults and youth: United States 2015–2016, Data Brief No. 288, October 2017
2. NPD Group: Plus size women's apparel market value forecast in the United States from 2015 until 2020. Emarketer, May 2017
3. Mintel Group Ltd.: Shopping for plus size teens' and women's clothing – U.S. Retailers shopped for plus size clothing. Analyst Insights, November 2012
4. O'Brien, R., Shelton, W. C.: Women's Measurements For Garment and Pattern Construction. US Department of Agriculture, No. 454 (1941)
5. Stampler, L.: The bizarre history of women's clothing sizes, October 2014. http://time.com/3532014/women-clothing-sizes-history/
6. LaBat, K.L.: Sizing standardization. Sizing in Clothing. Woodhead Publishing Limited, Cambridge (2007)

7. National Bureau of Standards: Body measurements for the sizing of women's patterns and apparel. Voluntary Product Standard CS 215-58. National Bureau of Standards, US Department of Commerce, Washington D.C (1958)
8. Ingraham, C.: The absurdity of women's clothing sizes, in one chart, August 2015. https://www.washingtonpost.com/news/wonk/wp/2015/08/11/the-absurdity-of-womens-clothing-sizes-in-one-chart/?noredirect=on&utm_term=.ca8fe038beeb
9. National Bureau of Standards: Body measurements for the sizing of women's patterns and apparel. Voluntary Product Standard PS 42-70. National Bureau of Standards, US Department of Commerce, Washington D.C (1970)
10. Bold Metrics: Apparel retailers start 2016 with 30% returns – almost a third of all holiday purchases made online will be returned (2015). http://blog.fashionmetric.com/apparel-retailers-start-2016-with-a-30-return-rate/
11. Ewen, L.: How the future of fit could spell the end of retail returns, July 2016. https://www.retaildive.com/news/how-the-future-of-fit-could-spell-the-end-of-retail-returns/421972/
12. Reagan, C.: That sweater you don't like is a trillion-dollar problem for retailers. These companies want to fix it, January 2019. https://www.cnbc.com/2019/01/10/growing-online-sales-means-more-returns-and-trash-for-landfills.html

Interdisciplinary Practice for Automobile Design

Extracting Contour Shape of Passenger Car Form in Rear View Based on Form Similarity Judgement by Young Chinese Consumers

Chunrong Liu(✉) and Minghua Zhang

Shanghai Jiao Tong University, Shanghai 200240, China
cheeronliu@sjtu.edu.cn

Abstract. It explores how to extract a contour shape to reveal underlying vital parts of passenger car form in rear view based on young Chinese consumers' form similarity judgement. By defining 21 character lines and their 75 segments for eighty representative forms selected from similarity judgement by young Chinese consumers, the slope values of all segments are calculated as an input for factor analysis and 13 common factors are extracted. The findings show that (1) the passenger car form in rear view can be extracted and presented as a reduced but recognizable contour shape with around forty-four percent loss of information on the form in rear view; and (2) the contour shape is mainly consisted of character lines defining such basic design parts as roof and side windows, the C-type contour of taillights, the U-type edges of sunken license plate area, the U-type boundary and the turning edge between the horizontal and vertical surfaces of trunk lid, the horizontal transitional edge on the upper rear bumper, and the contour of the corner area on rear bumper in rear view.

Keywords: Consumer research · Passenger car form in rear view · Contour shape extraction · Form similarity judgement

1 Introduction

Consumers choose products in terms of their own demands and preferences [1]. A product conveys a kind of value through its form and has an effect on the competition in marketplace [2]. The same is true for a passenger car. In Chinese passenger car market, young consumers have become the main force for car consumption [3] and propose a higher request of a passenger car form and its value, a key point in an intensely competitive market [4].

In the field of passenger car form research, the approach to character line is one of the main methods to define and analyze the passenger car form and has been used in some studies, for example, the descriptive model by extracting character lines from car form [5], the quantitative relationship between specific characteristic lines and sensitive image adjectives [6], the development of Toyota's car form summarized by defining and extracting character lines [7], and the evolution law and the trend of Volkswagen's car form summarized and predicted by extracting the representative character lines of Volkswagen's car form [8]. This approach is sometimes used for exploring forms in

C. S. Shin (Ed.): AHFE 2019, AISC 968, pp. 137–145, 2020.
https://doi.org/10.1007/978-3-030-20470-9_16

other types of vehicles, for example, freight locomotive's form [9] and the form of electric logistics light truck [10].

A car contains complex form information when observed and viewed from different angles [5]. Emphasized on analyzing the form information on passenger car form in rear view and based on character line definition method, an approach to extract the contour shape, the reduced but recognizable form expression, of passenger car form in rear view is completed in this study by quantitative cluster analysis and factor analysis methods with young Chinese consumers' similarity judgement on and data acquisition of passenger car forms in rear view.

2 Methods

2.1 Form Similarity Judgement and Cluster Analysis

Experimental Preparation. A total of 130 pictures of triple-compartment passenger car form in rear view with engine displacements from 1.5L to 2.4L are collected in Chinese mainland car market, involving in 31 brands such as Beijing Benz, Beijing-Hyundai, BYD, Dongfeng, Dongfeng-Honda, Dongfeng-Peugeot, Dongfeng- Renault, Dongfeng-Yueda-Kia, Dongfeng-Nissan, Dongfeng-Citroen, Qoros, GAC, GAC-Honda, GAC-Toyota, BMW-Brilliance, Chery, SAIC MG, SAIC Volkswagen, SAIC Roewe, SAIC-GM Buick, SAIC-GM Chevrolet, SAIC-GM-Wuling, FAW, FAW Audi, FAW-VW, FAW-Toyota, FAW-Mazda, Changan, Changan Ford, Changan-Mazda and Great Wall.

In order to reduce the influence of car body color, logo and other factors on the experiment, all pictures are converted to black and white mode and car bodies in pictures are uniformly placed on a white background while the brand logo and license plate are removed. Finally, 130 pictures are randomly marked with the serial number of v1, v2, …, v129, v130, respectively.

Form Similarity Judgement Experiment. In order to select the representative passenger car form samples, 30 subjects aged 18 to 30 including undergraduate, graduate and professionals from different industries, are invited to evaluate the similarity between 130 passenger car forms in rear view presented in printed pictures. During the experiment, 30 subjects are kindly asked to use an interactive grouping tool for grouping tasks [4] in a relatively quiet environment. When a subject completes his/her evaluation, the grouping tool automatically generates a piece of similarity matrix data. As result of the experiment, 30 valid similarity matrices are obtained and an averaged similarity matrix is calculated by averaging these similarity matrices.

Cluster Analysis. The averaged similarity matrix data is analyzed by hierarchical cluster analysis, plotting a dendrogram in cluster analysis result. According to the relevant cluster analysis principle [11], it is reasonable to classify 130 pictures of passenger car form in rear view into 7 categories by cutting the dendrogram at the appropriate position to set the number of categories.

Samples Screening. The K-means cluster analysis is executed by setting the category number to 7. The closer a sample is to the center of the category to which it belongs, the more it embodies information on forms of its corresponding category. Among the initial 130 form samples, 80 samples are retained as representatives for later experiments after those samples with relatively farther distance to the center of their corresponding categories are discarded.

2.2 Definition of Character Lines

The character lines of passenger car form are constructive lines with specific structural constraints and form connotations, which can provide a relatively complete description and expression of form. In the tomographic structure of the form characteristics, the character lines are on the basic class layer, that is, the "the least cost includes the most information" [5].

Edit Points and Segments on Character Lines. The character lines of passenger car form in rear view defined in this study consist of 21 descriptive lines that outline such parts as the roof, side window, rear bumper and wheel arch. The definition of character lines for passenger car form in rear view are shown in Fig. 1 and their labelling letters are listed in Table 1.

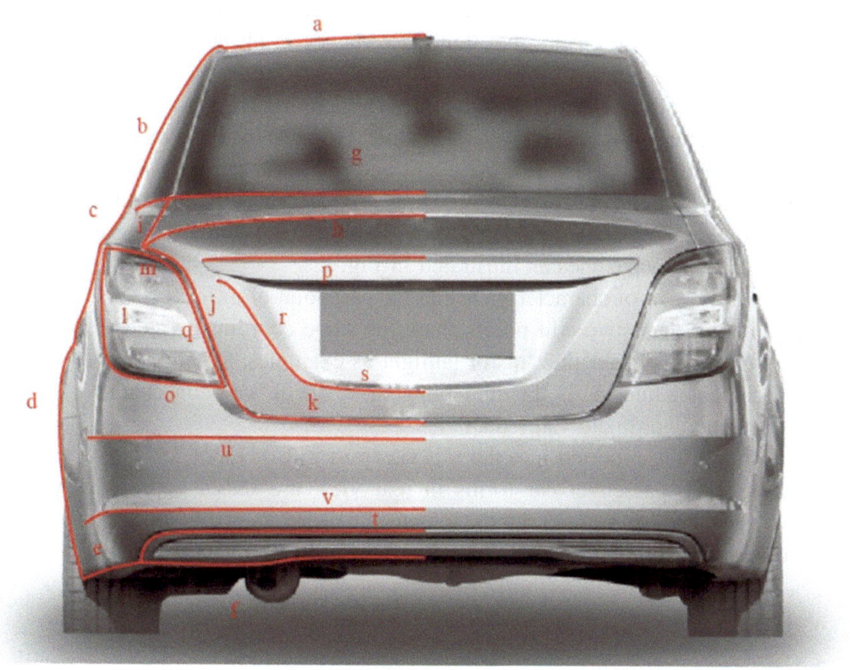

Fig. 1. Definition of character lines.

Table 1. Labelling character lines of different parts.

Parts	Character lines' labelling letters
Roof	a
Side window	b
Body shoulder	c
Wheel arch	d
Lower rear bumper	e, f, t, v, u
Taillight	m, q, l, o
License plate area	p, r, s
Trunk lid	g, h, i, j, k

The method to define edit points and segments on character lines and to obtain the coordinate value data of edit points are as follows.

Firstly, due to the symmetry of passenger car form in rear view, only character lines definition and data extraction of half of passenger car form are performed. Considering differences in whole contour and partial shape in passenger car forms in rear view, the versatility reflected by the configuration of character lines is discussed, and the simplification of the definition of character lines is maintained while a sufficient description of every form sample is ensured. Taking the taillight as an example, due to the large difference in taillight's shape, the character lines for taillight are more carefully defined as four outlines so as to ensure that taillight's shapes in all 80 form samples can be expressed accurately but concisely

Secondly, each character line is segmented by defining appropriate edit points with appropriate spacing on it. Taking character line 'a' for roof as an example, four edit points are set and marked as 'a1', 'a2', 'a3' and 'a4', respectively, and two adjacent edit points draw a segment. Thus, as shown in Fig. 2, four edit points divide character line 'a' into three segments, i.e., 'a12', 'a23' and 'a34'. Finally, 75 segments are included in 21 character lines as illustrated in Fig. 2.

Thirdly, for a pair of character lines that the end point of a character line is the start point of another, these two edit points share the same coordinate value. Taking character line 'o' and character line 'q' for taillight as examples, the end point of segment 'o78' on character line 'o' is the start point of segment 'q12' on character line 'q', having the same coordinate value data.

Fourthly, during imported into the top viewport in CAD tool for data acquisition, all pictures of 80 passenger car forms in rear view are scaled appropriately to the uniform height, and the left and right symmetry lines are aligned to the y axis and the x axis is used as the ground to align the bottom line of the wheel.

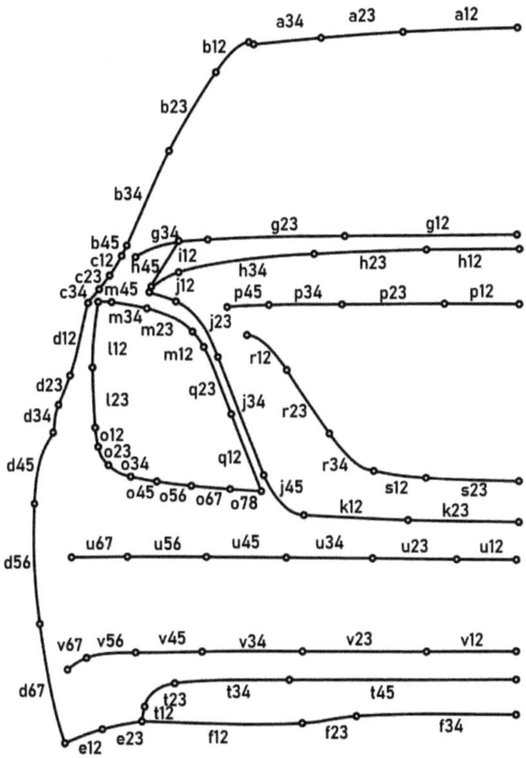

Fig. 2. Edit points and segments on character lines.

2.3 Coordinate Values of Edit Points

The slope of a segment reflects the degree of its own inclination. The calculation of the slope value of a segment needs to capture coordinate values of the start point and end point of that segment. During data acquisition, the (x, y) coordinate value of an edit point is displayed and then recorded when the cursor is put on it. In this way the coordinate value data of all start points and end points of 75 segments are sequentially recorded.

2.4 Slope Values of Segments

The slope of a segment can be calculated by the ratio of the difference between the end point' and the start point' y-coordinate values divided by that between the end point' and the start points' x-coordinate values. All slope values of 75 segments are obtained in a similar way. Taking segment 'a12' on character line 'a' as an example, by using the coordinate data of the start point, i.e., a1 (X_{a1}, Y_{a1}), and the coordinate data of the end point, i.e., a2 (X_{a2}, Y_{a2}), the slope value of segment ‘a12’ is calculated by $(Y_{a2} - Y_{a1})/(X_{a2} - X_{a1})$ and labelled as K_{a12}. Similarly, the slope values of other two segments on character line ‘a’ are calculated in turn and are labelled as K_{a23} and K_{a34}, respectively. Finally, the slope values of the segments of all 80 samples are calculated and recorded to obtain a complete piece of slope value data in a similar way.

2.5 Factor Analysis and the Extracted Contour Shape

Factor analysis is used to analyze the complete piece of slope value data for extracting contour shape of passenger car form in rear view. When factor analysis is performed, the principal component method and varimax rotation option are selected. By setting different eigenvalues for extracting factors, that is, eigenvalue is ‘greater than 1’, ‘greater than 2’, or ‘greater than 3’, factor analysis is executed repeatedly, resulting in the ‘total variance explained’ values of 80.704%, 56.286%, 36.245%, respectively.

Setting eigenvalue to be ‘greater than 2’ is used for common factors extraction in factor analysis to keep an approximate information reduction. As listed in Table 2, 13 common factors are extracted. Only when a segment has relatively large absolute value, greater than 0.5 set in this study, of the loading on one common factor but relatively small absolute value of the loading on the other common factors, it is a determinant of its corresponding common factor and is retained for drawing the contour shape. All those segments with above attribute of loadings are remained and as listed in Table 3, there are 47 segments retained which determine all common factors.

With the retained 47 key segments, the reduced contour shape extracted by factor analysis is illustrated as shown as the left figure in Fig. 3. The complete contour shape of the overall passenger car form in rear view is then restored by mirroring the extracted contour shape and illustrated as shown as the right figure in Fig. 3.

Table 2. 13 common factors extracted when eigenvalue is set to be ‘greater than 2’.

Rotated Component Matrix[a]													
	Component												
	1	2	3	4	5	6	7	8	9	10	11	12	13
a12	0.07	−0.20	0.15	0.07	−0.10	0.04	−0.08	0.02	−0.12	**0.51**	−0.09	−0.17	0.01
a23	0.07	0.03	0.15	0.05	0.10	−0.09	−0.02	0.09	−0.03	**0.83**	0.01	−0.03	−0.07
a34	0.13	0.13	−0.26	0.00	0.27	−0.02	−0.07	0.17	0.18	**0.65**	−0.09	0.13	−0.04
b12	0.00	0.01	−0.05	**0.82**	−0.04	0.03	0.10	−0.04	−0.01	0.16	0.09	−0.15	−0.02
b23	0.05	0.10	−0.01	**0.88**	−0.07	0.09	0.00	0.06	−0.06	0.10	0.00	−0.06	−0.05
b34	0.04	0.06	−0.02	**0.81**	0.09	0.09	−0.11	0.05	−0.09	−0.13	0.18	0.20	0.05
d34	−0.02	**0.62**	−0.01	0.10	0.03	−0.01	0.01	−0.03	0.01	−0.08	0.05	0.05	0.02
d67	−0.15	−0.10	−0.11	0.09	0.03	0.05	−0.14	0.33	0.09	0.05	−0.07	0.05	**0.60**
e12	−0.05	0.01	0.01	−0.11	0.01	0.05	−0.08	0.10	**0.85**	0.04	−0.03	−0.01	0.07
e23	−0.08	0.07	0.04	−0.05	0.01	0.10	−0.05	−0.04	**0.88**	−0.04	0.04	−0.03	0.01
f12	−0.06	−0.02	0.03	0.04	0.04	**0.71**	0.10	0.03	0.07	0.09	0.15	0.12	−0.17
g23	0.06	−0.37	0.04	0.13	−0.24	−0.25	0.03	0.19	0.12	0.27	**0.62**	−0.02	−0.04

(continued)

Table 2. (*continued*)

Rotated Component Matrix[a]

	Component												
	1	2	3	4	5	6	7	8	9	10	11	12	13
g34	0.13	−0.33	0.29	0.24	−0.18	−0.05	−0.01	−0.10	0.02	−0.09	**0.56**	0.16	−0.02
h12	−0.10	−0.22	0.18	0.05	−0.13	−0.29	0.15	0.01	0.05	−0.01	−0.02	**−0.59**	0.07
h23	−0.10	−0.41	0.14	0.00	0.00	−0.07	0.05	−0.09	0.14	0.13	−0.04	**−0.51**	0.21
j23	−0.06	0.09	−0.09	−0.01	−0.05	**−0.59**	0.00	−0.08	0.01	0.14	−0.03	0.20	0.01
j34	−0.02	−0.08	−0.05	0.06	0.01	**0.88**	−0.01	0.07	0.09	−0.07	−0.07	0.07	0.03
j45	−0.01	−0.07	−0.04	0.04	0.00	**0.87**	−0.02	0.07	0.09	−0.05	−0.05	0.07	0.02
k12	**0.55**	0.01	−0.01	−0.12	−0.05	−0.08	−0.24	0.07	−0.34	−0.02	0.20	−0.26	−0.31
k23	**0.61**	0.07	−0.12	−0.01	0.18	−0.02	−0.15	−0.19	0.07	0.18	0.05	−0.07	−0.30
l23	**0.63**	−0.12	0.14	−0.02	0.04	0.04	0.03	−0.04	0.03	−0.01	−0.02	0.40	−0.08
m12	0.00	−0.16	−0.02	0.01	**−0.57**	−0.05	−0.01	−0.06	0.11	−0.05	0.20	0.41	0.21
m23	0.15	0.06	−0.11	0.20	−0.06	0.08	0.04	**0.64**	0.07	0.09	0.19	−0.16	−0.18
m34	−0.01	0.06	−0.12	0.07	−0.14	0.07	−0.02	**0.90**	0.05	0.03	−0.03	−0.03	0.10
m45	0.01	0.00	0.03	0.02	−0.05	0.14	0.04	**0.83**	0.04	0.13	−0.07	−0.05	0.14
o12	0.13	**0.72**	0.12	0.01	−0.03	−0.10	−0.08	0.09	0.01	−0.04	−0.15	0.27	0.10
o23	0.11	**0.85**	0.10	−0.03	0.00	−0.02	0.03	−0.01	0.05	−0.01	−0.07	0.06	0.02
o34	0.14	**0.68**	0.14	−0.03	−0.05	−0.30	0.25	−0.09	0.03	0.11	0.01	0.00	−0.11
o45	0.12	0.08	**0.74**	−0.03	0.04	0.03	0.06	0.02	0.04	0.13	0.11	−0.04	−0.06
o67	−0.03	−0.03	0.41	0.03	0.09	0.03	**0.79**	0.02	0.07	0.10	0.06	0.03	0.01
o78	−0.05	0.11	0.10	0.01	0.01	−0.13	**0.89**	−0.07	0.02	0.00	0.03	−0.05	0.01
p34	0.00	0.05	−0.35	0.23	0.20	0.10	0.18	−0.05	−0.15	0.14	**0.64**	−0.13	0.04
r12	−0.07	−0.01	−0.03	−0.01	**0.79**	0.11	−0.05	−0.10	0.14	0.04	0.14	0.06	0.01
r23	0.10	0.01	−0.15	−0.01	**0.69**	−0.03	0.09	−0.04	0.01	0.08	−0.04	0.14	−0.08
r34	0.05	0.04	0.01	−0.01	**0.82**	0.03	0.00	0.03	−0.08	0.04	−0.03	0.06	0.10
s12	**0.62**	0.20	−0.04	−0.05	−0.01	−0.02	0.08	0.00	−0.21	−0.06	0.34	−0.19	−0.01
s23	**0.56**	0.22	−0.06	0.01	−0.03	0.00	0.04	0.02	0.00	−0.02	0.35	−0.13	0.07
t23	−0.05	0.09	−0.02	0.10	0.02	0.12	0.05	−0.01	**0.68**	0.01	−0.04	−0.08	0.11
t34	−0.11	**0.58**	−0.02	0.07	0.41	−0.02	0.06	0.12	0.12	0.00	−0.06	−0.20	−0.03
u12	**0.82**	−0.02	−0.02	−0.08	−0.10	0.03	−0.16	0.00	0.06	0.05	−0.12	0.03	0.02
u23	**0.73**	0.15	0.39	0.04	0.01	−0.09	0.02	0.15	−0.02	0.01	−0.10	0.03	0.10
u34	0.26	0.15	**0.73**	0.03	0.09	−0.10	0.25	0.07	0.06	−0.08	0.09	0.10	0.05
u45	**0.58**	−0.06	0.07	0.19	0.03	−0.02	0.19	0.16	−0.11	−0.08	−0.11	−0.05	0.08
u56	**0.79**	0.00	−0.02	0.11	0.04	0.02	−0.03	0.03	−0.03	0.02	−0.08	0.21	−0.01
u67	−0.10	0.03	0.10	−0.02	0.02	0.10	**0.61**	0.08	0.01	0.02	0.01	0.03	0.00
v34	0.08	−0.01	**−0.74**	0.00	0.06	−0.04	0.08	0.09	0.06	−0.09	0.21	0.03	0.06
v45	0.01	−0.03	**−0.69**	0.06	0.14	−0.02	−0.18	0.10	0.05	0.10	0.03	0.18	−0.10

Extraction Method: Principal Component Analysis.
Rotation Method: Varimax with Kaiser Normalization.
a. Rotation converged in 28 iterations.

Table 3. 47 segments determining 13 common factors.

Common factors	Retained segments
Common factor 1	k12, k23, l23, s12, s23, u12, u23, u45, u56
Common factor 2	d34, o12, o23, o34, t34
Common factor 3	o45, u34, v34, v45
Common factor 4	b12, b23, b34
Common factor 5	m12, r12, r23, r34
Common factor 6	f12, j12, j23, j34
Common factor 7	o67, o78, u67
Common factor 8	m23, m34, m45
Common factor 9	e12, e23, t23
Common factor 10	a12, a23, a34
Common factor 11	g23, g34, p34
Common factor 12	h12, h23
Common factor 13	d67

Fig. 3. The extracted and restored contour shape.

3 Conclusion

The contour shape for simplistically expressing passenger car form in rear view is extracted by quantitative analysis methods based on young Chinese consumers' perception of and similarity judgement on triple-compartment passenger car form in rear view from mainstream brands in Chinese passenger car market.

The result shows that the contour shape is described by 47 key segments on 21 character lines. It is mainly consisted of character lines defining such areas as roof and side window, the C-type contour of taillight, the U-type edges of sunken license plate area, the U-type boundary and the turning edge between the horizontal and vertical surfaces of trunk lid, the horizontal transitional edge corresponding to character line 'u' on the upper rear bumper, and the contour of the bottom corner area on rear bumper in rear view. Meanwhile, the restored complete contour shape of passenger car form in rear view can still be recognized after reduction with 43.714% loss of information on passenger car form in rear view.

In view of that it is extracted based on form samples selected by young consumers' form similarity judgement, the reduced contour shape implies to some extent consumers' perceptional characteristics of passenger car form in rear view.

Acknowledgement. The authors would like to extend thanks to Yi Jin for participating in part of the work in this study.

References

1. Nagamachi, M.: Kansei engineering: a new ergonomic consumer-oriented technology for product development. Int. J. Ind. Ergon. **15**, 3–11 (1995)
2. McCormack, J.P., Cagan, J.: Speaking the buick language: capturing, understanding, and exploring brand identify with shape grammars. Des. Stud. **1**, 1–29 (2004)
3. The White Paper of Chinese Individual Auto Brand in 2018. (in Chinese). https://www.d1ev.com/kol/61262. Accessed 1 Dec 2018
4. Liu, C.: Design Strategy Development for Product Innovation. Shanghai Jiao Tong University Press, Shanghai (2015). (in Chinese)
5. Zhao, D., Zhao, J.: Automobile form feature and feature line. Packag. Eng. **28**, 115–117 (2007). (in Chinese)
6. Hu, W., Zhao, J., Zhao, D.: Study on styling image of vehicle based on form feature lines. China Mech. Eng. **20**, 496–500 (2009). (in Chinese)
7. Chen, Z.: A feature-line-based research of TOYOTA brand automobile styling evolution. Master's thesis, Guangdong University of Technology, Guangdong (2013). (in Chinese)
8. Sun, D.: German Volkswagen's style evolution. Master's thesis, Jiangnan University, Jiangsu (2008). (in Chinese)
9. Wang, W.: Design of freight locomotive based on characteristic line. Master's thesis, Southwest Jiaotong University, Sichuan (2014). (in Chinese)
10. Deng, Y.: Research on molding design of electric logistic light truck based on morphological characteristics. Master's thesis, Taiyuan University of Technology, Shanxi (2018). (in Chinese)
11. Zhang, W.: Advanced Course of Statistical Analysis in SPSS. Higher Education Press, Beijing (2004). (in Chinese)

Research on Evaluation of CMF Based on Vision in Automobile Seat Design

Canqun He(✉), Xue Sun, and Weizhen Xiao

College of Mechanical and Electrical Engineering, Hohai University, 200 Jinling North Road, Changzhou 213022, China
hecq@163.com

Abstract. CMF design is the key to improve the quality of product design, which has been widely used in electronic consumer products, household appliances, product design for special users, interior design, and other aspects. This research is the CMF design of automobile seat in user's vision. Firstly, the main CMF elements of automobile seat and CMF combination scheme are investigated. Secondly, the most representative CMF combination scheme is defined by questionnaire and focus group discussion. Thirdly, with the help of the theory of fuzzy sets, the comprehensive level of users' satisfaction with the color, material and surface decoration of representative CMF combination schemes is counted, and the user preferences of CMF design of automobile seat based on vision are obtained. Finally, a new idea of CMF design of automobile seat is hoped to be provided.

Keywords: CMF · Automobile seat · Vision · Evaluation

1 Introduction

With the overall development of the automotive industry, it has brought about many improvements in automotive design. In terms of function, the noise reduction performance of automotive interior materials [1], the odor test of automotive interior materials [2] and other researches are all aimed to optimize the functional design of automotive interior. Of course, interior design has not only improved the function, it has gradually changed from functional design to overall atmosphere design, and now it gradually regards user perception as an important factor of automobile interior design [3, 4]. User perception is the basis of user experience influencing, and to a large extent, user perception of interior sensory experience determines the purchase behavior of users [5], so it is necessary to study the interior design based on user perception. As the main subsystem of automobile interior decoration system, the research of automobile seat system based on user perception evaluation is also very important.

CMF design is a deeper exploration under the subdivision of industrial design, and is the key to improve the quality of product design. It has universality and has been gradually applied to the design of products in various fields such as electronic consumer products, household appliances, special users' product design and interior

C. S. Shin (Ed.): AHFE 2019, AISC 968, pp. 146–158, 2020.
https://doi.org/10.1007/978-3-030-20470-9_17

design. CMF design methods are gradually enriched, including emotional context, market research, storytelling, trend iteration, character creation, and so on [6], which can improve the quality of product design. In addition, Nooree Na and Hyeon Jeong Suk, analyze CMF's C (Color) with different White colors, M (Material) with model texture, and F (Finish) with the gloss in the book "The Emotion Characteristics of White for Applications of Product Color Design" [7], reflecting The flexibility of CMF research.

Therefore, we choose to combine CMF design with automobile seat design on the basis of user perception. This paper makes an in-depth study of CMF design of automobile seats from visual perspective, hoping to provide new design ideas for the design of automobile seats, so as to optimize the design of automobile seats.

2 CMF Element Analysis of Automobile Seats

With the expansion of the automobile market, the online automobile market is gradually enriched, such as the Automobile Home, Easily Shared, SOHO Automobile, Pacific Automobile and so on. These platforms have brought great convenience to automobile users, and has gradually become the first choice for automobile buyers to purchase automobiles. Therefore, this study selected dozens of popular brands of automobile websites represented by the automobile home as the analysis domain of the CMF elements of this automobile seat. Through further analysis and arrangement, we determined that the texture and material itself should be taken as the M (material) elements of CMF elements of automobile seat, and the surface stitching and punching patterns of automobile seat should be taken as F (finish) elements.

2.1 Color Elements of CMF for Automobile Seat

The main color systems of automobile seats can be summarized as follows: black, red, grey, white, beige, brown and blue. In addition, there are a large number of color matching seats, which can be roughly summarized into two categories: the black-white-grey color matching seats and the black-white-grey matching other color seats. This paper summarizes the emotional characteristics of automobile seats with different color systems. The main color and basic emotional characteristics are shown in the following table:

As can be seen from Table 1, the colors of black, white and gray show the characteristics of high-end atmosphere. Beige shows obvious warmth and affinity, and red shows vitality and enthusiasm. The color and emotional characteristics of automobile seats are not unilateral. The depth of color, matching mode and shape of colors will have an impact on the emotional characteristics of colors in a certain seat. For example, in addition to the high-end atmospheric characteristics, white also has the characteristics of science and technology under certain conditions.

Table 1. Main classification and basic emotional characteristics of automobile seat color

Color Classification			Basic Emotional Characteristics
Series of Single Color	Black series		High-end, Luxury, Ambient
	White series		High-end, Sense of technology (concept automobile)
	Grey series		Ambient, Fashion
	Beige series	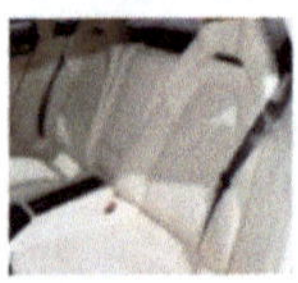	Warm
	Red series		Passion
	Brown series		Luxury (dark brown) Warm (light brown)
Series of Matching Color	Black-white-grey color matching		High-end, Ambient Fashion、Athletic
	Black-white-grey matching other color		Energetic

2.2 Material Elements of CMF for Automobile Seats

Material, as an important factor to be considered in the design of automobile seat, directly affects the quality and grade of automobile, and reflects the quality of automobile to a certain extent. At present, the materials of the automobile seat are mainly leather and fabrics, among which the leather includes dermis, microfiber leather, PU synthetic leather and PVC artificial leather, and the fabric includes cotton and linen fabric, wool fabric, blended fabric and chemical fiber [8]. In addition, there are some materials used in combination. Based on the overall visual effect of the automobile, the materials are roughly summarized as follows: pure leather, pure fabric, leather and fabric collocation, leather and velvet material collocation, and other materials collocation. As for what kind of leather and what kind of fabric to match, it is a deeper subdivision. Through a brief analysis, this paper summarizes the emotional characteristics of the basic materials of the main material matching types (Table 2).

Table 2. Main classification and basic emotional characteristics of automobile seat material

Color classification		Basic emotional characteristics
Single material	Pure leather (Leather, microfiber leather, PU, PVC)	High-end, Luxury
	Pure fabric (Cotton and linen fabrics, wool fabrics, blended fabrics and chemical fabrics)	Natural, Simple
Mixed material	Mix of different leathers (Such as dermis and leather)	High-end, Luxury
	Leather and fabric	Comfortable
	Leather and velvet	Advanced Unique
	Blending of other (non-cortical) materials	Individual, Novel

The analysis of M elements of automobile seat mentioned above should include not only the material itself, but also the surface texture. And the surface texture of automobile seat material is mainly embodied in the cortical material. The surface texture of leather material mainly includes lychee grain, BMW grain, Nappa grain, fine grain and so on. Among them, Lychee grain and fine grain are widely used in the automobile seat surface texture. BMW grain and Nappa grain are the texture of high-grade leather. BMW grain is generally used on some PU leather and microfiber leather, while Nappa grain itself is the texture of top-layer cowhide. Of course, some of them are designed to simulate the effect of leather and choose to apply Nappa grain on PU.

2.3 Surface Processing Elements of CMF for Automobile Seats

In this study, in addition to the surface texture mentioned above, we take the surface decoration effect of automobile seat design as the F element in the CMF design of automobile seat, including the stitching style, punching style and brand decoration. Firstly, stitching in CMF design of automobile seats can highlight the delicacy of hand-made and high-end added value. It integrates the delicacy of hand-made into industrial batch products skillfully, making it more artistic while meeting the functional requirements. In the analysis of automobile seat stitching style, we divide it into two levels. The first level is the stitching effect based on the overall feeling, which mainly includes general stitching and combination of general stitching and pattern stitching. Among them, pattern stitching refers to the use of stitching to draw a repetitive or regular pattern on the surface of the seat to decorate the it, while general stitching refers to the general overall stitching pattern between the stitches which does not constitute multiple repetitive patterns. The second level is the specific pattern of stitching, including pattern stitching, ordinary stitching, single and double stitching lines, two-color stitching, decorative stitching and flanging stitching. Different stitches have different decorative effects, as shown in Fig. 1.

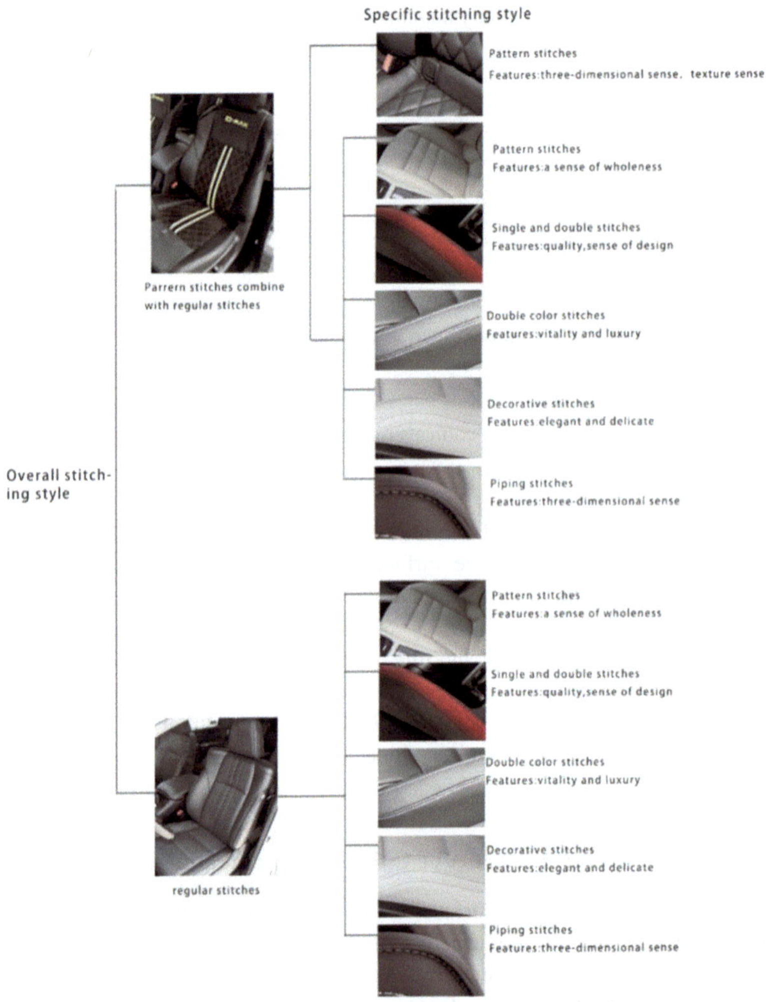

Fig. 1. Main classification and basic emotional characteristics of automobile seat stitching

For punching, it has two main functions: one is to have better function, to increase the ventilation and permeability of the seat; the other is to have a certain decorative effect. Especially with the improvement of punching technology, the punching style is getting increasingly abundant, and the decorative effect of the punching is correspondingly more and more obvious. These ample punching styles create a unique aesthetic for the leather seat. Similarly, the analysis of punching styles can be divided into two levels: whole and detail. The overall punching effect can be divided into detail punching and the main body punching, while the specific punching styles mainly include several categories: repetitive element punching, gradient element punching, penetration punching and geometric punching (as shown in Table 3).

Table 3. Main classification and basic emotional characteristics of automobile seat punching

Punching style		Basic Emotional Characteristics
Repetitive Punching	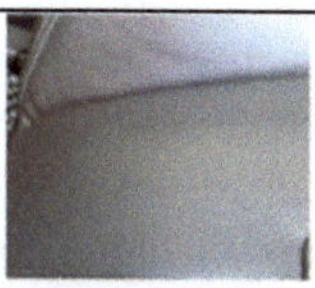	Sense of quality and decoration
Geometric Punching	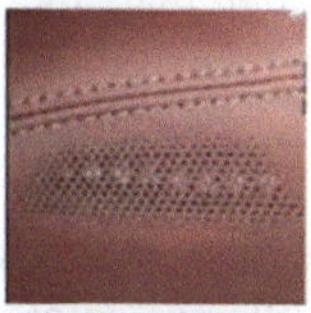	Sense of design
Gradient Punching	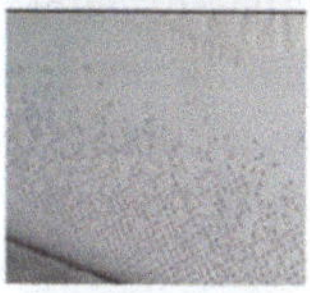	Sense of future
Penetration Punching		Sense of exquisiteness and energy

For brand decoration, although it is only the decoration of the details of the brand, it is also the main component of the automobile seat surface decoration effect. At the same time, it also has a certain role in brand promotion. At present, there are three main forms of brand decoration technology: embossing, embroidery, nameplate. Some brands apply the color splicing effect with brand characteristics to the design of the entire automobile seat as a brand decoration, such as the black strip stitching of the BMW M5 in white seats and so on.

3 Comprehensive Evaluation and Analysis of CMF Elements of Automobile Seat Based on Vision

According to the different nature of each element of CMF, we can know that some elements mainly act on human vision, such as color, texture stitches, and some elements mainly act on human touch, such as material and so on. Although material elements mainly affect human touch, they also affect human vision based on material characteristics. Therefore, in the comprehensive evaluation of automobile seat based on vision, we choose the color of automobile seat, seam stitching and the first level classification of automobile seat material for comprehensive evaluation.

3.1 Selection and Determination of CMF Combination Scheme

First, we need to screen CMF combinations to determine all CMF combinations. First, we chose the color red, the material leather and the texture general suture, then a CMF combination scheme based on the overall effect can be determined first, and then the combination forms of other colors, materials and textures can be determined in turn. In this way, 80 ideal CMF combination schemes can be determined, and 35 CMF combination schemes can be determined by preliminary screening combined with the actual situation of automobile seats on the market. After that, the user's preferences are understood through questionnaires, and then the focus group is used to discuss with those who have experience in using and practicing automobiles and new users of automobiles. So we can screen out the most representative combination scheme of automobile seats. We conducted a questionnaire survey by ranking preferences, and investigated the preferences of 224 people for different colors, materials and sewing styles of automobile seats. We assign samples one by one according to users'preferences. The one we like best is 8 points and the one we dislike least is 1 point. Then we make statistics. Taking color elements as an example, there are 178 valid data, and the statistics of each color are shown in Table 4. Figure 2 is the statistical results of the average values of material of the user preference questionnaire.

Table 4. Statistical table of the score of color elements in user preference questionnaire

Color classification	Mean value	Stand deviation
Black series	5.16	1.75
Grey series	5	2.27
White series	5.10	2.32
Beige series	4.66	2.64
Red series	5.01	1.69
Brown series	2.78	2.27
Black-white-grey matching	4.63	2.15
Black-white-grey matching other color	3.72	1.95

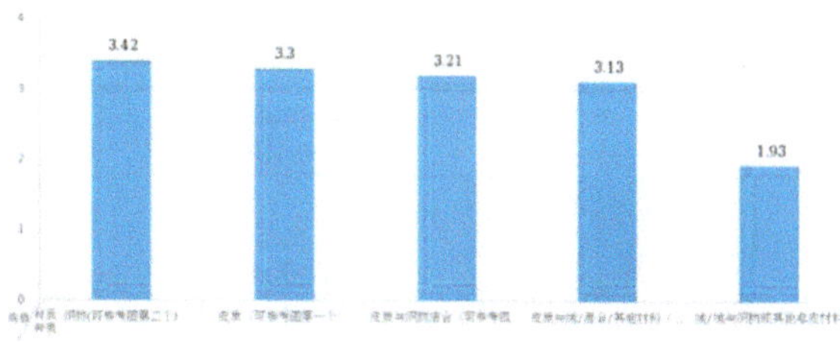

Fig. 2. Statistics of material elements in user preference questionnaire

Data statistics show that in terms of color, only black-white-grey and color matching seats and brown seats have lower average values, and the standard deviation between black-white-grey and color matching seats is small, so it can be considered that this scheme is not very popular with users. In terms of material quality, it is found that the average value of material collocation schemes between wool, fabric or other non-leather materials is lower, and the average value of other schemes is similar and higher, so it can be considered that this scheme is not very popular with users. In the overall effect of sewing, the number of users who like general sewing is slightly higher than the number of people who like pattern sewing, but there is no significant difference. In the survey of users' preferences about specific stitching and punching patterns, the average values of each stitching pattern and punching are similar. We can consider more the significance of the design effect of details to seat design.

On the basis of the research results, through focus group discussion, the CMF combination of visual-based representative automobile seats is determined. In the discussion, our core task is to select a representative CMF combination of automobile seats. Firstly, 35 combination schemes are shown, and the group members give some suggestions about which schemes are representative, which are not feasible and which are similar. Through recording, collating and summarizing, 16 representative vision-based CMF schemes for automobile seats are finally determined (Table 5).

3.2 Comprehensive Evaluation of Representative CMF Combination Schemes

With the help of fuzzy evaluation method [9, 10], users can evaluate the importance of CMF three elements to determine the weight of the three elements in the CMF combination. Combining weight and CMF elements, the comprehensive evaluation of CMF combination is obtained. Fuzzy comprehensive evaluation method is a comprehensive evaluation method based on fuzzy mathematics. According to the theory of membership degree of fuzzy mathematics, the comprehensive evaluation method transforms qualitative evaluation into quantitative evaluation, that is to say, it makes a general evaluation of things or objects restricted by many factors by using fuzzy mathematics.

First of all, we surveyed 224 people to find out the importance of color, material and surface decoration for automobile seat design. The importance of user selection mainly has five dimensions, which are very unimportant, unimportant, indifferent, important and very important. The following table is a survey of the importance of CMF elements (Table 6).

Table 5. Sixteen representative CMF combination schemes for automobile seats

Representative CMF portfolio	Typical case	Representative CMF portfolio	Typical case
1 red Pure leather General stitching		2 Red Pure leather Pattern stitching	
3 Red Leather and velvet patchwork General stitching		4 white Pure leather Pattern stitching	
5 white Single material General stitching		6 beige Pure leather General suture	
7 beige Pure fabric General stitching		8 Black Grey Monochrome Pure leather General suture	
9 Black Grey Monochrome Pure Leather Pattern stitching		10 black Leather - velvet splicing material Pattern stitching	
11 black Leather - velvet splicing materal General stitching	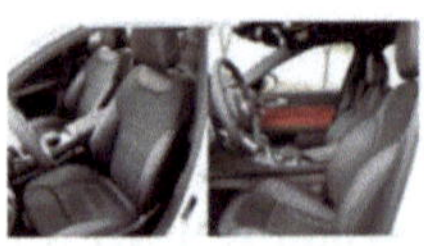	12 black and grey color matching leather and fabric materials	
13 Black, White and Grey Color matching Pure fabric		14 Black Grey Monochrome Pure Fabric General stitching	
15 Black, White and Grey Color matching Pure leather General stitching		16 brown Pure leather General stitching	

Table 6. Statistical table of importance survey of CMF elements

Elements	Very unimportant	Unimportant	Indifferent	Important	Very important
C(color)	8.48%	4.02%	22.32%	55.36%	9.82%
M(material)	0%	16.38%	8.47%	47.46%	27.68%
F(surface decoration)	5.65%	5.08%	24.29%	51.41%	13.56%

The weight is calculated by quantitative statistical method.

Firstly, the percentage of each index can be calculated by statistics of the survey results, and then the limit is 60%. If the importance is greater than 60%, the index is effective, and it is found that all three indicators are effective. Then the very important assignment is 1, the unimportant assignment is 2, the indifferent assignment is 3, the more important assignment is 4, and the important assignment is 5. Then the important and above weights are: important: 4/(4 + 5) = 0.4444; very important: 5/(4 + 5) = 0.5556.

Then the weight of color is:

$$(55.36 * 0.4444 + 9.82 * 0.5556)/\{(55.36 * 0.4444 + 9.82 * 0.5556) \\ + (47.46 * 0.4444 + 27.68 * 0.5556) + (51.47 * 0.4444 + 13.56 * 0.5556)\} = 0.3102$$

The weight of material is:

$$(47.46 * 0.4444 + 27.68 * 0.5556)/\{(55.36 * 0.4444 + 9.82 * 0.5556) + \\ (47.46 * 0.4444 + 27.68 * 0.5556) + (51.47 * 0.4444 + 13.56 * 0.5556)\} = 0.3763$$

The weight of surface decoration is:

$$(51.47 * 0.4444 + 13.56 * 0.5556)/\{(55.36 * 0.4444 + 9.82 * 0.5556) + \\ (47.46 * 0.4444 + 27.68 * 0.5556) + (51.47 * 0.4444 + 13.56 * 0.5556)\} = 0.3135$$

Therefore, the weight of the three elements is 0.3102 for color, 0.3763 for material and 0.3135 for surface decoration.

After that, we extracted 16 samples and asked 224 users to judge whether the color, material and surface decoration of the sample can satisfy them. We divided the degree of satisfaction into five levels. With the help of the method of fuzzy evaluation, we took CMF as the factor set of fuzzy evaluation, satisfaction as the evaluation set, and the weight set of CMF as the weight set. The mathematical model of fuzzy evaluation is established. In the process of establishing the model, we should first make clear the correspondence between each index and evaluation. Taking the first sample as an example, the following table is the statistics of the evaluation of its various elements (Table 7):

Table 7. CMF user satisfaction statistics of sample 1

Elements	Very dissatisfied	Dissatisfied	Indifferent	Satisfied	Very satisfied
C(color)	6.07%	9.82%	25.45%	48.21%	9.82%
M(material)	0%	17.51%	11.3%	46.89%	24.29%
F(surface decoration)	3.39%	7.91%	27.12%	48.02%	13.56%

Thus we can establish a fuzzy judgment matrix:

$$P_1 = \begin{bmatrix} 0.067 & 0.0982 & 0.2545 & 0.4821 & 0.0982 \\ 0 & 0.1751 & 0.113 & 0.4689 & 0.2429 \\ 0.0339 & 0.0791 & 0.2712 & 0.4802 & 0.1356 \end{bmatrix}$$

From the previous analysis, we can see the weight ω:

$$\omega = (\,0.3102 \quad 0.3763 \quad 0.3135\,)$$

Then we can get a comprehensive evaluation **B**:

$$\mathrm{B} = \mathrm{P} \cdot \omega \tag{1}$$

Then it can be obtained by fuzzy calculation:

$$B_1 = P_1 \cdot \omega = (\,0.1 \quad 0.2 \quad 0.2 \quad 0.3 \quad 0.2\,)$$

From this, we can know that for the first sample, 30% of the people think this set of plans is very unsatisfactory or unsatisfactory, 50% think this set of plans is general or satisfactory, and 20% think this set of plans is very satisfactory.

Similarly, we can calculate the comprehensive evaluation B of the other 15 schemes. They are:

$$B_2 = (\,0.1 \quad 0.1 \quad 0.2 \quad 0.3 \quad 0.3\,)$$
$$B_3 = (\,0.1 \quad 0.2 \quad 0.2 \quad 0.3 \quad 0.2\,)$$
$$B_4 = (\,0.1 \quad 0.2 \quad 0.2 \quad 0.3 \quad 0.2\,)$$
$$B_5 = (\,0.1 \quad 0.2 \quad 0.2 \quad 0.3 \quad 0.2\,)$$
$$B_6 = (\,0.1 \quad 0.1 \quad 0.2 \quad 0.4 \quad 0.2\,)$$
$$B_7 = (\,0.1 \quad 0.2 \quad 0.2 \quad 0.3 \quad 0.2\,)$$
$$B_8 = (\,0.1 \quad 0.2 \quad 0.2 \quad 0.3 \quad 0.2\,)$$
$$B_9 = (\,0.1 \quad 0.2 \quad 0.2 \quad 0.3 \quad 0.2\,)$$
$$B_{10} = (\,0.1 \quad 0.1 \quad 0.3 \quad 0.3 \quad 0.2\,)$$
$$B_{11} = (\,0.1 \quad 0.2 \quad 0.2 \quad 0.3 \quad 0.2\,)$$
$$B_{12} = (\,0.1 \quad 0.1 \quad 0.2 \quad 0.3 \quad 0.3\,)$$
$$B_{13} = (\,0.1 \quad 0.1 \quad 0.2 \quad 0.3 \quad 0.3\,)$$
$$B_{14} = (\,0.1 \quad 0.2 \quad 0.2 \quad 0.3 \quad 0.2\,)$$
$$B_{15} = (\,0.1 \quad 0.1 \quad 0.2 \quad 0.3 \quad 0.3\,)$$
$$B_{16} = (\,0.1 \quad 0.1 \quad 0.2 \quad 0.3 \quad 0.3\,)$$

We can see that 30% of the people in the second, twelve, thirteen, fifteen and sixteen programs are satisfied with them, that is, a relatively large number of people are satisfied with them, 60% of the people in the sixth and tenth programs and a relatively large number of people think that they are generally or satisfied with them. These plans, namely 2, 6, 12, 13, 15 and 16, also have a relatively small number, that is, 20% of the unsatisfactory population.

4 Conclusion

In the process of researching on CMF of automobile seat, with the help of fuzzy evaluation method, I investigated the comprehensive evaluation of CMF elements of automobile seat on the visual level. Through the research, I determined that the CMF combination styles of automobile seats with high customer satisfaction. There are mainly red pattern seats, Beige pure leather seats, Black-gray color matching leather and fabric seats, black-white-grey color matching pure fabric seats, black-white-grey color matching pure leather seats and brown pure leather seats. It can be seen that most users still like the design of automobile seats with some decorative effect, rather than simple and pure color and simple sewing. At the same time, most users are conservative, some too neoteric designs and the use of a large number of decorative details have not been recognized by more users.

The automobile seat can be a kind of individual product or a part of the automobile interior system. In this paper, the automobile seat can be analyzed and studied independently. Then we can get a design scheme or idea which is universally accepted of the CMF design of automobile seat. At the same time, this study can also provide ideas for the design of other automobile parts, and then a design scheme from details to the whole for the automobile interior is proposed.

References

1. Zhao, M.M., Du, Z.F.: Research status and development trend of sound absorption and insulation performance of automobile interior materials. J. Ind. Text. **27**, 6–8, 16 (2009)
2. Tang, B.B., Guo, G., Xia, J.: Olfactory experience evaluation and device of automobile interior materials odor. J. China Mech. Eng. **28**, 206–214 (2017)
3. Bao, H.: Research on design trend of automotive interior trim. J. Des. **11**, 58–59 (2017)
4. Pang, L.Q.: On the "people - oriented" automotive interior design. J. Automot. Pract. Technol. **20**, 179–181 (2018)
5. Chen, Y.W., Yu, S.L.: Research on automotive interior design based on user perception. J. Sci. Technol. Innov. **6**, 49–51 (2018)
6. Zhao, W.: CMF design process and method in industrial design. J. Sci. Technol. Econ. Market. **9**, 1–2 (2018)
7. Na, N., Suk, H.J.: The emotion characteristics of white for applications of product color design. Int. J. Des. **8**(2), 61–70

8. Du, Q.: Development and application of automobile interior fabric. J. Zhejiang Text. Cloth. Vocat. Tech. Coll. **8**, 22–25 (2009)
9. Shen, Z., Zhang, B.: Research on DNA analysis of product design based on fuzzy theory. J. Sci. Technol. Prog. Countermeasures. **29**, 74–78 (2012)
10. Zhao, Y.Y., Bian, F., Li, J.T.: Study on material intention of products based on Kansei engineering. J. Mech. Des. **8**, 117–121 (2015)

Study on the Design Elements of Shape and Interface of Modern Agricultural Machinery

Yitong Liu(✉) and Huabin Wang

School of Design, South China University of Technology,
Guangzhou 510006, China
1006532167@qq.com, hbwang@scut.edu.cn

Abstract. To improve the interfacial configuration and the core competitiveness of domestic agricultural machinery equipment. Using the concept of design science, this dissertation summarizes the methods of shape and interface design of modern agricultural machinery, mainly exploring from the modeling, color and enterprise VI design. In conclusion, the application of design science in the equipment of agricultural machinery can effectively improve the ergonomic effect of the interfacial form, enhance the aesthetic quality and additional value of the products.

Keywords: Agricultural machinery · Appearance design · Design method

1 Introduction

With the growth of the awareness of innovation in China, the field of industrial design is attracting more and more attention. Production and sales of any product are inseparable from the blessing of design. As a great country whose prosperity depends highly on agriculture, the manufacture of agricultural machinery equipment plays a significant role in China's national agricultural development. China has been making efforts to develop advanced manufacturing technology during these years. However, the form of domestic agricultural equipment is still poorly designed, falling behind that of other developed countries. Therefore, how to systematically summarize and conclude the modeling design methods of modern agricultural machinery and improve its scientific and technological content at the same time is a problem worth further exploration.

This study was funded by the Guangdong Provincial Social Science Research Fund Project GD17XYS18.

C. S. Shin (Ed.): AHFE 2019, AISC 968, pp. 159–170, 2020.
https://doi.org/10.1007/978-3-030-20470-9_18

2 Analysis of Main Design Elements of Agricultural Machinery Equipment

According to the use of agricultural machinery, it can be classified as field management machinery, post-harvest processing machinery, agricultural conveying machinery and animal husbandry and aquaculture machinery [1]. According to the method of system analysis, the structure of agricultural machinery can be classified as function host, control system, running system and functional implementation system. According to the overall form, it can be classified as internal structure and external man-machine interface. This article studies the shape and interface design elements of agricultural machinery. The main research object is the shape element, followed by color collocation and enterprise VI design.

The shape of product determines its overall style and impression of users, which is the key point of agricultural machinery equipment design. Appropriate color collocation can enrich the modeling level of the product and also make the equipment adapt better to the environment when scientifically used. And VI design is a significant way to form brand DNA of agricultural machinery equipment and improve core competitiveness of enterprises.

3 Shape Design Method of Agricultural Machinery Equipment

The shape innovation is the most direct embodiment of product innovation [2], and it is also the primary method for enterprises to survive and develop in the fierce market competition [3]. Good appearance design can not only visually enhance the overall image of agricultural machinery products, but also close the distance between products and users. Due to the huge gap of design development between China and foreign countries in the past few decades, it is still difficult to catch up with the design level of other countries in a short period of time. However, only by improving the appearance design and perfecting the system design method on the basis of existing technical level, can the competitiveness of domestic agricultural machinery equipment be promoted.

3.1 Harmony and Unity of the Overall Shape of the Product

The design and development of products are determined by many factors, and it is not advisable to separate any of them [4]. Similarly, agricultural machinery consists of multiple parts. During the design process, only when the coordination and balance among different parts be achieved, can the harmony of the overall shape be maintained. Taking the model design of a series of rotary tiller in Guanglian as an example, the rotary tiller consists of rotating cutter teeth, manual operation room, crawler belt and chassis. In the design of some parts, such as the chassis design, its shape should conform to the overall direction of the fuselage, and be harmony with other parts of the fuselage. The track shape of the rotary tiller is generally forward-tilting on the basis of the quadrilateral, and has a certain impact force visually, so this trend should be followed when designing the chassis shape. Additionally, according to the basic type of the track, the box is cut and chamfered to provide a sense of smooth as well as mechanical strength. The center of gravity of the rotary cultivator body is in front, giving an unsteady feeling, but after assembling the tail rotary knives, the body will reach equilibrium.

In the case of model derivation, the lateral derivation method can be used to derive the overall shape by drawing the side view of the fuselage in order to seek the best solution (Fig. 1).

Fig. 1. A series of rotary tiller in Guanglian.

The unity of the styling trend is reflected in the representative products of some well-known agricultural machinery brands, and in order to avoid the heavy visual experience of the body of the equipment, the upward trend is usually adopted (Fig. 2).

Fig. 2. Comparison of modeling trends of major brands of agricultural machinery products.

3.2 The Philosophy of Establishing and Breaking the Product Model

On the basis of styling consistency, if the fuselage consists of unchanging blocks, it will make the device appear dull, sluggish and bulky. Appropriate use of common modeling methods such as shape decrease and cutting can break the original frame and make the form more flexible. The following picture shows the dump truck design of Japanese machinery brand Komatsu. Based on typical square shape, appropriate chamfering and remodeling changed the original bulky box and bring more design details to the product. At the same time, the inclined angle caused by shape cutting of the bottom of the bucket can increase the tipping angle, which is more convenient for the dumping operation. Japanese brand CANYCOM also use the same design method, which adopts the form cutting to convey the light and flexible style. The color of red makes the agricultural machinery products subvert the traditional image of clumsy and heavy, presenting a more modern visual effect while displaying the mechanical power.

In the design process, it is not arbitrary to change the shape, aesthetic principle should be used appropriately to assist this task (Fig. 3).

Fig. 3. Komatsu and Canycom tipping bucket design.

3.3 Relationship Between the Virtual and the Real Shows the Modeling Level

The virtual and the real, the hidden and the exposed, they are the actual performance of the modeling level. The flat surface without any change will make the body appear dull and unwieldy and lack the input of design, resulting in the cheap sense of the product. The relationship between virtual and real can be reflected in the front and back contrast of the shape block. Appropriate segmentation or stamping on the same plane can not only enrich the visual effect of the modeling, but also ensure the solidity of the panel surface and make it not easy to be bent. And the relation of hidden and exposed decided the breathable effect of product body. If all the mechanical structures of agricultural machinery products are covered by differently shaped cabinets, the products presented must be heavy and clumsy, and cannot reflect the unique beauty brought by the machinery. The mechanical structure of the agricultural machinery is exposed without affecting the safety factor and the function of the equipment, and the mechanical aesthetics with full power can be displayed. Some products adopt metal mesh design on the surface of the chassis, and replace the material of the side part with the mesh metal plate. When the transparency of the chassis is properly enhanced, the convenience of maintenance can be facilitated. If necessary, the metal mesh can be removed to check and adjust the internal structure of the chassis (Table 1).

Table 1. Products comparison in modeling level of various brands.

Brand / model number	Product picture	Block stamping	Block segmentation	Metal mesh cover	Exposed mechanical parts	Color distinction
New Holland T7		√	√	√	√	√
Claas Arion-640		√	√			√
New Holland Bigbaler		√	√		√	√
John Deere 2C-354		√	√		√	√
John Deere 6J-2104		√	√	√		√
John Deere 6B-1404		√	√		√	√

3.4 The Application of the Contrast and Harmony Between the Big and the Small

In the modeling process of industrial products, the relationship between the large and the small is also widely used, such as large and small shapes, high and low, thick and thin, long and short [5]. The basic blocks can be divided and get the contrast between different parts to enrich the product shape. In the visual illusion produced by the product design, the large objects seem to be far and the small objects be closed, so the contrast between the blocks can also enhance the overall virtual and real relationship. Additionally, it can bring tension and impact to the body, which is a styling effect that cannot be achieved by a single block. The primary and secondary relationship is very important. The first priority is to maintain its complementary relationship [6]. The main body, the secondary body and the subsidiary body [6] should be clarified between the faces of the plurality of body blocks, and coordinate with each other through reasonable allocation of proportion, so as to achieve a stable overall shape.

3.5 The Application of Analytic Hierarchy Process in the Summary of Design Methods

After the above discussion on the basic design methods of agricultural machinery and equipment, the main design points and design principles are summarized as follows (Fig. 4):

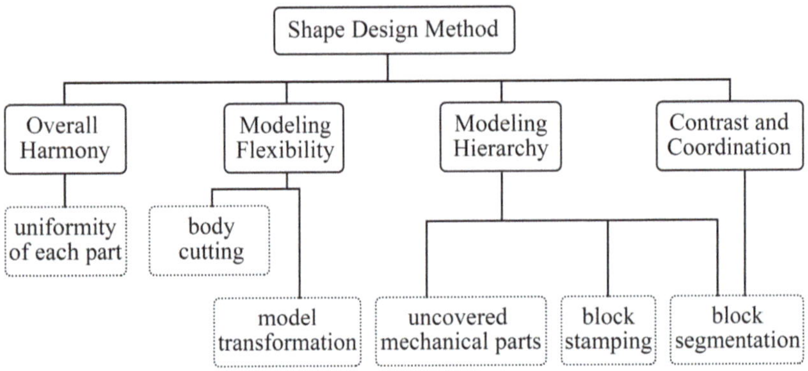

Fig. 4. The summary of shape design principles of agricultural machinery

Taking some representative products of different brands at home and abroad as examples, the Analytic Hierarchy Process (AHP) can be used to decide the weight among design principles and rate these representative products respectively according to the design points summarized before. The score is contrasted and verified to sort the representative items and finally test the application of the design principles to actual products (Fig. 5 and Table 2).

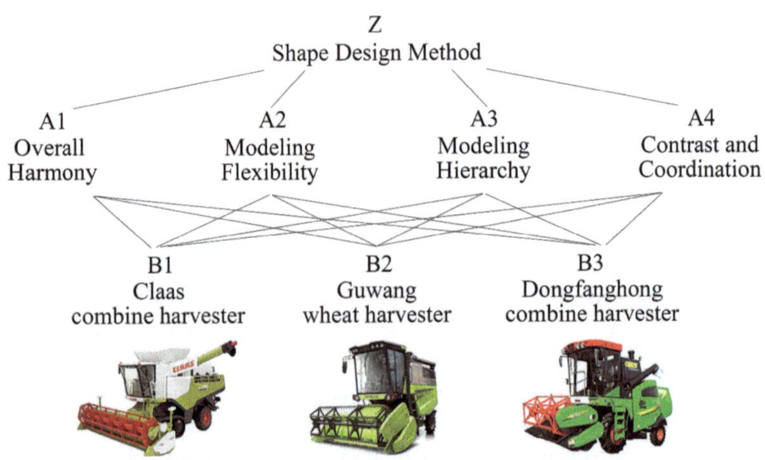

Fig. 5. Schematic diagram of analytic hierarchy process

Table 2. Judgment matrix Z-A

Z	A1	A2	A3	A4
A1	1	5	3	5
A2	1/5	1	2/3	1
A3	1/3	3/2	1	3/2
A4	1/5	1	2/3	1

Establish a weight table for the design principle:
The judgment matrix of the above total target Z is:

$$A = \begin{bmatrix} 1 & 5 & 3 & 5 \\ 1/5 & 1 & 2/3 & 1 \\ 1/3 & 3/2 & 1 & 3/2 \\ 1/5 & 1 & 2/3 & 1 \end{bmatrix} \tag{1}$$

Add by column and normalize:

$$A = \begin{bmatrix} 0.577 & 0.588 & 0.563 & 0.588 \\ 0.115 & 0.118 & 0.125 & 0.118 \\ 0.193 & 0.176 & 0.187 & 0.176 \\ 0.115 & 0.118 & 0.125 & 0.118 \end{bmatrix} \tag{2}$$

Find the vector row and normalize it to get the eigenvector W = (0.579, 0.119, 0.183, 0.119).

Make a consistency check on the above judgments:

Maximum eigenvalue of matrix A: λ max = 4.0014, CI = 0.00047, CR = 0.00052 < 0.1, meet the requirements for consistency.

For the product examples, determine the relative weight between product B1, B2, and B3 (Tables 3, 4, 5 and 6).

Table 3. Judgment matrix A1-B

A1	B1	B2	B3
B1	1	1/2	4
B2	2	1	7
B3	1/4	1/7	1

Table 4. Judgment matrix A2-B

A2	B1	B2	B3
B1	1	1/3	1/4
B2	3	1	1/2
B3	4	2	1

Table 5. Judgment matrix A3-B

A3	B1	B2	B3
B1	1	1/3	1/2
B2	3	1	2
B3	2	1/2	1

Table 6. Judgment matrix A4-B

A4	B1	B2	B3
B1	1	1/2	5
B2	2	1	7
B3	1/5	1/7	1

Compare the matrices A1, A2, A3, and A4 for batch inspection (Table 7).

Table 7. Matrix batch inspection

	A1	A2	A3	A4
W1	0.315	0.123	0.164	0.334
W2	0.602	0.320	0.539	0.590
W3	0.083	0.557	0.297	0.076
λmax	3.0020	3.0183	3.0092	3.0142
CI	0.001	0.009	0.005	0.007
CR	0.0017	0.0158	0.0079	0.0122

CR1, CR2, CR3, and CR4 are all less than 0.1, and the matrices A1, A2, A3, and A4 pass the consistency test.

The total level of the hierarchy CR = 0.0059 < 0.1, passed the consistency test.

The weights of B1, B2, and B3 for the total target are (0.267, 0.555, 0.178) respectively, so the weights of the schemes are: B2 > B1 > B3.

4 Color and Decorative Design of Agricultural Machinery

4.1 Color Application in Agricultural Machinery Equipment

The color of the surface of the product is generally divided into two major categories: primary color and auxiliary color. The main color often determines the basic color of the product, and the auxiliary color can play the role of highlighting the main color, which can have good effects in product aesthetical features [7]. In the color design of agricultural machinery equipment, the method of matching the colors black, white and gray with one or two main colors is often adopted to improve the overall modeling level of the product. The main color is mainly used in monochrome, similar colors, contrast colors, complementary colors, etc. (Fig. 6).

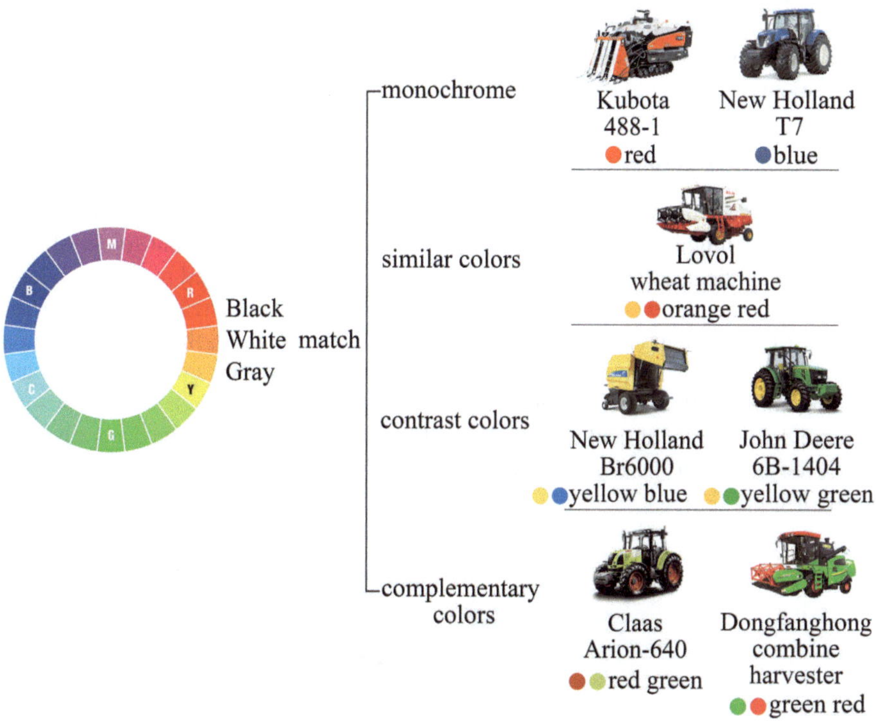

Fig. 6. Common color matching of agricultural machinery equipment

Not paying attention to the hierarchical structure of product styling, it is very low-level to unify the material and color of the block surface. This can be reflected very clearly through the simple comparison of the appearance of Chinese and foreign agricultural machinery. Foreign agricultural machinery usually pays great attention to the restoration of the color of the parts and materials, from a steel plate to a screw, try to restore its true color without affecting the color of the subject. This color distribution method can highlight the overall agricultural machine modeling style and enhance the sense of science and technology of agricultural machinery equipment. Product detail design is not only reflected in the excellence of each component material, color matching is also an important factor to improve the harmony and balance of the overall vision.

Domestically produced agricultural machinery, in view of cost saving or time considerations and lack of design awareness, usually adopts a uniform painting method on the same piece. Using this method, the agricultural machinery equipment that is finally presented will reveal a sense of cheapness in the overall color of the design, and it cannot reflect the power of the machine and its unique charm.

For the use of color, it is necessary to combine product characteristics and product image that conveyed to consumers. If a large area of black is used in the color design of the farm transport vehicle, the product itself will reveal a heavy feeling. This does not meet the consumer's expectations for the field transporter. The light and flexible shape can stimulate the user's desire to purchase. Therefore, color design should be considered in combination with various aspects such as using environment and functional requirements.

4.2 Decorative Design of Agricultural Machinery

Agricultural machinery equipment is usually decorated with decorative strips of various patterns to highlight brand characteristics or to convey certain content to consumers. The design of the decorative belt will affect the style of the integrated product to a large extent. For example, the sharper decorative color blocks will increase the impact and strength of the agricultural machinery products. On the contrary, the sleek decorative design will soften the image of the agricultural products. The use of brighter and jumping colors will show a flexible and lightweight effect, while the heavier colors can increase the power of agricultural products. Therefore, the design of the decorative belt should be based on the definition of the style of the agricultural machinery and the needs of users (Fig. 7).

Fig. 7. Decorative strip design of various agricultural machinery brands

5 Enterprise VI Design

Although many domestic enterprises have clearly realized the driving effect of product design and the improvement it brings in core competitiveness, most of them still do not form brand awareness, the products designed have not formed a series, and lack product identification and brand DNA. Aiming at the design of continuity of agricultural machinery product modeling, firstly, the existing product system of the brand should be analyzed, and the important characteristics of the representative brand should be obtained, which can be continued in the subsequent design, so as to achieve the purpose of identification, solidification and inheritance through the systematic characterization of brand modeling genes [8].

As mentioned in the previous article, the Japanese machinery brand Komatsu Group has a high degree of recognition for its agricultural machinery products. Its agricultural machinery equipment basically uses bright yellow as the main color, the decorative belt of the fuselage and the product logo all use blue-yellow color collision and black-white color matching. In addition, the manual operation room usually uses black frame design with transparent glass, and the tracked design basically uses the inverted trapezoidal shape. From product shape to color matching, they all have unified characteristics, which is very helpful for consumers to form brand impression and promote the sales of products (Fig. 8).

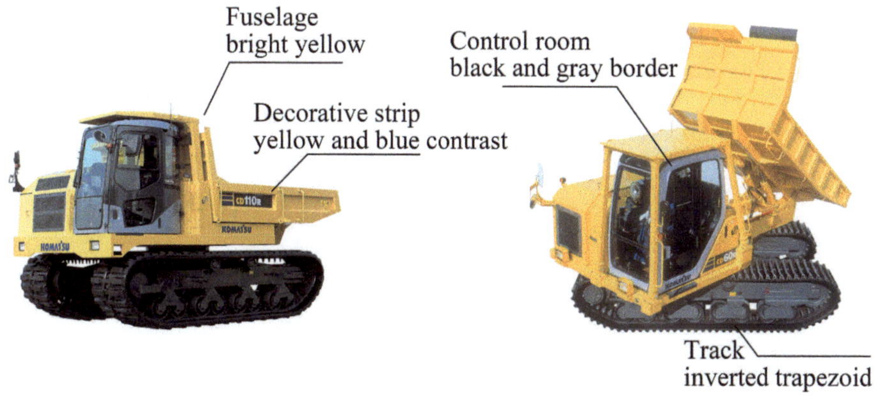

Fig. 8. Komatsu agricultural machinery products

6 Conclusion

In this article, three design elements of shape, color and enterprise VI recognition of agricultural machinery equipment are analyzed and summarized by using the concept and basic principle of design science.

In the aspect of product shaping, through the analysis and comparison of a large number of examples of agricultural machinery product design, this article sums up the unification of the overall product shape, the philosophy of establishing and breaking the product model, the hierarchical relationship between the virtual and the real, the contrast and harmony between the big and the small, and their practical application effects in product design, and carries out the weight comparative analysis of the design principles and design examples by using the analytic hierarchy process. In the aspect of color application, the influence of common color matching methods and detail treatment on the overall design level of products is analyzed. In the part of enterprise VI identification, the importance of brand design DNA in improving enterprise identification and product core competitiveness is demonstrated through the analysis of brand examples.

References

1. Ministry of Agriculture of the People's Republic of China, NY/T, 1640-2008: Agricultural Machinery Classification. China Standard Press, Beijing (2008)
2. Li, Y.: Product Innovation. China Textile Publishing House, Beijing (2004)
3. Pan, P., Yang, S.: Research status and tendency of product form innovative design and its evaluation system. J. Mach. Des. **29**(5), 01–05 (2012)
4. Andreasen, M.M.: Modelling - the language of the designer. J. Eng. Des. **5**, 103–115 (2007)
5. Hu, L., Lou, Y.: Introduction of Industrial Product Design. Higher Education Press, Beijing (2004)
6. Hannah, G.G.: Design Element - Royna Reed Costaro and Visual Relationship. Shanghai People's Fine Arts Publishing House, Shanghai (2013)
7. Wang, W., Kou, R., Yu, S., Yang, X.: Research on design decisions model for requirement groups driven product main color design. Mod. Manuf. Eng. **04**, 36–41 (2014)
8. Wang, S.: Agricultural machinery modeling design based on brand experience. Packag. Eng. **38**(2), 105–109 (2017)

Interdisciplinary Design and Education

Exploring a New Future in Collaborative Design Processes in Education

Joyce Thomas[1(✉)], Jerrod Windham[1], Suresh Sethi[2], and Manish Arora[3]

[1] Auburn University, Auburn, AL, USA
jkt0022@auburn.edu
[2] University of Illinois at Urbana-Champaign, Champaign, IL, USA
[3] Indian Institute of Science, Bengaluru, India

Abstract. The design process… what is it? How do we teach it in industrial design programs within the US? What is the role of users and how does the design process influence collaboration? There are many similarities yet also differences in how we approach teaching at undergraduate and graduate levels between universities. This paper discusses the experience of design educators from two universities in the US exploring assistive and medical device design projects and begins to explore where do we go from here.

Keywords: Design process · Discovery · Self · Problem finding/solving

1 Introduction

In Universities within the United States, undergraduate and master's industrial design students have explored assistive and medical device design from the perspective of the end-user who needs these devices to support their quality of life. These projects that focus on form design and usability typically run 7–15 weeks at the undergraduate level, and a semester or longer at the master's level.

Educators seek to prepare students for practice in a product development process that is undergoing changes with improvements in technologies that speed up the entire development process. Within undergraduate industrial design programs, this often means that functional improvements are frequently "imagined" and less defined in product solutions that pay attention to unmet user needs. At the master's level, there is focus both envisioning new solutions and demonstrating technical feasibility of the same with the help of rapid prototyping and open source modular electronics. The time and research level between the approaches can vary dramatically, as do the learning and innovation outcomes. Both levels of design education are including the process of self-discovery in different ways to help better understand users in problem-finding and problem-solving.

C. S. Shin (Ed.): AHFE 2019, AISC 968, pp. 173–184, 2020.
https://doi.org/10.1007/978-3-030-20470-9_19

2 Design Process Background

Design process has been studied since the late 1960s when the development of a new engineering 'design methods movement' began through a series of conferences and design methodology books [1]. These studies began as a deconstruction of how designers and engineers work, and the stages they go through while designing a product.

Scientific (analytic) methods were described in terms of patterns of problem-solving behaviors to discover *what exists*. Design (constructive) methods were concerned with how things ought to be and were employed in *inventing things which do not yet exist* [2, 3]. In the early 1970s, this definition was expanded to characterize design problems as "wicked" (social and cultural problems that are difficult and perhaps impossible to solve), while describing science and engineering problems as "tame" [4].

Within the 1980s, a new generation of models and strategies for executing design projects was being developed, particularly in engineering design [5]. Design studies began moving away from early views that prescribed the "one right way to do design (such as form follows function ideology)", instead advocating incorporating a variety of design processes [6].

The traditional methods of a linear product development process, where design, engineering, and manufacturing development were completed sequentially, gave way in the early 90s to the emergence of "concurrent" models of product planning and development [7, 8], where project teams composed of designers, engineers, manufacturing experts, product planners and others work together from the start to ensure design intent (styling, features, functionality) follows through to production. Thus, in practice, designers began working in back and forth iterative processes, developing designs "while they collaborate with others- a nonlinear practice that takes advantage of multiple resources and perspectives" [6].

An engineering design study found novice designers used a 'backward reasoning' process of trial and error (design, prototype, test, repeat) on average 23 times within a design project. Experienced designers using a 'forward reasoning' process evaluating decisions prior to implementation averaged 3 iterations of development [9], "expert designers do not work from 'intuition' but have recognisable and appropriate strategies for dealing with their ill-defined problems, as research in understanding design expertise has shown." [10].

The design process followed by industrial designers today is more descriptive than in the early prescriptions that were derived in the late 60s and 70s. There is fluidity in this process because each design is a wicked design process, meaning that every single design scenario is going to be different and therefore every aspect of research that is done towards designing things is going to be quite different. "Design is a complex, multifaceted phenomenon, involving: *people*, a developing *product*, a *process* involving a multitude of activities and procedures" [11].

Current practice in design process continues to shift, moving towards human-centered methods that incorporate asking questions about how people live, work and engage with products, developing a deep understanding of people and discovering a variety of methods to bring them into the design process of new products.

"Education of future designers… is one of the most important tools for bringing research to practice" [12]. One operational model of industrial design education process used in Australia by 66 students each completing individual sponsored projects is broadly described in 5 steps: 1-Task Clarification, 2-Concept Generation, 3-Evaluation and Refinement, 4-Detail Design of Preferred Concept, 5-Communication of Results [13]. Another model used within a school in the US over the course of 5 years engaging nearly 300 students is described as an iterative process including Discover, Unpack, Define, Decide, Develop, Deliver [14, 15].

In industrial design education, new product development generally emphasizes the industrial design process, working in the fuzzy front end and finding ways to step away from preconceived notions that reflect the student designer's bias and not the actual user. These projects focus primarily on form design and usability, while also trying to consider manufacturing, and commercialization opportunities.

The following sections discuss the design processes used in developing medical/assistive products at two American universities.

3 Undergraduate Industrial Design Process at Auburn University

In the spring of 2008, Auburn undergraduate Industrial Design (ID) began an interdisciplinary association with the cross-campus Center for Disability Research and Policy Studies, which has since been integrated into The Department of Special Education, Rehabilitation & Counseling (SERC). Students from both programs collaborated on teams to address the needs of a local individual with some type of physical disability to conceptually design an assistive technology device. Over the years, the teams collaborating with Auburn ID have also included students in Alabama State University's Department of Prosthetics and Orthotics and Georgia Institute of Technology's Electrical and Computer Engineering Program.

Prior to the start of each spring's project, the two core faculty members (one ID and one from the collaborating team) worked together to uncover volunteer participants, apply for Institutional Review Board (IRB) approval, and review the goals of the collaboration.

The exact make-up of each team has varied from year-to-year, depending on the number of students in each program and the number of volunteer participants. Interdisciplinary teams usually consist of 4–6 students working with one volunteer participant who is referred to as *client-user*.

This semester-long ($\sim$15 weeks) studio project is positioned in spring of the 3rd year (approximately 16 months after ID students are accepted into the program) and immediately following a Design Methods course taught the previous semester.

3.1 Discovery

Rather than start with a specific product design brief defined by faculty, the design process begins with a kick-off meeting connecting client-users with their student teams, timeline, meeting schedule, proposed deliverables, as well as examples of assistive technology. The aim has been to inspire and shape expectations as student teams and client-users get to know one another in an informal/relaxed setting. Positive language is encouraged when discussing disability (e.g. challenge vs. problem) and focuses on opportunities rather than limitations.

The overarching goal of the meeting is *discovery*. Teams are directed to uncover a minimum of three challenges/potential opportunities for assistive technology to address issues associated with the client-user's disability, empowering the student teams to generate their own design brief. They start with some variation of the following prompt:

1. *When I am at home, I wish I could* ____________.
2. *When I am at work, I wish I could* ____________.
3. *When I am at school, I wish I could* ____________.
4. *During recreation, I wish I could* ____________.

3.2 Life-Enhancing Research, Design Opportunities and Evaluation

Student teams compile information relevant to the client-user's disability including: population number of those with similar disabilities, medical understanding of the disability, and existing assistive technology. Combining their *discovery* and *research*, teams are guided by faculty to define three *design opportunities* that are carefully chosen and revised through *evaluation*. Is there already a solution to the challenge uncovered? If so, can the existing solution be improved? Is it feasible to successfully address the challenge and fabricate a working prototype within the context of a three-month timeframe? What is the potential impact if successfully completed, both for the individual user as well as others with a similar disability? While we do not immediately eliminate high-tech opportunities, low-tech solutions are often chosen as they are more feasible to complete within the context and constraints of the studio collaboration.

3.3 Human-Centered Research

When possible, teams visit the client-user to observe them in the contextual environment identified in each of their 3 design opportunities. ID students document their observations for each through illustrated storyboards.

Students observe and, if possible, participate in the client-user defined activity. This aspect is critical to helping the students begin to understand that they are not designing for themselves; they do not 'imagine' their user as often happens in ID education [16]. As a particularly illustrative example, a client-user with cerebral palsy reported struggling to move herself from her wheelchair to her bed. The team visited her home to observe the maneuver and try it themselves using only their arms. This direct contextual interaction led to additional questions and a better understanding of the challenge that is documented in the form of illustrated storyboards (Fig. 1).

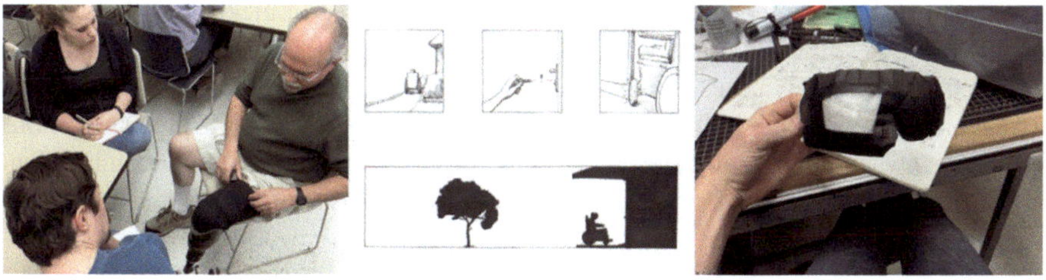

Fig. 1. Discovery, storyboarding, and empathy modeling

3.4 Empathy Modeling

The first *design* challenge for each team is to create a way to simulate their client-user's disability experience; this opportunity for self-discovery in the design students helps them gain understanding of their own (dis)abilities as they try to experience another's (dis)abilities. In some instances, this is easy such as maneuvering a common journey in a wheelchair. Others are not as straightforward. This exercise was first integrated into the collaboration in 2013.

When working with one low-vision client-user, the team created an empathy device simulating blindness in her left eye by completely covering the left lens of safety glasses with black tape. The right lens was covered with multiple layers of crinkled plastic wrap, relying on the client-user's verbal description of the visual distortion she experienced with her right eye.

Figure 1 shows three steps in the design process used within this course.

3.5 Concept Generation and Client-User Feedback

From this point onward, there is very little difference in the teams' approach to designing than on typical undergraduate ID projects, other than the continued participation by the client-user at key evaluation points. ID students generate many sketches and develop a minimum of three fully-thought-out concepts for each of the three opportunities. Drawings and annotations are crafted to communicate in a manner easy for the client-user to understand. Teams present their design opportunities to the client-users and solicit feedback. Students are encouraged to have drawing tools ready in order to quickly sketch alternatives and additions based on the feedback (Fig. 2).

Fig. 2. Concept generation review, rough mock-up, final model

Students then refine sketches and determine how best to quickly test the concept, possibly only testing important details. They fabricate rough and quick mock-ups using inexpensive and easily accessible materials such as foam core, duct tape and PVC pipe (one team crafted a telescoping measuring cup using two sizes of PVC pipe and a store-bought dishwashing brush handle). The prototypes are tested by the team members, and in some instances, they are tested directly by the client-user.

3.6 Refinement and Delivery

When possible, teams fabricate fully operational prototypes that are given to the client-user. However, where a prototype needs further testing, they are not given to the client-user at the conclusion of the project (e.g. products that support the user's weight or have some other safety concern).

Over the course of 10 years, this interdisciplinary collaboration between students and client-users has been impactful. Their interaction from *discovery* through *feedback*, is a key educational aspect to the project that makes it stand apart from other school projects. Unlike many university design exercises that include a hypothetical design brief and abstract market research, the challenge is real, tangible, and right in front of them. Due to the relationship students build with the client-user, it is also personal.

4 MFA Industrial Design Process at University of Illinois at Urbana-Champaign (UIUC)

At the core of the MFA Industrial Design program at UIUC, is the idea of human-centered design; user experience, emotions, storytelling, are common industrial design tools. As designers, we create products embedded with sensorial qualities, aesthetically appealing forms appropriate to function. The MFA Industrial Design program extends beyond the framework of form and function. It is not about inventing things but much more about interpreting the user experience and behavior within the realms of technology, symbolic values, culture, and society. The attempt is to find simple answers; breaking into new realms of experiences and new ways of designing are not easy. The course is built around a studio learning environment, with a combination of lectures, practical workshops, and seminars.

Design students are expected to provide the products that will connect people to the wave of innovation transforming the society. Looking within the context gives the student answers and design solutions. Students search for the design opportunity through investigation of values and qualities that users hold essential in everyday life. How do designers start the design process, given that they want to come up with a yet non-existent product? At the outset of many projects, designers are not even aware of what they should know or what they are looking for. The design projects during the studios are "a starting point for a journey of exploration; the designer sets off to explore, to discover something new" [17].

Designing a novel product is founded principally on the culture of technology and business. Business demands a product that is original, genuine, surprising and beautiful. "The designer has the capacity for envisaging a non-present reality, analyzing it and modeling it externally" [18]. A 'designer's' most important method of working through the problems of existence is using the images she or he invents, judges, and manipulates. A work of 'design' is not an illustration of the thought of its maker, but rather the final manifestation of that thinking itself [19]. A designer's experience, thoughts, and emotions are all manifested in the designed object. Lived experiences resulting from direct perception appear to be a critical factor in the generation of design ideas [20]. These experiences are, by definition, stored in the unconscious and become manifested through the ideas they generate. Visual, non-verbal thinking is fundamental to how designers work [21] and much of the knowledge involved is experience-based, tacit and embedded within the act of designing [22].

Students of MFA Industrial Design at UIUC chose areas from everyday context to create products. In the semester (16 weeks) long design studio project, the students were concerned with these questions: What kind of products and services can they build? What do they want to preserve and how or for whom would they preserve it? What is replacing that which must go? Designers have a dialogue with their drawings while thinking about the product form and its structure. By understanding and expressing their innate ideas, designers are able to develop a concrete product form.

An example of one project is illustrated by a healthcare device developed by K. Chou. The prompt for this project challenged the students to understand the current landscape and to look at an existing device or an object that augments our bodies and fulfils the needs of the user. Possible starting points included wearable technology/ wearable identity, invisible interface/visible material, abstract forms and body in motion, wearable emotions (merging of real and artificial), physical proximity and social distance (a digital reality). Using principles and exploration techniques of a human centered design approach, these graduate students researched the user experience, analyzed the data and conceptualized their ideas.

Chou's own culture and personal experiences influence the way she interprets, connects and develops products relevant to the context. Just data from research cannot supply the solution. From her project, it is clear that experience and prior knowledge, having a background in both fine arts and natural sciences, is what she drew from in order to clarify and reason out her solution. Her design process includes *observation, forming ideas, reasoning, exploration and prototyping.*

4.1 Observation

Direct *observation* is made explicit through a storyboard. Her drawings are rationally deduced results (Fig. 3). Her design process continues through *forming ideas* (Fig. 4), *reasoning* (Figs. 5 and 6), *exploration* (Fig. 7), and *prototyping* for testing and validating the solution (Fig. 8).

4.2 Forming Ideas

Fig. 3. Storyboard of cardiac emergencies

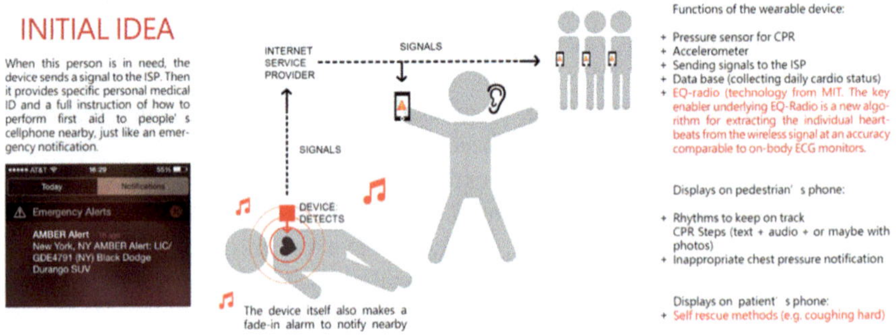

Fig. 4. Concept exploration

4.3 Reasoning

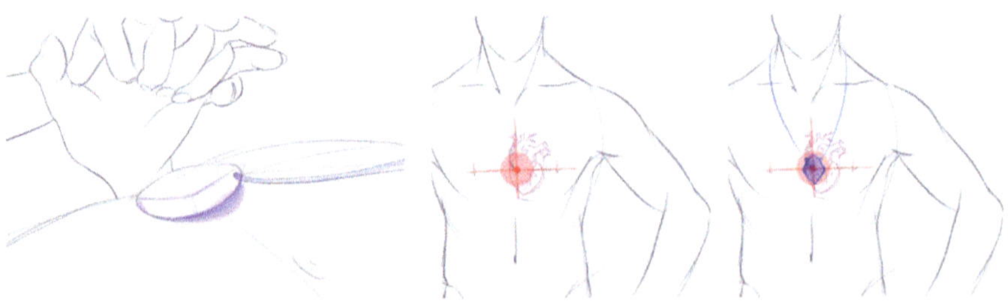

Fig. 5. Correct Spot and Pressure for CPR

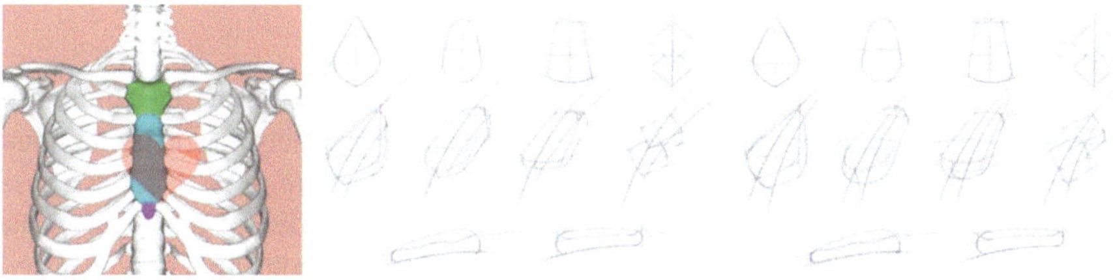

Fig. 6. How Should the device fit to the sternum?

4.4 Exploration

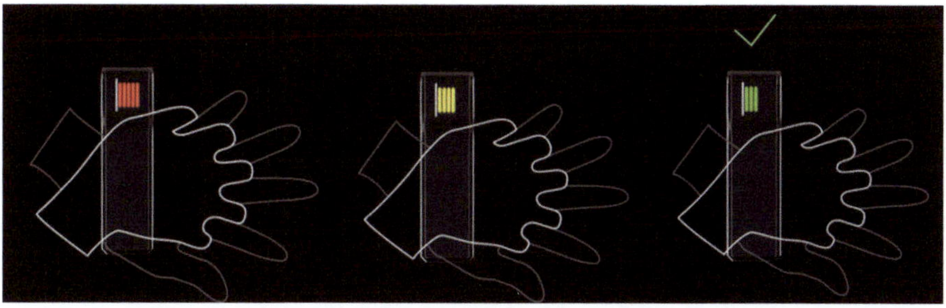

Fig. 7. Light showing levels of pressure for carrying out CPR

4.5 Prototyping

Design should be understood as an entity embedded in daily life, coming from one's interaction with the world. In the design process, designers tend to mitigate the influence of experience through strict adherence to rituals and procedures and by appealing to logic and reason. However, this personal prism gives design its life – it is no longer a set of materials, processes, and price. The designer will discover that designing is not set apart from daily life but springs forth from visual impressions of one's own experience. In this act of exploration, there is a discovery of self, which brings out the meaning and constructs new confidence.

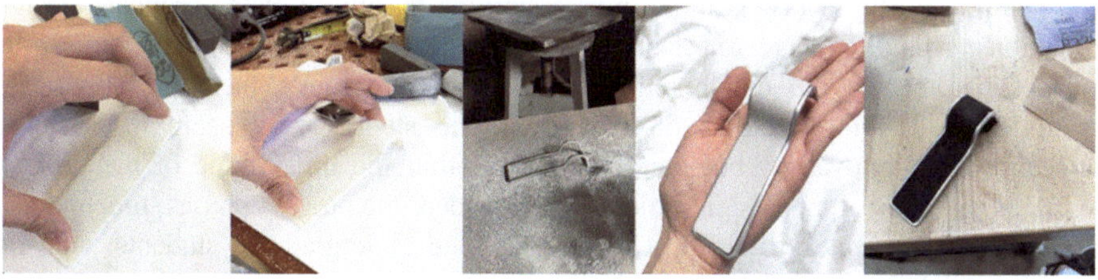

Fig. 8. Model making

5 Discussion

All three design educators and authors of this paper each have a strong background in professional practice, especially in the area of consumer products. Their professional experiences in the collaboration required between disciplines (e.g. engineering, business, manufacturing) to bring new products to life with their design intent intact has had powerful influences on their approaches to teaching at both undergraduate and graduate levels. We recognize that the ability to *visualize* is a significant contribution by industrial designers in the product development and problem-solving process. Designers use a variety of drawing tools/skills to help find problems and then to understand, identify and define them. These methods of visualization not only help us communicate with a wider audience of client-users and other collaborators, but they are truly significant to helping the designer begin to parse out their own observations and research, and then to discover ways to solve problems.

On comparison of the undergraduate projects and graduate projects in ID discussed in this paper, the ways of conducting the design studios is similar – the idea for students of *discovering the self* and engaging with their surroundings to find the right problems to solve. Specific product prompts are often more common in undergraduate projects (where the entire studio is engaged in problem-solving in a particular product area e.g. lighting or seating). At the graduate level, the project prompt is frequently broad enough that students may seek out problems to solve in very diverse areas, and the expectation quality of research is greater from the MFA students. This allows them the freedom to search for their own direction, making them more confident.

In design education, defining the start point of a student project can have a significant impact on both the design and learning outcomes. How specific or open-ended is the project brief, criteria and constraints? Are all the students addressing the same issue?

The undergraduate course at Auburn discussed here provides a challenge that is real, unlike many university design exercises that include a hypothetical design brief. The specific challenges are not defined at the start of the project, which provides the student teams the opportunity to define the brief, criteria, and constraints in direct interaction with the client-user. Students are problem-finding and -solving with their person-with-disabilities volunteer, and so most of the designs are very unique (separate teams are not seeking to solve the same problem). The idea of *self* is often put into direct contrast to the client-user. This helps them explore both shared and not-shared experiences. One outcome is that these projects help establish an understanding and foundation for how to shape desired results of future problem identification and research.

The role of the faculty is more in assessing the feasibility of the proposed design directions. This often leads to selecting solutions that are low-tech, simple, and can be fully fabricated by the collaborating team in the program's facilities. High-tech, complicated design directions discovered by the student may seem to be more appealing and innovative, but not achievable within the time frame of the project. While the genesis of the design project is open ended, the process phases, timeline and deliverables are acutely defined. This provides the undergraduate students with the structure needed to complete the project.

Both undergraduate and graduate ID education involve a process of self-discovery. This discovery of the self at the beginning helps them be able to understand the user better as they launch into a project. Without understanding their own biases, students can't separate themselves from the design and concentrate more purely on the user needs, wants, and desires. While this self-discovery is actually *part of the design process* in ID education (both grad and under grad) it often is not considered a formal part of the design process.

Learning is about multiple views – when it is for a single owner (self), there is only one view. Within design critiques (also not formally included in the design process), there are multiple views that help push student designers towards thinking beyond themselves and can offer the potential to discover new opportunities within the framework of the project.

At the undergraduate level the educational focus is on the designer's toolkit, helping them understand '*how*' to use the tools, and '*how*' to design. At the master's level, students are looking for more of what they haven't learned in their undergraduate degree or professional practice. This is the timeframe when educators really push them to think more deeply about and asking '*why*' you do things – it relies on an advanced level of maturity within the students. At the master's level in other art fields such as music, students seek to hone and polish their techniques, skills, and maturity in performance – to become a master at their instrument. In the field of ID, we seldom become a master in one specific area of product development because we are jumping from product to product and often completely different product categories. Instead we seek to train masters in how to find the correct problems to solve.

There is a level of ambiguity in this. The designer is an individual. Many people would say that is a challenge when designing for others. It is also what may be a key. What is the role of the individual in design education? It is not about designing for yourself, but about understanding yourself and discovering the right problems to solve, searching for the simple answers.

6 Conclusion

In the construction of this paper and the conversations around our topics we have come to understand these are some of the approaches we take in our current journey as educators. There are many similarities in how we approach teaching at undergraduate and graduate levels between Universities in the US. This is what we have been doing, but is it where we need to go? We propose that this discussion should be much broader than what we have touched on here.

References

1. Cross, N.: Science and design methodology: a review. Res. Eng. Des. **5**, 63–69 (1993)
2. Gregory, S.A.: Design science. In: Gregory, S.A. (ed.) The Design Method. Butterworth, London (1966)
3. Simon, H.A.: The Sciences of the Artificial. Mass., MIT Press, Cambridge (1969)

4. Rittel, H.W.J., Webber, M.: Dilemmas in a general theory of planning. Pol. Sci. **4**, 155–169 (1973)
5. Cross, N.: A history of design methodology. In: DeVries, M.J., et al. (eds.) Design Methodology and Relationships with Science, pp. 15–27. Springer, Dordrecht (1993)
6. Henderson, K.: The role of material objects in the design process: a comparison of two design cultures and how they contend with automation. Sci. Technol. Hum. Values **23**(2), 139–174 (1998)
7. Andreasen, M.M.: Design methodology. J. Eng. Des. **2**(4), 321–335 (1991)
8. Pugh, S.: Total Design: Integrated Methods for Successful Product Engineering. Wokingham, UK, Addison-Wesley (1991)
9. Ahmed, S., Wallace, K., Blessing, T.M.: Understanding the differences between how novice and experienced designers approach design tasks. Res. Eng. Des. **14**, 1–11 (2003)
10. Cross, N.: Expertise in Professional Design. In: Ericsson, K.A., Hoffman, R., Kozbelt, A., Williams, A.M. (eds.) Cambridge Handbook of Expertise and Expert Performance, 2nd edn, pp. 327–388. Cambridge University Press, Cambridge (2018)
11. Blessing, L.T., Chakrabarti, A.: DRM, a design research methodology. Springer, London (2009)
12. Telenko, C., Sosa, R., Wood, K.L.: Changing conversations and perceptions: the research and practice of design science. In: Chakrabarti, A., Lindemann, U. (eds.) Impact of Design Research on Industrial Practice, pp. 281–309. Springer, Cham (2016)
13. Lewis, W.P., Bonollo, E.: An analysis of professional skills in design: implications for education and research. Des. Stud. **23**(4), 385–406 (2002)
14. Shin, C., Thomas, J.: Exploring two design processes: fast and slow. In: Chung, W., Shin, C. (eds) Advances in Affective and Pleasurable Design. Advances in Intelligent Systems and Computing, AHFE 2017, vol 585. Springer, Cham pp 3–15 (2017)
15. Thomas, J., Strickfaden, M.: From industrial design education to practice: creating discipline through design sprints. In: Chung, M.W., Shin, C.S. (eds.) Advances in Interdisciplinary Practice in Industrial Design, AHFE 2018, AISC 790, pp. 111–121 (2019). https://doi.org/10.1007/978-3-319-94601-6_13
16. Strickfaden, M., Heylighen, A.: Who are they? Student voices about the 'other'. In: Include 2009. Helen Hamlyn Centre RCA, London (2009)
17. Cross, N.: Design Thinking Understanding How Designers Think and Work. Bloomsbury Publishing Pte, London (2016)
18. Archer, L.B.: The nature of research into design and design education. In: Archer, B., Baynes, K., Roberts, P.H. (1992). http://idater.lboro.ac.uk/wp-content/uploads/AR_LP2_Bruce_Archer.pdf. UK
19. Arnheim, R.: Visual thinking. In: Kepes, G. (eds.) Education of Vision, pp. 1–14. Studio Vista Ltd, London (1965)
20. Sethi, S.: Every design tells a story. In: ICERI2015 Proceedings, pp. 6529–6535. IATED Publications: Seville (2015). https://library.iated.org/publications/ICERI2015
21. Cross, N.: Designerly ways of knowing. Des. Stud. **3**(4), 221–227 (1982)
22. Schon, D.A.: The Reflective Practitioner. Basic Books, New York (1983)

Industrial Design Participation in Project/Matrix Management

Randall Bartlett(✉)

School of Industrial and Graphic Design, College of Architecture, Design and Construction, Auburn University, Auburn, USA
bartlrn@auburn.edu

Abstract. Team management often referred to as concurrent design/engineering and matrix management is now embracing this practice as a proven methodology. Empowerment, the main ingredient is given to the team to make decisions collectively, thus eliminating total approval from the next level of management. Allowing the team to make decisions minimizes delays and moves the product to market faster. The industrial designer's role in team management is different from most other members in that they are dealing with an area that is often viewed as subjective. Most businesses are not acutely aware of the importance of design. Industrial designers are often considered to be just visual people and not actually concerned with practical elements. On the contrary, the industrial designer is very practical, their problem-solving methods and a holistic view is a tremendous asset in business today, especially in design thinking. Design cannot be avoided, no matter what the company or organization produces.

Keywords: Organizational · Virtual · Management

1 Introduction

The project management direction in consumer product manufacturers today is team management. Team management often referred to in the past and recently as concurrent design/engineering and matrix management is now embracing this practice as a proven methodology. The main ingredient is that of empowerment given to the team to make decisions collectively, thus eliminating the total approval from the next level of management. The concept is based on the team making decisions in order to minimize delays and hopefully bring the product to the market fast.

In particular, this paper addressed the role of the industrial designer as a key member that participates in the decision process. The industrial designer's role in team management is different from most other members in that they are dealing with an area that is often viewed as being subjective. Most businesses are not acutely aware of the importance of design. Industrial designers, viewed by peers and even other designers, are often considered to be just visual people and not actually concerned with practical elements. On the contrary, the industrial designer is very practical, their problem-solving methods and the holistic view is a tremendous asset in business today,

C. S. Shin (Ed.): AHFE 2019, AISC 968, pp. 185–192, 2020.
https://doi.org/10.1007/978-3-030-20470-9_20

especially in design thinking. Design cannot be avoided, no matter what the company or organization produces.

The industrial designer is an integral participant in the team management concept. The importance of industrial design in team management will be discussed in terms of team management participation.

In organizations today, especially consumer product manufacturers, project/matrix management has been the preference for making management decisions. Hierarchical type of management, as we know it today, is being replaced by project/matrix management concepts. Why is this type of team management more important now than it has been in the past? One reason is that the change in economics, society, culture, environment, technology, politics, and international affairs is taking place at an accelerating rate. And change is having a dramatic impact not only on individuals but also on organizations. Task forces, departmental teams, cross-functional teams, and project teams are replacing the cumbersome hierarchical organization structure of the past in many organizations [1].

The basic concept behind project/matrix management is the attempt to reduce levels of management to a flatter hierarchy. With flattened levels of management the decisions are participatory, thus the results will be at an elevated confidence level. Information and communication are passed in a horizontal direction rather than a vertical format. If a request is given by upper-management it doesn't trickle down through the chains of command like the hierarchal management concept but is received by the entire team at the same time. Receiving information from the same source decreases fallacious communication. The team members are responsible for the direction and with their involvement, they typically stay focused on following through with the team's decisions. This enables a company to push decision making into the teams rather than consuming the time of top management. They need time to concentrate on the future and not the crisis of the minute [2].

In general, matrix management is especially effective when numerous complex factors are involved in decision-making and when the environment is dynamic. Matrix organizations can react quickly. The personal rewards, motivation, and excitement of participants can be high. Since the backgrounds of the participants are diverse, the quality of decisions and ease of implementation are improved [3].

2 History

Team management, even though the current trend of organizations is not a recent management concept. Edward Deming finds team management roots in the work, a statistics professor at New York University in the 1940s. At the end of World War II, Deming helped set the stage for Japan's remarkable recovery by advocating the participation of small work teams in the task of planning and guiding efficient production methods [4].

Although the Japanese regarded Deming with reverence, he was largely unknown to business leaders in the United States. In fact, he worked with the Japanese for almost 30 years before he gained recognition in his own country. In recent years, U.S. companies have turned their attention to Deming and embraced his philosophy. In

America today, corporations are looking to the success of Japan for management techniques that might be used in their organizations. The major elements of project/matrix management are the things that Deming proposed to Japan. Our corporations had the source in their backyard but neglected to utilize the techniques. Team management stayed foreign to the American companies in general and did not show any emergence until the 1960s.

Team or project management emerged in an unobtrusive manner, starting in the early 1960s. The term project or program management was used to describe a type of structure that already existed in various forms. No one can claim to have invented project management; its beginnings are often cited as the ballistic missile program of the space program of the United States. As early as 1961 the line-staff concept was in the obsolescence stage and the growing trend toward the use of a "functional teamwork" approach to the organization [5].

3 Team Work-Based

With the success of Japan, Deming's knowledge and the seeming failure of a hierarchical management method, the efforts have been aimed at team management. Perhaps in the part, management felt that they could not get to a decision by involving several people, as in a team. However, by involving the people who will be directly affected has proved to be a convincing method. This not to say that power is in numbers but if you have the right people and enough of them, then the decision should reflect a higher level of accuracy. An accurate decision is critical in business and the reaction time to incorrect decisions is also critical. These decisions can be accomplished by the team more efficiently and confidently that if only one individual is responsible. One of the attributes of team management is the team is held responsible for their decisions, right or wrong they have to endure with them. Each team member knowing that he or she is corporately responsible will challenge and evaluate the suggestions of the other members, which elevate the certainty of the proposed decision.

Agreement in the team's decisions is people cooperating to meet a common goal, which is referred to as teamwork. This includes all types of people doing work that calls for joint effort and exchange of information, ideas, and opinions. In teamwork, productivity is increased through synergy: the magic that appears when member generates new ways of getting things done and that special spirit for making them happen [6]. Teamwork is exactly what has to take place to have an effective group of people solving the problems with the right solutions. The key word is work. In some organizations, the work is often busy work, which is primarily the act of doing something with very little accomplishments. This plagues our industries in an unknown form and can continue virtually unnoticed by management. I like to refer to this as project-based team members and not product based members. Let me explain the differences of the project-based member compared to a product-based member.

4 Project vs. Product

A project-based member is one who mainly projects oriented and not product or production focused. This is not to say that the person is not a good performer or contributor but their sights are not aimed at turning the project into a product. This happens not by plan but rather by default. A product based team member understands the importance of converting the project into a product and the reality that in due time the project will ultimately become a product. The market response time to new product introduction is crucial in business today. The team management concept can't afford too many project-based members or that is what the results will be, a project. It is the responsibility of the team members to keep a check on their progress in order to avoid the project only syndrome. As stated, this is not the original intent of the team embers but we can all and probably have been guilty of being project people.

The question that should be asked is, how do we avoid the trap of just working on a project and not bringing it to closure? There needs to be a check and balance system set up in the teams. Team members need to ask themselves what progress is being made to closing this project and bringing it to market. It could be upper management is causing the delays or a number of other reasons may be keeping the task in the project mode. The charge of the team is a total team evaluation to identify reasons why the project is not progressing toward a product. Once the reason or reasons are identified then action should be taken. By no means should this become a project within a project? There are enough sub-projects within the team's list of items to complete than adding tasks, which expand the problem. A key to staying focused is communication within the team. Each person has to feel that they are equal and contributing member. Synergy is generated as people work together to meet common goals.

The correct balance of team members is essential in determining solutions and directions that face our companies today. Since effective teams, those correctly balanced, are highly interactive, teamwork depends heavily on the interpersonal skills of the members. In a team setting, this personal interaction takes on special importance because the number of relationships among members is sharply increased [7]. Sometimes, this does not run so smooth. The team member is dealing with more people than before, thus the interaction is increased resulting in greater chances for conflict. This will require members to work together skillfully and cooperatively. These skills include listening, dealing with interpersonal conflict, negotiating, and influencing.

The question that should be should be addressed is, do the team members collectively have the skills as listed? Several books and articles have been published to prepare and suggest ways of developing the skills required to become a good team member. A concern that I have is, we spend many man-hours away from being productive by teaching skills that we should probably already possess. Each team member should comprehend and employ the characteristics of a good team participant. This may become evident to some individuals after they have been involved with the team for a while and have observed those members with good team skills.

5 Industrial Design Participation

With the understanding of the project/matrix management concept and the requirements of the participants, each individual team player can react to the charge of the particular team. Each individual, with their own area of expertise and talents, support and solidify the team. Even though there are several team members that could be used as a model there is one in particular that are typically trained to be used in team settings. This particular member of focus is the industrial designer. Industrial designers may not be the exemplary players but their education is based on solving design problems and in many cases in a team setting. Problem solving in product design is an interactive process involving many possible ideas, probable solutions and people from different areas.

One of the skills that the industrial designers bring to the team is that they are trained to ask the questions the engineers seldom think of. Design challenges assumptions, while engineering focuses on how to make something work. Design concentrates on how to making it usable. Industrial designers see the product from the viewpoint of all users: buyers, manufacturers, services, sellers, and end-users. They focus on incentives to buy, ease of learning, and o overall satisfaction with the products [8].

This is not to say that industrial designer does it all. Industrial designers are not engineers, chemist, or marketing or manufacturing specialist. Nor is this to say that engineers, for example, are capable of coming up with creative design ideas. They do, all the time, what it does say is that industrial design is the crossover discipline between marketing, sales, customers, engineering, and manufacturing. It is the logical place where all these disciplines meet in the product development process [9]. The designer's solution is often supported by employees inside the corporation and with outside people, too. Interaction with marketing managers, mechanical engineers, production workers, industrial engineers, suppliers, and consumers is a vital part of the initial design process. Involvement and support form the above-mentioned participants solidify the solutions, giving the designer credibility.

The design process encompasses research and feedback with various individuals throughout the stages of the process. A good design receives total support from the interactive individuals that contributed to the design efforts. Teamwork is established almost by default during the development of the design process. The designer's effort or solutions do not stop at a concept; in reality, this is the beginning of the project is to become a producible and marketable product. Industrial designers seek and severance the support from the contributing individuals.

One common requirement that keeps surfacing in project/team/matrix management concepts is participatory. This seems to be a key ingredient in making a successful team along with other important factors. Participation is essential in making decisions that are accurate and timely. The industrial designer is participatory oriented by education and primarily because of the steps that were taken in the design process requires interaction with diverse individuals.

In the plan, with the team participation of the industrial designer, team management concepts should benefit from having an industrial designer as part of the management team. A problem exists with this; most managers view the area that industrial design covers are subjective. There is truth to this assumption, however; a design is only partly subjective. A crude definition of design is, to plan, to show the way. Visual references may be used to plan or to show the way but planning may be in the form of setting directions, research, developing possible solutions and other problem-solving techniques. Unfair as it may seem the industrial designer still has to prove their importance and sometimes need for the company. With the industrial designers moving into management positions, being effective team players and contributing to the organization's success will begin to change this fallacy. We cannot expect our managers of today or tomorrow to become design experts, we must find the means to enable them to recognize good design so that they will be in a position to promote a more comprehensive design policy [10].

6 Design Awareness

A majority of the companies today are not acutely aware of the importance of design much less good design, especially from the management side. Industrial designers, viewed by their peers and even other designers, are often considered to be just visual people and not actually concerned with practical elements. Good design has to be recognized and perceived more than just a visual entente. The recognition will become clarified when the results of the participatory industrial designer are placed on a team, which is not functioning very effectively. The industrial designer can become the catalyst for the team, keeping it balanced and educating the team members about good design.

Corporate management needs to learn how to engage, understand and work with professional designers if we are to make good design a higher priority in our business firms and in our everyday life [11]. This needs to take place in the education process of our new managers and companies will have to place greater emphasis on design. The question that has occurred to many is, why has it taken so long to bring good design into the management decision-making process? One answer is in the narrow focus of management education, which traditionally has either ignored design or viewed it as a value concept beyond the legitimate scope of business education. This narrow view of design, especially the importance of an industrial designer, is changing in business management. It has not totally arrived in every companies plan but the front-runners in the industry today are starting to notice the importance of good design. Business leaders are becoming more sensitive to the ways a corporation projects itself to the public, whether in graphics and advertisements or in products and services, the subject of design is emerging as an important concern [12]. With design awareness becoming more evident in our industries a certain degree of management chaos is created.

Even though awareness of design is what the designer may desire for the company, design decisions such as aesthetics and human factors should be made by the industrial designer. In a lot of companies, everyone wants to be a designer. This attitude is good but the team should function as a team and leave the decision about design to the designer. Also, the designer should not make engineering or purchasing decisions. Only recommendations should be made concerning another individuals area of expertise. The designer will have to be sensitive to the other team member's suggestions concerning design. I see this as part of educating the team about design. An obvious problem is most people view design as purely subjective, recognizing only the cosmetic side of design. Design is much more than putting a pretty cover on a product. This subjective, but with the industrial designer working in project teams the real issues concerning good design will be exposed.

Industrial design optimizes the function, value, and appearance of products and systems for the benefit of both the user and manufacturer. Its concern is not simply the external packaging or aesthetics of a product, but the product itself as functional, manufacturable and marketable [13]. Designing is an attitude of the mind and must permeate management and more than a painting for the president's dining room or mural in the cafeteria. Lack of design understanding and awareness in our society is due to our worldwide educational system [14]. Our education system lacks the appreciation for design or as mentioned by Schutte, we do not have a real understanding of design. We are slowly being educated of design with the participation of the industrial designer in the business-world but this does not reach the education system, as we know it today. Design is often viewed as an art rather than a functional element. This does to only exist in our businesses as stated earlier but it's the consciences of the consumers. There are positive changes on the horizon toward design and the awareness is surfacing in the recent flood of foreign products in our market. If the U. S. can't raise the level of design awareness perhaps Japan can. This can be traced back to the team management concept roots employed by Deming.

Good design can be justified on the grounds of improving effectiveness in large organizations. Good design saves money and time. Design enhances communication between people and simplifies manufacturing. Conversely, the lack of design often causes design by default, which often means disorder, confusion, and lack of currency [15]. Disorder and confusion cannot be tolerated in the business today nor can it exist in team/matrix management. In time, the industrial designer's philosophy and methods should bridge gaps in the teams they participate in.

Design awareness and importance may not be the key to effective project or matrix management teams, but it does elevate the concern for quality. This will began to create a team philosophy, a common direction or at least an agreement among the members, to build the much needed individual relationships. Understanding the importance of good design is essential to building a productive team that is aware of the user's needs.

7 Conclusion

Project/Matrix management is reaching and taking told in our corporations today. The concept of moving the project from marketing to design to engineering to manufacturing is becoming a method of the past. Now representatives from these departments are being assembled in teams to virtually manage the project from concept to production.

This type of management concept is very exciting for companies, especially the way they have to position their products in today's competitive markets. The marketing demands are becoming increasingly complex. New products are often the only way some companies can survive. Our manufacturers, in some cases, are being forced to secure goods overseas in order to be a player in today market. This is primarily because our reaction times to product development and product introductions have been sluggish. Missed product introduction dates can cost a company a major reduction in the market share; this can be fatal.

American manufacturers no longer can afford to make product development mistakes. They are under such incredible time and budgetary pressures that there is literally no "doing it over". In many cases, they do it right the first time or go out of business [16]. Team management concepts are charged with doing it right the first time. This attitude is typically the main attribute in team management because of the number of individuals that are involved in making decisions. As stated earlier, the team approach brings skills together which are typically complimentary. Exactly which skills and how they fit together will depend on the product under consideration as well as the individuals.

The industrial designer is trained to consider things in a holistic context. Their skills and ability to solve problems allows them to be involved with almost all project teams. Management must first understand the importance of good design before the industrial designer will be selected on project design teams. Ironically, this may not happen until the industrial designer has shown evidence of their contributions as a team member.

References

1. Dinsmore, P.C. (ed.): Human Factors in Project Management. AMACOM, New York (1990)
2. Dinsmore, P.C. (ed.): The Handbook of Project Management. AMACOM, New York (1993)
3. Harkins, J.R.: Is design doing its job? Mach. Des., 53–58 (1994)
4. Kerzner, H., Cleland, D.I.: Project/Matrix Management Policy, and Strategy. Van Nostrand Reinhold, New York (1985)
5. Okay, M.: Managing Product Design. Wiley, New York (1984)
6. Schutte, T.F. (ed.): The Art of Design Management. Tiffany & Company, New York (1975)
7. Wynn, R., Guditus, C.W.: Team Management: Leadership by Consensus. Merrill Publishing Company, Columbus (1984)

Moving from Dependency, Wastefulness and Risk Toward Sustainability and Resilience Through Designed Systems

Carlton R. Lay(✉)

Auburn University, Auburn, USA
crl0030@auburn.edu

Abstract. Technology, automation, and their related systems, have been put in place to add efficiency and convenience to our lives, allowing modern cultures to live prosperous and "easy" lives compared to most cultures throughout history. Our lives of ease have many desirable attributes but, we have simultaneously developed a high level of dependency and a lack of resilience in the face of many common challenges. Through intentional and strategic design, our society could be nudged into a philosophical shift toward a life of resilience, independence, and greater humanness.

Keywords: Resiliency · Sustainability · Technology · Design · Systems design · Infrastructure

1 Introduction

> Behold, this was the guilt of your sister Sodom: she and her daughters had pride, excess of food, and prosperous ease, but did not aid the poor and needy. They were haughty and did an abomination before me. So I removed them, when I saw it.
>
> - Ezekiel 16: 49-50.

There are both positive and negative things to say about the prosperous and relatively "easy" lives we live today in modern culture. Technology, automation, and the many systems that have been put in place add efficiency and convenience to our lives. Our lives of ease have many desirable attributes but, we have developed a high level of dependency and a lack of resilience in the face of many common challenges. Our dependency is on fragile systems and infrastructure that eventually will fail. A desire for a life of "prosperous ease" and the desire for convenience has ushered a systemic and often unrecognized dependency on fragile systems into modern cultures. Even when our systems run smoothly, our dependencies have been proven to be a risk to the planet's environment and resources, as well as to the maintenance of knowledge. When the systems fail, our dependencies are a risk to culture, economic stability, and to life itself.

There currently appears to be a growing understanding of the true nature of the risks we live under in our modern society and an unrecognized convergence of differing cultural movements. At minimum, there is a handful of counter cultural

C. S. Shin (Ed.): AHFE 2019, AISC 968, pp. 193–204, 2020.
https://doi.org/10.1007/978-3-030-20470-9_21

rejections of particular areas of risk and dependency by a growing population taking the form of movements and sub-cultural trends. A few of these movements include the organic and non-GMO food movement, the off-grid movement, the home school movement, the tiny house movement, the prepper movement, the urban gardening trend, the non-plastic food and water storage trend, water filter usage trend, and the movement for the legalization of medical marijuana as an alternative to big pharma.

The unique convergence of each of these typically widely divergent movements is toward a greater level of personal resilience and an independence from or a response to prevailing societal systems or methods of production. There are, perhaps, varying motivations behind these desires for alternatives to the prevailing systems as there are widely differing beliefs and philosophies between these movements. These motivations likely include environmental, moral, spiritual, physical and mental health, financial, desire for control and freedom from other's control.

2 Drivers of Dependency and Lack of Resiliency

The rise of wealth inequality in the US and other developed nations, and the systems in place that provide for it increase a variety of risks, benefits a small percentage of the population within those countries, and decreases our resilience to a variety of challenges. The Guardian writes that Bill Gates, Warren Buffet and Jeff Bezos, the three wealthiest businessmen in the United States, hold as much wealth as the entire bottom half of the US population [1]. According to the Harvard Business Review, globalization has provided significant income growth for two categories of individuals – the global 1% and the middle class of Asia, [2] from where almost 90% of the next billion people to enter the middle class will emerge [3].

Globally, the middle class is expanding based on Asia's numbers alone, but in developed counties, the middle class has stagnated or declined [1]. Even those that argue the economic benefits globalization and transnational corporations recognize the risk involved in allowing power and influence to grow as much as it does. Kordos and Vojtovic write in *Transnational corporations in the global world economic environment* that "the oligopolistic nature of international corporations can undermine market forces and allow unreasonable profits." These TNCs attempt to retain influence, by supporting authoritarian regimes. Examples of transnational corporation involvement in world politics include the ITT (International Telephone & Telegraph) financial support of the Nazi regime in 1933. [4], and the overthrow of the Chilean government in the 1973 [5], The technical support that IBM provided the Nazi regime during WW2 that some argue make it complicit in the Holocaust [6] United Fruit Companies economic and political control over several countries in Central America [7], The Anglo-Iranian Oil Company (BP) involvement with the CIA and UK government in the overthrow of the Iranian government in 1953 [8] Monsanto's recent influence over US policy [9, 10] among many more.

The systems of production, data collection, commerce, labor, and finance and others are in place or are operating in a particular way for the distinct advantage of influential individuals. With the current systems in place and operating as they do, motivation and incentive for increased resiliency is minimal.

Whether the simultaneous resistance to these prevailing societal systems and the current societal status quo by these counter cultural movements is happening in isolation from one another or coming into existence as an unrecognized but connected larger societal shift in mindset is not yet determined. Through intentional and strategic design, a connection between these counter cultural movements could be fostered and our society nudged into a philosophical shift toward a life of resilience, less dependence on questionable organizations, and greater humanness.

3 Our Dependencies

An individual or societal dependency is potentially created when a technology is developed that increases efficiency, increases production capabilities, or lowers cost relative to the prevailing system or method of production or technology with which the new product will compete. Once the shift from old to new technology dominance occurs, the dependency on the new technology begins. The shift in usage of any particular system does not constitute a new risk until the old technology has faded from mainstream production and consumption. Once a system, product, or technology has been removed from production and use for more than a generation, the knowledge of the previous technology will likely begin to disappear, the capability to manufacture or produce the product is diminished and skill sets required to bring the previous prevailing system back on line may be lost.

An example of technological dependence and loss of important knowledge is the Inuit hunter's recent use and growing dependence on GPS navigation. Rather than developing their historically strong way finding skills in the dangerous landscape on which these people live and hunt, this new dependence on GPS technologies has led to a growth in accidents during hunting season. According to Claudio Aporta, an anthropologist from Carleton University, in writing about the Inuit people, "A unique talent that has distinguished a people for centuries may evaporate in a generation" [11].

GPS may be considered a form of automation and indeed, the key to much innovation and technology is automation of processes that humans use to accomplish a given task. Starting a fire, flying a plane, tracking a location, moving heavy objects, are all done more efficiently through automation. A 2013 article by *Nicholas Carr in The Atlantic* explains that the performance and capability of individuals who depend on automation on a regular basis to perform their jobs or tasks within their job has deteriorated and, as software, rather than hardware, is becoming the dominant method of automation, the deterioration is being found in a wider variety of careers and professions including professions requiring significant training and knowledge. The skills, capability and knowledge of doctors, bankers, architects, and lawyers as well as non-profession related activities such as shopping, cooking, socializing, navigating, and even child rearing are all being reduced. Many software programs take on intellectual work—observing and sensing, analyzing and judging, even making decisions—that until recently was considered the preserve of humans. Humans are being expected, less and less to translate information and observations into knowledge [12].

Alternative methods of accomplishing many tasks such as storing and preserving food without electricity, cleaning clothes by hand, keeping warm or cool without electricity or communication and navigation without the use of smart devices and GPS are disappearing or have already disappeared. [13].

Technological development is no doubt an important and significant building block of the modern society, economy, and the life we enjoy, but often the consequences to a new technology are never anticipated. The individuals or groups involved in technological development are often too involved in the tech itself to consider the downsides and the sponsors of technologies have often discouraged "thinking" about the potential risks. A short list of examples of both old and new technologies that have been developed/are being developed with either intentional or unintentional ignorance of unintended consequences include nuclear weapons, nuclear power, drones, mini drones, artificial intelligence, autonomous vehicles, GPS, social media, among others.

Companies involved in technological development are understandably focused on their bottom line, moving forward with a profitable product, or achieving an end goal. Rarely do companies have a team dedicated or give focused effort to the analyzation of the potential overall negative societal impacts and risk factors of the products they develop. Whether it is a material choice that produces more waste and pollution, an algorithm that has designed in bias, or data collection that puts consumers at risk of identity theft (Facebook, Google, Apple, credit bureaus, etc.), the default assumption is often that increased efficiency and technological development is beneficial progress, in and of itself.

However, in 2016, a coalition was formed to bring together companies and partners to create a net positive approach to business and product development, through a common set of principles and practices. This coalition, called the Net Positive Project, has 11 companies involved as of Jan 2019 that recognize the need for responsible production and operation. According to the Net Positive Project website the mission is "To create a world where companies drive financial success and create "net positive" impacts by putting more back into society, the environment, and the global economy than they take out." [14] It is encouraging to see that at least a handful of companies are beginning to see their potential for having either a negative or positive impact on society, environment, and economy whether intended or not. Indeed, without intention and specific actions for a positive impact, a negative impact is likely.

On the other end of the spectrum, in 2018, many employees of technology development corporations such as Google & Microsoft began speaking out on the ethical questionability of the projects they were being tasked to work on. According to a 2018 article by Vox, significant numbers of employees at Google have raised concern over the ethical and moral decisions of taking on particular military contracts involving the creation of artificial intelligence technology designed to analyze drone video footage in order to identify human targets for drone strikes. [15].

Why do we, as a society, continue to allow more and more of our lives to be tracked, analyzed, categorized and governed by large corporations and allow increased levels of automation through hardware and software into ethically questionable areas? For Profit? Control? Do we assume that any technological advancement is "progress"? Are we addicted to convenience? Have we, individually and societally, become dependent on the products and services of these companies? Or perhaps, as blue's singer/songwriter Keb Mo puts it, we are just "Victims of Comfort".

As described above, greater automation for the purpose of increased efficiency in tasks such as way finding, identification of individuals, controlling a home environment, and many others lead toward potential risks such as loss of privacy, loss of knowledge, and lack of resiliency. Similarly, ease of acquisition of goods and services in modern cultures leads to potential risks such as irresponsibility, entitlement, lack of work ethic, and a lack of resiliency. The ease in which we acquire entertainment, the ease in which we acquire perceived basic necessities or "human rights" such as internet [16], cable television, streaming media, gasoline, as well as the ease in which we acquire true basic necessities such as water, food, and shelter lead to these risks. A few brief examples of irresponsible behavior developed from ease of acquisition include the taking advantage of the welfare system, the stealing of property, waste of food and other resources, the use of inexpensive products as "disposable". In all these instances, the products or services involved are often abused, neglected, wasted or disposed of.

Writing about the dependency created and the abuse of the welfare system, Carrie Lukas, a contributor at Forbes writes about the abuse and exploitation of aid and assistance programs. Not only does the abuse create result in inefficient use of public funding, but more significantly "it enables a dependency that might be temporarily attractive, but limits someone's life prospects and chances for long-term fulfillment." [17].

Modern culture's relationship to food and food production is complicated and constantly in flux. A study conducted by the US Department of Agriculture has found that 150,000 tons of food is wasted in US households every day, and that the waste is happening not only at the individual level but at the retail level as well. At the time of the study, in 2012 only four of the largest 10 grocery chains in the US had specific food waste reduction commitments" [18]. Developments in agriculture have provided such ease of access to certain foods that the value of these foods is not understood. The movement of individuals re-thinking their relationship with food is often based on the increasing mistrust of "big agriculture" whose aim is to increase efficiencies regardless of potential negative impact [19]. Surveys and reports over the past several years reveal a trend that this re-thinking the relationship with food is resulting in a strong increase in at-home food gardening. According to a 2014 report by the National Gardening Association, the increase in millennials choosing to plant gardens for the purpose of food is greater than any other segment of the population. [20].

4 How Should We then Live?

Soon, we must all face the choice between what is right and what is easy.

- Dumbledore in *Harry Potter and The Goblet of Fire*

To free ourselves from, or lessen our exposure to, the dependencies and the irresponsible behavior that result from a variety of conveniences, there must be a creative solution, a desirable alternative offered. Just as millennials are making choices to grow their own food, opportunity to create a cultural shift is ripe and can be accomplished with strategically designed products and systems. As our addiction is severe, the barrier to entry must be low and the reward must be sufficient to begin weaning ourselves away from the ease of convenient choices and toward responsible and even augmented living through knowledge, action, and independence. Personal, familial, and communal resiliency can be achieved.

Although attention is given to the risks that our dependencies create, not all dependencies should be avoided nor are bad in and of themselves. Not only do we have families, loved ones, beating hearts, oxygen and even a higher power that we may depend on, we truly do enjoy many of the conveniences of modern life. Ideally, well designed products and systems will provide competitive experiences and conveniences with less systemic risk than our current options in the marketplace. Alternative options will, no doubt, require a learning curve, a budget, and an effort on the part of the user, but what product worth having does not require these things. In fact, the products that require a bit more from us, the ones that we put forth the effort to "know" or understand, are usually the more personally valued and treasured objects we own. The pour over coffee maker, the home brewing kit, the new heavily featured smart device, the high end bicycle, the beautifully crafted pocketknife, the musical instrument of choice.

Again, through informed and strategically designed products and systems, the cultural philosophy of dependency and ease could be successfully nudged toward one of resiliency, sustainability, and appreciation. The task will be to create a desire in consumers to do more than merely consume, but to have a stake and a knowledge in the things they need and enjoy, and by doing so, enjoy them all the more.

5 Specific Systems of Convenience and Their Risks

Outlined below are 5 areas that have been systematized or automated for the convenience and large scale delivery of goods and/or services. The systemization of each of these areas has led to what is now an individual and societal dependency, weakness, and risk to environment, life, liberty, and knowledge. In each of these areas the conveniences currently enjoyed as well as the current associated risks with each convenience are briefly described.

Food Production and Delivery. More advanced logistics provides for Just in Time (JIT) delivery proving convenient for retailers in that they do not have to warehouse large quantities of food beyond what is on the shelves. Retailers can provide fresher foods with deliveries as often as daily. The risk exists in that there is typically only a one to two-day supply of food at any store under normal circumstances causing shortages before significant weather events and other situations where delivery can be delayed.

Home Delivery has been increasing in popularity due in part to the busy lives of modern citizens. Convenience and personal efficiency is increased (or perceived as increased) by being able to order groceries on-line and not even having to travel to receive them. If this develops into the primary method of food acquisition, then the food supply risk during weather or other events increases further.

The technical advances of GMO's, hormones, and chemical insecticides have allowed insecticide resistant plants, drought tolerant plants, disease resistant plants, and plants with larger fruits and vegetables. Associated risks include potential negative health consequences to humans from all of these advances. Colony Collapse Disorder

(CCD) of pollinators and the population decline of Monarch butterflies have also been widely blamed on particular chemical insecticides and herbicides. Improper use, over spraying or improper delivery of herbicides and pesticides has caused crops of neighboring farms to be destroyed.

Significant percentages of food production supplied by small areas (California) – This lack of diversification has little benefit but is a risk. When there are problems within a region that produce over 90% of much of our countries produce, then shortages will likely occur. Common problems in California include drought, fire, and landslides, others could include insects and other pests, invasive species, and disease that can easily spread in a region.

Potable Water/Municipal Water Treatment. Municipal water treatment systems are depended upon and most often successfully provide clean tap water for drinking, cooking, cleaning. When there is a problem, it can go undetected, ignored, or hidden for long periods of time and cause serious long term or short term health issues. If there is a contamination it becomes systemic and is potentially a city wide contamination, the Flint Michigan water crisis being a recent example that has led to many municipalities discovering their own water problems.

Because of the ease of access, there is much Inefficient usage and waste as mentioned before. This inefficient usage is found in farming methods, toilets, leaks, sprinkler systems and more.

Decisions about water contents for both intentional and unintentional additives and contaminants are made at bureaucratic levels of government, rather than at the individual, family or community. Levels of pesticides, pharmaceuticals, chlorine, lead, fluoride and other intentional and unintentional additives deemed "safe" are determined at the government level.

Trash Disposal/Sanitation. Regular trash pick-up and delivery to a landfill is a system that is immensely time saving convenient as many individuals don't have the capacity to haul the amount of trash we produce to a landfill for regulated and more healthful disposal. As it is quite convenient to roll our provided trash bins to the curb for pickup we are not intentional nor conscious about the volume of trash we generate. Though the majority of the volume and weight of trash can be recycled or composted, a high percentage finds its way into lakes, rivers and eventually the ocean. A recent study by Great Britain's Royal Statistical Society found that only about 9% of all plastic that has been generated has been recycled and over 6 billion metric tons of plastic has become waste in either landfills or pollution in the natural environment [21].

Toilets and plumbing systems are perhaps the innovations that have provided the most significant increase in societal health as our personal waste never need come in contact with others before it has been treated. What is widely unknown is that the water supply to many homes is actually dependent on pump stations. The water towers that you can see in most cities and towns are used to supply water to homes and businesses through hydrostatic pressure due to gravity. This allows water to be supplied for short periods even when there is a loss of power. As the water level in those towers is lowered, water from the local reservoir is pumped into the water tower.

Hot water, also greatly improving sanitation, is usually dependent on electricity. Gas (natural or propane) hot water heaters that have a tank and that have a pilot light are the one type of hot water that are not dependent on electricity to operate. The hot water heaters that are growing in popularity and in efficiency are the instant type, both gas and electric, both of which have a "brain" that require electricity to operate. Tank Based systems that maintain the temperature of all the water in the tank can supply hot water for a short period of time until the water stored in the tank cools.

Food Storage, Refrigeration, Cooking. Refrigeration of food and drugs allows significantly longer life of particular food and pharmaceutical products. As it is so convenient to refrigerate our food, we refrigerate many things that may not need refrigeration at all depending on how they are acquired. Most of us have never learned much knowledge of earlier generations, including how to safely store foods without electrical refrigeration. One example is egg safety and storage. Washing eggs after they have been laid removes the bloom, also called the cuticle, and allows bacteria through the pores in the eggshell and into the egg. "Washing eggs in *cool* water actually creates a vacuum, pulling unwanted bacteria inside even faster" [22]. Eggs that have not had this layer removed by washing, are safe at room temperature for up to 2 weeks. Once an egg has been washed, they must be refrigerated. The US requires all commercially sold eggs to be washed and refrigerated while other countries require that eggs not be washed [23].

Much of the food we purchase has been processed in some way, to either change its characteristics or to extend its shelf life. Modern methods of food preservation include processes such as chemical washes, irradiation, and food additives. Traditional methods of food preservation include drying, salting, smoking, and fermentation. Within the past century, food preservation has transformed from a process typically done in the home to a process done at the industrial level [24]. Knowledge of the traditional methods of preservation is being lost and the nutritional content of much of modern processed foods has been altered and lessened.

Climate Control and Clean Air. Forced Air Heating and Cooling for quick/accurate/ and programmable temperature setting at the touch of a button, sound of one's voice, or via smart algorithmic learning are all possibilities today. We don't even have to lift a finger, literally, to have and maintain a completely comfortable home climate. Of course risks involved are again related to electricity which is required for the modern heating and cooling system.

The smarter and more connected the systems are, the more vulnerable they are to privacy and security risks such as tampering and hacking. The visibility of these risks has increased with examples of the nefarious possibilities seen in television shows and movies. In 2018 episodes of both Madam Secretary, shown on CBS, and Mr. Robot, on USA Network, show in length the risk of smart home hacking.

Because homes are so easily heated and cooled with modern systems, newer homes are typically not built with the considerations that allow more efficient use of resources or make use of natural resources, solar radiation, the landscape, climate considerations, etc. that can help to heat and cool homes without any additional input. Alternate heating methods that were once standard, such as fireplaces or wood burning stoves are often no longer built into the home, and if they are included, are more decorative than they are functional as an effective heating system.

Because of the effectiveness of electric central and forced air systems, homes can be and are designed around street layouts rather than solar consideration that allow for passive heating in the winter and less interior warming in the summer. This makes the comfort of a home almost completely dependent on electrical systems unless the home happens to be on a street angled toward the proper direction with the windows properly placed to allow for these natural heating and cooling effects. This lack of or ignoring of knowledge by builders, architects, and city planners creates a significant rise in heating and cooling costs of homes, energy and resource usage, and therefore an increase of pollution from whatever fuel is used to create the electricity in any given region.

6 The Influence of Electricity

As has been described, electricity has been the primary driver of how the architecture and the products we use today were/are developed and operate. Electricity has been and incredible benefit to all individuals and societies that have access to it. It has shaped our products, systems, and lives so thoroughly that any lengthy disruption to the electrical grid or our supply of electricity causes serious problems. An awareness and response to this lack of resiliency to the dependency on electricity and the means to produce it can be traced back to the early to mid. 20th century at least.

In 1929 Buckminster Fuller introduced his Dymaxion House in which he "imagined the disappearance of infrastructural networks" with the conceptual design of a home that contained its own fuel source, septic tanks, and water filtration [25]. In 1970, architect Michael Reynolds proposed his 'Earthship' concept that allowed those who dwell within to remove their dependency on the modern electrical grid and other networks that he claimed were responsible for "urban decline" [25].

The environmental risks of electricity production and the risk to the supply chain of electrification has encouraged individuals and counter cultural movements to conceptualize, design and construct alternative ways of living. In Reynold's movement of self-sufficiency, individuals and groups are encouraged to be the primary constructors of as much of their living environment as feasible. [T]he inhabitants of autonomous homes were expected to be totally (or almost entirely) devoted to the construction, repair, modification and maintenance of makeshift systems" [25]. These "makeshift systems" are the same type of systems that can be found on hundreds of "DIY" or "Life Hack" YouTube videos today. These "solutions" are typically "made" without a larger system in mind, without strategic thinking, and are not built with user experience in mind. These are the systems and products of "makers" and "DIYers" not designers.

In addition to the above categories that revolve around electricity and/or severe short term risk and resiliency issues, there are several more areas that have evolved into the current modern systems that potentially create long term risks and resilience issues. These additional categories include Modern Medicine, Early Childhood Education, Protection & Defense, Transportation, Communication, and there are likely many more that have not been mentioned or have yet to be considered or recognized as a risk.

In all 10 of the areas mentioned above, modern day methods and conveniences have or are in the process of leading us to forget previous knowledge that could be valuable to retain and may again be necessary.

Each area of dependency and therefore risk and lack of resiliency is already currently being addressed by one counter-cultural movement or another, but is being addressed without a systems thinking or design thinking cohesiveness. As mentioned above, strategic design can foster the connection between these counter cultural movements to aid in a philosophical shift of our society toward a life of independence, resilience, and greater humanness. A systems thinking approach could potentially incorporate many or all of the above areas into a new and different but enhanced system of living. A way of life that incorporates knowledge of self, knowledge of particular skill sets, environmental consciousness, moral & spiritual freedom, improved health, financial independence, culminating in the opportunity for resilience, responsibility, sustainability, and increased control over one's own life choices.

7 A Design Studio Opportunity

A paper alone is not a sufficient engine for paradigm shift. However, the creation of designed "solutions" could act as stepping stones for individual resilience within each of the categories above in the form of a product or system. These physical "solutions" have the potential to be a unified series of products and systems that encourage the persons already questioning the benefit of the modern choices within their particular area of concern to take a particular step not only within their area but also within adjacent areas. The "series" or "collection" or "family" aspect of the products, their designs and an accompanying marketing & distribution strategy could encourage an interest in adjacent areas that, when combined, create a "system of living" that competes with the lives of "ease" and risk that the population currently live. Each of these products or systems could have benefits not only when society's larger systems fail but immediate benefits to life today within or along with the convenient and modern world in which we find ourselves.

Over the past few semesters, several third-year industrial design studios have been given a design brief and project related to several of the above categories of technological advancement and associated lack of resiliency. A few of the categories that have been explored, resulting in designed solutions, include personal food waste management, food preservation, food production, emergency or short term heating, emergency or short term cooling, and alternative medicine.

These individual "solutions" are as simple as a single tool and as complex as a closed loop food production ecosystem. Quality, cost, beauty, simplicity for the user, low barrier to entry, recognizable and experiential benefit, empowerment of the user, must all be considered in order to design and develop a successful line of products and systems. The solutions themselves will also require analyzing for any additional implications or risks that they may bring forth. Over time, the series of products/systems developed will ideally create the connection between the disparate movements into a cohesive paradigm shift in the way individuals, communities, and societies think, behave, and live. With additional research, development, funding and promotion, the most successful designs could be a catalyst toward augmented, responsible, sustainable and resilient lives.

References

1. Collins, C.: The wealth of America's three richest families grew by 6000% since 1982, (2018). https://www.theguardian.com/commentisfree/2018/oct/31/us-wealthiest-families-dynasties-governed-by-rich
2. Milanovic, B.: Why the Global 1% and the Asian Middle Class Have Gained the Most from Globalization (2016). https://hbr.org/2016/05/why-the-global-1-and-the-asian-middle-class-have-gained-the-most-from-globalization
3. Kharas, H.: The unprecedented expansion of the global middle class (2017)
4. https://www.brookings.edu/research/the-unprecedented-expansion-of-the-global-middle-class-2/ (2017)
5. Sampson, A.: The Sovereign State of ITT, Hodder and Stoughton. ISBN 0-340-17195-2 (1973)
6. The Pinochet File: How U.S. Politicians, Banks and Corporations Aided Chilean Coup, Dictatorship. Archived 2015-09-12 at the Wayback Machine. Democracy Now! (2013)
7. Burkeman, O.: IBM 'dealt directly with Holocaust organizers'. The Guardian (2002). guardian.co.uk.. Accessed 31 July 2017
8. Livingstone, G.: America's Backyard: The United States and Latin America from the Monroe Doctrine to the War on Terror. Zed Books Ltd. – via Google Books (2013)
9. The National Security Archive: CIA Confirms Role in 1953 Iran Coup (2013). nsarchive2.gwu.edu.. Accessed 21 Aug 2018
10. Baum Hedlund: Monsanto Government Influence Has Fueled Unrivaled Corporate Power (2017). https://www.baumhedlundlaw.com/monsanto-government-influence/
11. Sager, J.: Global Research. Monsanto Controls Both the White House and the US Congress (2018). https://www.globalresearch.ca/monsanto-controls-both-the-white-house-and-the-us-congress/5336422
12. https://www.theatlantic.com/magazine/archive/2013/11/the-great-forgetting/309516/
13. Carr, N.: All Can Be Lost: The Risk of Putting Our Knowledge in the Hands of Machines. The Atlantic (2013). https://www.theatlantic.com/magazine/archive/2013/11/the-great-forgetting/309516/
14. https://www.lowtechmagazine.com/2018/12/keeping-some-of-the-lights-on-redefining-energy-security.html
15. https://www.netpositiveproject.org
16. Campbell, A.F.: How tech employees are pushing Silicon Valley to put ethics before profit (2018). https://www.vox.com/technology/2018/10/18/17989482/google-amazon-employee-ethics-contracts
17. A Human Right: Everyone Connected. A Human Right. http://ahumanright.org/blog/2012/01/vint-cerf-internet-access-is-a-human-right-2/
18. Lukas, C.: Yes America, Your Fellow Citizens DO Abuse Federal Aid Programs (2012). https://www.forbes.com/sites/carrielukas/2012/10/02/yes-america-your-fellow-citizens-do-abuse-federal-aid-programs/#5beefc3a68a2
19. https://www.theguardian.com/environment/2018/apr/18/americans-waste-food-fruit-vegetables-study
20. The Center for Food Integrity: A Dangerous Disconnect: New Research IDs Food and Ag Trust Gaps (2017). http://www.foodintegrity.org/research/consumer-trust-research/current-research/
21. https://garden.org/special/pdf/2014-NGA-Garden-to-Table.pdf
22. Parker, L.: Planet or Plastic? https://news.nationalgeographic.com/2017/07/plastic-produced-recycling-waste-ocean-trash-debris-environment/. National Geographic (2018)

23. Winger, J.: The Prairie Homestead. Eggs: To Wash or Not to Wash? https://www.theprairiehomestead.com/2011/10/eggs-to-wash-or-not-to-wash.html, https://www.betterhensandgardens.com/fresh-eggs-washed-unwashed/
24. Weaver, C., Dwyer, J., Fulgoni, V., King, J., Leveille, G., MacDonald, R., Ordovas, J., Schnakenberg, D.: Processed foods: contributions to nutrition. Am. J. Clin. Nutr. **99**(6), 1525–1542 (2014). https://doi.org/10.3945/ajcn.114.089284
25. Coutard, O., Shove, E.: Infrastructures in Practice, The Dynamics of Demand in Networked Societies, pp. 58–61. Routledge, Abingdon (2018)

Designing Under the Influence: Exploring the Motivations and Obstacles of Young Design Students

Benjamin Bush(✉)

School of Industrial and Graphic Design, College of Architecture, Design, and Construction, Auburn University, Auburn, AL, USA
bushbem@auburn.edu

Abstract. Industrial design education takes its cues from the profession of Industrial Design. A century old discipline blending human science, engineering, and art. For better and worse, the design profession's flexibly and obscurity make it difficult for design students to find their way. This paper will analyze the motivations and obstacles that design students face during their academic career. Both internal motivations and external obstacles will be viewed through the lens of the influential people and the presence of existing systems. Concluding the discussion of influential people and existing systems will be three key implementations that can aid design students in maintaining their motivations and outpacing their obstacles.

Keywords: Industrial design · Design education · Motivations · Obstacles

1 Introduction

The last thing Mom says before we hang up the phone is, "Sweety, do your best." I fake a chuckle and say, "I will". I look like I have my act together, but on the contrary I'm a mere 3 months from graduation and don't know if I can live up to all the expectations. Like many college design students, I wanted to get the perfect internship, land the perfect job, and negotiate the perfect salary all while seamlessly managing a social life. It's an ambitious goal, and standing in the way are struggles, uncertainties, and the fear of inadequacy.

But I was just a college student – there is so much more to look forward to! Industrial designers have the unique ability to make a wide impact on society, but with that freedom also comes endless choices. This first seems like an advantage, but transforms into an overabundance of options that can cause even the best decision makers to falter. And when students need to make decisions they consult the sources that have the most influence. Where do these influences, both positive and negative, come from? What forces are pushing these students forward and encouraging them to be the very best? Inversely, what are the obstacles facing these students that keep them from realizing their goals?

Design students operate in a different environment than most. Rather than traditional testing, students are graded on semester-long projects. Meanwhile, the scope of

C. S. Shin (Ed.): AHFE 2019, AISC 968, pp. 205–213, 2020.
https://doi.org/10.1007/978-3-030-20470-9_22

their projects can be astronomically different from the one before. Instead of following a set curriculum, design students are tasked with the responsibility of designing for the future. To do this, they have to first construct and evaluate multiple possible futures by examining the intersection of human science, engineering, business, and art. Both sides of their brain are forced to ride in tandem. In terms of design education, the practice isn't that old. In 1919 the Bauhaus emerged in Weimar, Germany making Industrial Design a distinct discipline. It was described as a "utopian craft guild combining architecture, sculpture, and painting into a single creative expression" [1].

Perhaps there has always been and continues to be an issue with identity. This world of design is constantly in flux from the trends of the artistic influences to the technological advancements of the manufacturing sector. For a discipline still defining itself, it's no wonder young design students struggle to find their place.

This paper will analyze the motivations and obstacles that design students face during their academic career. Both internal motivations and external obstacles will be viewed through the lens of the people that influence students and the systems they exist within. The influencers will be unpacked through the subtopics of the individual his or herself, the family, the instructor, their friends, and social networks. The systems explored will cover the subtopics of high school education, professional competition and hiring tendencies, and award giving organizations. Concluding the discussion of the aforementioned topics will be three key implementations that aid design students in maintaining their motivations and outpacing their obstacles.

2 People

2.1 The Individual

Many of a person's internal motivations are shaped and significantly impacted by external sources. It could be reasonable to skip straight to the external sources that impact the design student, but there is one factor that should be analyzed first: their talent and interest. If a student recognizes that he or she has a natural proficiency, he or she may seek to develop those interests through practice. Dean Simonton describes this realization where, "you all of a sudden find yourself different. You see yourself as different. You have different goals. And these diversifying experiences can take a lot of different forms," and these diversifying influences, "tend to lead to creative genius" [2].

Specified talent and interest at a young age are uncommon. But if talent is rare, then identifying and pursuing an interest at a young age is almost unheard of. Angela Duckworth notes that she, "[doesn't] think most young people need encouragement to follow their passion. Most would do exactly that - in a heartbeat - if only they had a passion in the first place" [3]. Most young people don't know what they want to do for the rest of their lives. This is a good thing. Students need to be given the freedom to explore different career paths before they invest the next decade pursuing a specific vocation.

2.2 The Family

Of all the factors that impact a design student, the family is notably the strongest. Industrial Design is still a relatively unknown discipline despite its emergence in the social lexicon in the last few years. The general public, while owning many consumer electronics fail to consider that someone plays a major role in the shaping and interfacing of their favorite products. This lack of public knowledge consequently extends to the mysterious design school environment. Design school, in its inherent structure, greatly differs from the high school educational model. For this reason, families may struggle to comprehend the challenges most design students encounter – for many the blending of engineering, art, human sciences, and business is difficult to comprehend.

Yet, more essential to a student's success than their parents' understanding is their parents' support. College students, especially design students, who receive support from their family tend to be more accomplished. Encouraging, well informed parents who grant their children the opportunity to explore multiple interests and professions give their children excellent momentum as they journey through college. However, university itself is but a fraction of a young person's life – the majority of influence and direction that a young student receives from their family happens while under their parents' roof. Support in college is great, but support and guidance while coming of age is of essential.

Households of artistic, engineering, or even design backgrounds aren't necessarily guaranteed to produce design savants. However, exposure to that type of thinking does help. Nico Muhly, the youngest person to ever have a commission from the Metropolitan Opera, was the child of both a painter and a documentary-filmmaker. His parents were supportive of his musical interests, but more importantly they "found the good middle point." What set Nico up to succeed wasn't simply that his parents were artists, but it was that they worked at "creating a household in which ideas were spoken about". Nico expresses that "the real luxury of my childhood was not necessarily being surrounded by art in that way, but by people who read and thought about a million things…" [4].

2.3 The Instructor

The role of the instructor in terms of motivation or discouragement of students is well documented. A poor instructor not only frustrates, he or she simultaneously mars a person's interests in a certain topic. Adversely, great teachers evolve good students into great students. Angela Duckworth says that great teachers can actually have more influence than opportunity or inherit talent. A "great coach or teacher" may matter more than anything about the individual [2]. A good professor balances nurturing a student with challenging them to achieve their highest potential. If design students pursue instructors that are known for developing clever and hardworking young professionals, students will surely set themselves up to succeed.

2.4 Friends

Friends are an interesting part of the equation in the university setting. Yes, the university serves as a place to prepare for a potential profession, but higher education by default provides students with opportunities to interact and grow socially. College friends are an incredibly important choice. As the old proverb goes, "iron sharpens iron" [5]. Friends have a way of persuading each other: those that prioritize social functions too much can lead even a highly dedicated student astray, while dedicated friends can help keep a student on track and bring out the best in them. This is especially true of peers that are a part of the same cohort. In an article examining the Architectural studio, Boyer and Mitgang noticed a special comradery between students who "stay up late, are never home, spend all their time in studio, and belong to a clique of other architecture students… here in this earliest phase of becoming an architect we see kernels of architects larger values. Such as the principle of peer review…" [6]. By investing time in positive relationships during school, students arm themselves with an arsenal of comrades they can call on once they have gone out into the professional world.

2.5 Social Media

The most controversial and misunderstood influence on the college design student is social media. With over 1 billion users [7] and ⅔ within the age range of 18–34 [8], it is undeniable that social media is incredibly pervasive. Though these forms of community prove to be wildly popular, whether or not they are beneficial must be examined.

Because rates of tech and screen addiction are on the rise, there has been an increasing concern for how patrons manage their digital hygiene. Gizmodo editor Kashmir Hill remarks that social media, like many connected services, is "so ubiquitous and fundamental to our lives that [its] offerings have replaced core functions of our brains" [9]. It is interesting to note that computational giants like Bill Gates and Steve Jobs severely limit their children's technology usage [10]. Even the tech companies themselves are aware of the potential damage: Apple has introduced 'Screen Time' as a means for users to better understand and manage the time they spend interacting with a smartphone [11].

Unfortunately, much damage has already occurred. Studies from Nottingham Trent University and Swansea University show that social media like Instagram is addictive. It causes an over dependency of connectedness and results in heightened forms of anxiety. It leads "to greater feelings of social isolation", and leads users to unfairly compare themselves to others. The aggregation of digital 'followers' doesn't inherently produce a better social life. Alice G. Walton says, "there seems to be a cap on the number of friends a person's brain can handle, and it takes actual social interaction (not virtual) to keep up these friendships" [12].

A publication from the University of Pennsylvania [13] simply states that "(w)hen you look at other people's lives, particularly on Instagram, it's easy to conclude that everyone else's life is cooler or better than yours." This significantly impacts students who are in the design curriculum. It takes years to develop the visual communication skills of an accomplished designer, yet when they open Instagram design students see the best work that the best designers have to offer they instantly feel inadequate.

Rather than seeing the mistakes and the messy parts of the process, students only see the obsessively crafted final product. Students get the impression that they have to produce the highest level of visual output to gain acceptance in the profession. There seems to be a fear of being identified as "not there yet" and rather than face public ridicule, students hesitate to share their work, even in safe spaces like the design studio. Fearful of judgement from "followers" or "friends", students remember the mantra that's been whispered in their ears: one cannot fail what one never attempts.

However, social media isn't all bad– it exposes design students to a wide world of creatives who share many things in common. Ed Glaeser says that having access to a large and interactive community "plays to our ability to learn from one another"... "Ideas colliding, on purpose and by accident. Also, there's competition"... "and with that competition comes strong incentives to create" [2]. Social media also serves as a significant source of aesthetic inspiration. Because there is a low barrier to entry, there is also freedom to post without worrying about whether an archaic system believes the content is permissible or is impermissible.

3 Systems

3.1 High Schools

Most high schools and their prevailing pedagogy don't prepare students for creative industries. These schools focus on giving students a base set of skills necessary for successful careers or to pursue higher education. Unfortunately, design students carry the mindset and ideology of that system into the design classroom, where that system has no place. While test taking, the memorization and recollection of facts, is the preferred measurement of quality for the high school system, this form of assessment doesn't translate well into the studio. Design schools rarely give tests and design prompts almost never have a purely right or wrong answer.

Siloed learning environments also impede future design students. The design world is project-based and requires designers to recruit the skills of other talented professionals in order to execute a large goal. Group work and project-based learning activities are rare at most schools, and when tests are the primary source of assessment "failure" is viewed as a student's worst possible outcome. In design school peer collaboration is essential. Early, frequent failure can provide incredible insights and accelerate the pace of a project. High schools don't set out to inhibit students from pursuing creative careers, it's just that old systems and hard to define career paths don't gel very well.

Not all high schools are created equally – though rare, certain schools are uniquely geared to prepare students for the creative profession. These specialty schools grant students the opportunity to identify their interests and begin to craft a passion that could drive an individual throughout their entire creative career. In terms of hard skills, like visual communication, these specialized schools give their students an early start on the "10,000 h needed for an individual to become an expert on a topic" [14]. Students who receive such design skills (drawing, form-giving, and computer modeling) are, as a result, potentially years ahead of their classmates.

3.2 Professional Domains

Bookending the other side of a student's college career stands the professional world. The professional domain can be an interesting motivator or obstacle depending on the interests of the design student. One such obstacle can be location: certain career opportunities occur more or less based on the part of the county a student lives in. For instance, many tech jobs are located on the West coast while more traditional design jobs are housed in the Midwest. Regardless of the type of design a student chooses to pursue, the profession as a whole is competitive.

Frequently, skillful students find job placement while a portion of graduating seniors end up finding employment outside of the design profession. While the scope of design extends into the health, governmental, interfaces, systems, and service fields, the bulk of industrial design still resides in manufacturing. A healthy consumer market allows many students to enter the design field while a slower economy can quell even the most talented of young designers.

3.3 Design Employers' Tendencies

Industrial Design is a problem-solving endeavor, and acts as the intersection of engineering, humanities, business, and art. Designers have always been proud of and vocal about the superiority of its process but there still lies a very human problem: everyone loves talent, natural giftedness. Chia-Jung Tsay simply states that, "we do love naturals" There have been many cases where capable students who have superior ideas, more work experience, and heightened teamwork abilities get overlooked simply because of their lack of visualization skills. Chia's research, "pulls back the curtain on our ambivalence toward talent and effort. What we say we care about may not correspond with what - down deep - we actually believe to be more valuable. It's a little like saying we don't care at all about physical attractiveness in a romantic partner and then, when it comes to actually choosing whom to date, picking the cute guy over the nice one" [3]. There is nothing malicious about this approach. It is simply human nature: we enjoy things that are attractive and effortless.

3.4 Awarding Bodies

In conjunction with the human tendency to admire surface over substance, there lies the design awards. By no means do students live or die by their acceptance or rejection from design competitions. However, success in this arena can boost students' confidence while rejections leave them questioning their own propensity to create good design.

Consider this, in the early 2000's Kobe Bryant, Adidas, and Audi teamed up to make an automotive inspired shoe. The result of their interaction led to the adidas Kobe ONE and subsequently the Kobe TWO. The Kobe TWO was described as having a "clean design and subtle branding" and was awarded an IDSA silver in the

IDSA IDEA awards of 2002 [15]. Not only did the shoe perform poorly in the marketplace, but this is the last shoe that Adidas made for the 'Black Mamba'. The embarrassment was so great that Kobe Bryant chose to not produce a shoe at all for the 2002–2003 season [16]. The following year he signed with Nike.

Winning awards is good and can assist in professional development but should never be the goal in creating quality design. Competitions have their own agenda and are judged by humans entitled to their own opinions and biases. Just because a design doesn't receive an award doesn't mean that the design is inherently bad. Oppositely, just because a design receives an award doesn't mean that it will be successful in the 'real world'.

As we wrap up the internal and external forces that come from the systems like high school education and professional tendencies it's wise to consider adopting the mindset of controlling what one can control and letting the rest go. Because these systems are slow to change, it benefits young designers to rely upon and develop personal and professional relationships that will encourage them to thrive. If young designers surround themselves with encouraging voices, consult the wisdom of concerned mentors, and stop comparing themselves to every post on social media, they will develop into the well-rounded change agents that the design profession needs.

4 Conclusions

Identifying motivations and obstacles, both internal and external, are good practice for any design student. More importantly, they must decide how they will respond to these influences. In order to overcome these obstacles, design students may increase their success while promoting a healthy departmental culture by implementing the following three practices.

4.1 Destigmatize Failure

We have to destigmatize failure. The design studio should be a space where new and controversial ideas can grow and flourish. It should act as a place of transparency where all participants are given permission to try and fail without judgment. Professors need to strike a delicate balance between demanding greatness though competition and having students invest in one another through collaboration. Academia is not the career field. The classroom should serve as their training ground, a habitat for challenging oneself more and more. Failure is a pathway to success, not the enemy of it.

4.2 Team Selection

The next step is to choose the correct people to surround oneself with. The relationships students build and maintain with others will have the most profound impact on their creative future. Learning from experienced mentors and peer feedback is essential to development, so it is imperative students select that group carefully. One cannot please everyone, and more importantly not everyone has the authority to give feedback. Students must choose their influencers thoughtfully.

We are influenced by what they see, what our minds take captive. All stimuli, including those from social media need to be viewed through a critical lens. Students need to make an active choice to follow designers who have integrity and are transparent in their design process. They need to make the effort to locate and champion people who change the world with their creations [17] rather than those who produce excellent renders of boujee headphones.

In person relationships are easy to address, but connections through social media are murkier. Digital connections have a duality: social media can be used for influence and inspiration, but it can also be an escape or entertainment. The average person scrolls through 5 miles of social media content in a year [18]. Considering its addictive nature and ability to impact the psychology of a user, digital connections need additional caution.

All humans have a mere 24 h in a day. Therefore, if one is investing time they must consider the return on that investment. Design students need to create an agenda and be mindful of where their attention is directed. The design student should treat digital connections as a growing tool, while also seeking opportunities to interact, in person, with these connections.

4.3 Develop Grit

The last step is more of an attitude, an attitude that must be adopted for students to achieve true success in and beyond the classroom. There exists a common misunderstanding that the magical pixie dust called 'creativity' is what makes a good designer. Nothing could be farther from the truth. The abundance of natural talent, or lack thereof, has little bearing on the future success of a design student. Great design students are marked by the presence of grit – a tenacity to try, fail, learn, and try again. Angela Duckworth states that, "no matter the domain, the highly successful had a kind of ferocious determination that played out in two ways. First, these exemplars were unusually resilient and hardworking. Second, they knew in a very, very deep way what it was they wanted. They not only had determination, they had direction" [3]. The greats became great because they possess the perseverance to get back up, tune out the noise, and try again.

References

1. The Bauhaus (1919–1933). https://www.metmuseum.org/toah/hd/bauh/hd_bauh.htm
2. Where Does Creativity Come From (and Why Do Schools Kill It Off)? http://freakonomics.com/podcast/creativity-2/
3. Duckworth, A.: Grit: The Power of Passion and Perseverance. Scribner, New York (2016)
4. How to Be Creative. http://freakonomics.com/podcast/creativity-1/
5. Proverbs 27:17. https://www.biblegateway.com/passage/?search=Proverbs+27%3A17&version=NIV
6. Boyer, E.L., Mitgang, L.D.: Building community: a new future for architectural education and practice, 2nd Printing edn. A Special Report. Jossey-Bass Inc Pub. (1996)
7. Instagram. https://techcrunch.com/2018/06/20/instagram-1-billion-users/

8. Percentage of U.S. Adults That Use Instagram. https://www.statista.com/statistics/246199/share-of-us-internet-users-who-use-instagram-by-age-group/
9. Life Without the Tech Giants. https://gizmodo.com/life-without-the-tech-giants-1830258056
10. Why Steve Jobs and Bill Gates Both Severely Limited Their Kids' Tech Use. https://www.inc.com/jessica-stillman/why-steve-jobs-bill-gates-both-severely-limited-their-kids-tech-use.html
11. Screen Time. https://www.apple.com/newsroom/2018/06/ios-12-introduces-new-features-to-reduce-interruptions-and-manage-screen-time/
12. Ways Social Media Affects Our Mental Health. https://www.forbes.com/sites/alicegwalton/2017/06/30/a-run-down-of-social-medias-effects-on-our-mental-health/#4d026db62e5a
13. Hunt, M.G., Marx, R., Lipson, C., Young, J.: No more FOMO: limiting social media decreases loneliness and depression. J. Soc. Clin. Psychol. **37**(10), 751–768 (2018)
14. Gladwell, M.: Outliers: The Story of Success. Back Bay Books, New York (2011)
15. KOBE Two. http://www.idsa.org/awards/idea/computer-equipment/kobe-two
16. The Evolution of Kobe Bryant's Signature Sneakers. https://www.complex.com/sneakers/2016/04/evolution-of-kobe-bryant-signature-sneaker
17. My invention that made peace with lions. https://www.ted.com/talks/richard_turere_a_peace_treaty_with_the_lions
18. Kurtz, A.J.: Perfect Isn't Better. https://99u.adobe.com/videos/59398/adam-j-kurtz-perfection-is-a-myth

Lessons from the Archive: Still Relevant 50 Years Later

David Smith(✉)

Auburn University, Auburn, AL, USA
drs0049@auburn.edu

Abstract. Human interaction and sustainability represent huge concepts impacting design practice in multiple disciplines. Because of this impact on design practice, the conversations around these concepts is critical to the profession. The discipline of Industrial design in the making of objects has a rich history of applying these guiding principles, while by comparison graphic design in the making of communications has more recently started to address some of these concepts. This paper highlights bits of conversation from present to 50 years ago about the role of human interaction and sustainability in industrial and graphic design. For the contemporary designer it is important to recognize and embrace these ideas, so that the next generation of designers do not have to rediscover them.

Keywords: Human-centered design · Sustainability · Graphic design · Design thinking · Ethical responsibility · Ethical design · Design futures

1 Introduction

Whether it be peer organizations or academia, those involved in discussions about design practice will often tackle some broad concepts. In these discussions, the idea of a designer's ethical responsibilities will eventually come up. It can be hard to wrap your arms around a subject that may not be explicitly expressed. This is not new, as long as design practices have existed, designers have discussed how to move forward with their work in a responsible way. This paper examines select high points of that conversation over that last 50 years from the perspective of both industrial design and graphic design.

One way to encapsulate this area of a designer's ethical responsibility would be to say, "the impact of one's work on people and the world." Within that statement you have the two very robust subjects of human-centered design and sustainability.

Some say that all encountered problems can be worked out by human-centered design. They propose that embracing human-centered design means believing that all problems, even the seemingly intractable ones like poverty, gender inequality, and clean water, are solvable [1]. The above statement uses the idea of human-centered in its broadest sense. Don Norman gives another description of human-centered design from his book, *The Design of Everyday Things*, "… an approach that puts human needs, capabilities, and behavior first, then designs to accommodate those needs, capabilities, and ways of behaving" [2]. This definition is also very broad, for example,

C. S. Shin (Ed.): AHFE 2019, AISC 968, pp. 214–219, 2020.
https://doi.org/10.1007/978-3-030-20470-9_23

to address human needs is to encompass essential needs such as food, shelter, health, and safety. It also implies usefulness, well-being and longevity—it's human centered. While the concepts of sustainability and human-centered design overlap, we will separate them sometimes in historical highlights.

2 Recent Examples

In a 2017 report by the National Endowment for the Arts, Industrial Design is identified as a key ingredient in the future success of small and medium-sized firms. The report states, "Industrial design is not a new field, but what is relatively new in the past decade is the recognition of industrial design as a means to significantly increase a company's value through design thinking" [3]. The report found industrial design using design thinking as a key factor in solving business problems. Design thinking is the process at the root of human-centered design, and a hot topic for companies wanting to innovate. Design thinking revolves around a deep interest in developing an understanding of the people for whom the products or services are designed [4]. This deep interest is much more than describing a user persona or identifying a target market. It is to come from true empathy of a person's needs, and not colored by assumptions or personal opinions. According to the report, industrial design utilizes design thinking—at least for the last 10 years.

AIGA is the professional organization for graphic design. The leadership of AIGA published *Design Futures* on their website in 2018. *Design Futures* describes seven trends related to graphic design, many of which have been caused or exacerbated by rapid technological advancements. The article suggests that designers and design educators think about the implications of these trends and adapt their practice accordingly. Trend number four is titled *Core Values Matter*. One of the main points of this trend is the interweaving of commerce and social responsibility. One of the challenges of incorporating core values into a practice is the urge to categorize some projects as "design for good" and other projects as "not for good." The article suggests that educators particularly should resist this compartmentalization of projects and teach students to understand that all design should be accountable for producing positive social and environmental consequences [5]. The implication here is that for many (if not most) in the graphic design discipline, there is a lack of inherent accountability for positive or negative social and environmental consequences resulting from their practice.

The final example from recent history is the First Things First Manifesto of 2014. The Manifesto first appeared in 1964, was updated in 2000, and revised again on its 50th anniversary in 2014 by visual artist Cole Peters. The First Things First Manifesto is a signed statement by primarily designers (but inclusive of other visual artists) that expresses disappointment with the state of their industry and its effects on cultures and societies across the world. The 2014 version expands the scope to include those working in digital environments and the technology industry. It is also a call for refocusing of priorities to shift away from profit over people business models, towards the exploration and production of humble, meaningful work and beneficial cultural impact [6].

From the above recent examples, one might surmise that graphic design more than industrial design is still maturing and searching to establish how ethical responsibility and a human centered approach can be integrated into practice.

3 Aging Voices

The thirty year period just prior to our last decade saw design visionaries extolling the ethical use of design, and a burgeoning worldwide concern for sustainability. These factors coalesced into a heightened industry awareness of ethical responsibility.

Born in Austria, Victor Papanek is an American architect, industrial designer, teacher and author that applied the principles of socially responsible design in collaborative projects with concerns such as UNESCO and the World Health Organization. He considered design as a political tool in developing countries and peripheral communities in Europe and the US. The groundbreaking ideas and uncompromising critique of contemporary design culture contained in his book, *Design for the Real World,* initially divided the design community. Ultimately, however, the book was a huge success; translated into twenty-three languages it remains one of the most widely read design books to date [7]. In the preface of the book, Papanek takes a shot at industrial design, naming it one of the most harmful and phony of professions. He then turns to advertising design (a subset of graphic design), and names it phonier than industrial design. To summarize his thinking about advertising design Papanek writes, "Advertising design, in persuading people to buy things they don't need, with money they don't have, in order to impress others who don't care, is probably the phoniest field in existence today." This very harsh critique is supported by Papanek's view that design in this age of mass production demands high social responsibility from the designer. Because everything must be planned and designed, he believed that design had become the most powerful tool that mankind uses to shape the environment and society. [8]. Papanek had a clear vision of the potential of design to benefit people and the world. In the Sixties when the book was originally written, he found the design professions wanting.

Another important design influence at the turn of the century is Bruce Mau, who started his career as a graphic designer but evolved into an interdisciplinary designer. He currently operates a practice named Massive Change Network. Mau was a pioneer in design thinking and has evolved his own design thinking methodology to inspire innovative solutions to challenges in any field or environment and on any scale [9]. In his optimistic book, *Massive Change* He boldly asks the question, "Now that we can do anything, what will we do?" The book addresses eleven subjects impacting the future, from energy to politics, in which Mau ties all together with the concept of a design economy. He highlights various aspects of each subject area with short articles describing the latest research and hopeful potential. In explaining the design economy concept, Mau specifically mentions the industrial and graphic design disciplines saying, "... we began to explore systems of exchange, or design 'economies.' Instead of looking at product design, we looked at the economics of movement. Instead of isolating graphic design, we considered the economics of information" [10].

The green movement of the Seventies in America finally started to have widespread impact on the design community in the Eighties and early Nineties influencing architecture, product design, packaging design and printing. Ecological oriented themes in design books and professional publications placed sustainability issues into the consciousness of all design professionals. One such book, *Green Design* by Dorothy Mackenzie, addresses sustainability in multiple design disciplines. She deftly summarizes the shift in thinking during that period. During the eighties the recognition of good design was on the rise around the world. However, that rise took a downturn at the end of the decade. Focus on form and fashion equated to superficial glitziness that was targeted to those wanting the social status that comes with expensive and unnecessary material things. Design was devalued—seen as elitist and irrelevant by the public. Mackenzie writes, "Designers have participated fully in the disposable society, creating new styles with increasing frequency, and therefore necessarily building in obsolescence." This practice led to criticism of the design profession from environmental groups because of the huge waste of natural resources and failing to use their skills in more useful ways. As global interest in environmental issues grew, businesses were paying attention and began looking for ways to be part of the solution. This shift of priorities provided opportunity for design to have a stronger role in environmental considerations, as well as other ethical concerns. [11].

For the first time, graphic design holistically embraced using recycled paper, starting its journey as a discipline into sustainability.

Design writer Rick Poynor's poignant article about the First Things First Manifesto 2000 update in *Émigré 51* quotes artist Johanna Drucker that the process of exposing the underlying ideological basis of commercial culture boils down to a simple question that we need to ask, and keep on asking, "In whose interest and to what ends? Who gains by this construction of reality, by this representation of this condition as 'natural'?" [12]. This question should serve as a guide for accountability of the social and environmental consequences from one's practice.

4 Lessons from the Archive

In the early Sixties, the Industrial Design Forum, an organization of Auburn University industrial design students began publishing an annual named *Auburn Design*[1]. This publication which contained industry articles, student work and promotion of the industrial design program at Auburn lasted about 20 years. Articles were written by faculty, students and industry guests. Many of these articles touch on or feature the ideas of human centered design and sustainability.

Page 35 from the 1976 Volume 16/17 of *Auburn Design* was a letter to prospective students. The text is solely devoted to issues of human need and sustainability. Following is the entire text:

> "A Career for You. Are you concerned about your environment, consumer product safety, control of sensorial pollution, barrier-free design for the handicapped, or better housing and

[1] At the time of this writing the published issues of *Auburn Design* are not officially archived.

transportation? That is, would you like to induce changes for the humanization of technology on spaceship earth? If you care about these fundamental ecological problems of our technological society, you might consider a professional career in Industrial Design. The Industrial Designer's concern for the human being and his opportunity to participate innovatively in the shaping of man's future attracts creative young men and women to this profession" [13].

The text speaks to the focus of the program on design responsibility over business or aesthetic concerns, and reveals the integration of this philosophy into the core of the curriculum during this time period.

A second example is an article from Volume 14. The name of the article is "Where is the Measure of All Things?" written by the Auburn Industrial Design program director Walter Schaer. The overall theme is that man is not the measure of all things, but that all things natural and man-made are parts of a dynamic planetarial organism on which man himself is only a component. Schaer points out multiple examples of technological achievements that turned into limitations, and warns readers not to trust in slogans such as "growth is progress" and "bigger is better." He makes a case for rethinking how we approach technology:

"A restructuring process of our technical environment demands a full understanding not only concerning foreseeable 'consequences of use' and the elimination of the 'constraints of use,' but also the formulation of new 'objectives of use.' We must recognize that questions of use of science and technology are always moral and planning questions. That is, they contain ethical value judgements, which can only be successful if they coincide with man's biological limitations. We must learn how to manipulate and design the scientific and technological forces. This vital process is design in its broadest connotation and it is a social activity exercised by everybody" [14].

It is easy to find many other examples of thoughtful discussion about human centered design and sustainability within the pages of the *Auburn Design* annuals.

5 Conclusion

It is about the same time that the early *Auburn Design* annuals were being published that the original 1964 First Things First Manifesto was released. The manifesto was written out of frustration with the state of design. Frustration that most design practice was centered on profit over human need, and with no regard for human or environmental consequences. Ken Garland, the writer of the manifesto studied design at the Central School of Arts and Crafts in London in the early 1950s, and for six years was art editor of *Design* magazine, official mouthpiece of the Council of Industrial Design. Garland was in a good position to get a clear overview of the design scene at that time. It is striking to see the stark contrast of practice between the faculty and students in the Auburn Industrial Design program and that of the general design industry. Even more striking is that *Auburn Design* in the Sixties and Seventies speaks of ethical design in such an integrated natural way while fifty years later most of the design industry is still struggling with the concept, and in 2014 felt the need to revise the First Things First Manifesto yet again.

Industrial design as a discipline seems to have a more established footing than graphic design. It is credited with using design thinking as an integral part of its design process. Also integral to industrial design is the use of human factors, which automatically makes it sensitive to human needs. The question is, does industrial design as a discipline take the concept of human need beyond the specific project specifications to consider the unintended consequences of design choices? On the Industrial Designers Society of America website, it is easy to find information on professionalism, process and business issues, but not so much on issues of design ethics and sustainability.

The AIGA website for graphic design contains plenty of discussion about design ethics and responsibility. Are these discussions on the website a reflection of the discipline as a whole or is it the leadership of the profession trying to wake up the rest of the graphic design community. Interactive design has brought design thinking closer to being a part of the bigger graphic design process but is still a long way off from affecting the industry in a holistic way.

Planned obsolescence shows up in the form of replacing a perfectly good smart phone every couple of years. Advertisements still abound trying to persuade people to buy things they don't need, with money they don't have, to impress people who don't care. Perhaps it's time to relearn some lessons from the archive.

References

1. IDEO.org: The Field Guide to Human-Centered Design. IDEO.org, San Francisco (2015)
2. Norman, D.: The Design of Everyday Things. Basic Books, New York (2013)
3. Sheppard, H., Spearman, C., Carter, K.: Industrial Design Report May 2017 rev3.pdf. National Endowment for the Arts (2017)
4. Dam, R., Siang, T.: What is Design Thinking and Why Is It So Popular? https://www.interaction-design.org/literature/article/what-is-design-thinking-and-why-is-it-so-popular
5. Davis, M.: Introduction to Design Futures. https://www.aiga.org/aiga-design-futures/introduction-to-design-futures/?mc_cid=da65c6927b&mc_eid=0860ab517f
6. Peters, C.: First Things First (2014). https://dpya.org/wiki/images/e/e2/First_Things_First_2014.pdf
7. Papanek Foundation – About. https://papanek.org/about/
8. Papanek, V.: Design for the Real World. Academy Chicago Publishers, Chicago (1985)
9. Mau, B., Leonard, J.: Massive Change. Phaidon Press, London (2010)
10. Mau, B.: Bruce Mau's Massive Change Network. http://www.massivechangenetwork.com/bruce-mau/
11. Mackenzie, D.: Design for the Environment. Rizzoli, New York (1991)
12. Poynor, R.: Emigre Essays First Things First Revisited. https://www.emigre.com/Essays/Magazine/FirstThingsFirstRevisited
13. Schaer, W.: Auburn Design 1976, vol. 16/17. Industrial Design Forum, Auburn (1976)
14. Schaer, W.: Auburn Design 1972, vol. 14. Industrial Design Forum, Auburn (1972)

Collaboration Models for Teaching Design Within Specialized Contexts

Shea Tillman(✉)

School of Industrial and Graphic Design, Auburn University, 207 Wallace Hall, Auburn, AL 36849, USA
tillmts@auburn.edu

Abstract. While there exists a wealth of human-centered design projects in which students can successfully render their creative skills while producing a compelling portfolio piece, it is often problematic for faculty to frame and organize a product design collaboration obscured within a space of limited experience or access. This paper will compare and contrast three collaboration models employed in teaching design to undergraduate students on projects that required an in-depth understanding of unfamiliar contexts. Strengths and weaknesses observed through each of these approaches will be outlined, as well as the learning and design outcomes from projects focused on assistive technologies, commercial construction, and the healthcare sector. Faculty and team leaders will gain insights in how to structure their teaching collaborations to maximize student interactions, learning, and design results on specialized projects.

Keywords: Human-centered design · Industrial design · Interdisciplinary · Collaboration

1 Teaching Design: Collaboration with Specialization

Critiques of student aptitude upon entering the workplace have traditionally pointed to a disparity in higher education between theory and practice. This divide between understanding the foundations of a discipline in school and the experiential expertise that comes with repeated applications to 'real-world' problems is legitimate, and while recent criticisms are often manifested towards disciplines within liberal arts [1], they also hold true of professional programs such as industrial design. The profession of industrial design is oriented towards identifying and solving problems from a human perspective, rather than that of technical invention or a market venture. Since visual aesthetics is often a prominent component product design within the human perspective, industrial design education has traditionally placed an emphasis on form-giving and visual communication as required learning outcomes for a designer to be successful in the profession. Industrial design studio coursework and training, faculty hires, and student admission are each influenced by the demonstration of abilities to generate visual solutions that are appropriate. To this end, design faculty often develop student projects to teach an isolated principle or to expand student creativity toward a thoughtful ideal. As a student progresses through a design program, the variables and

C. S. Shin (Ed.): AHFE 2019, AISC 968, pp. 220–229, 2020.
https://doi.org/10.1007/978-3-030-20470-9_24

complexity of these projects should increase to the degree that visual aesthetics become an outgrowth of solving a more focused, often specialized design problem. The disconnect to professional practice for an industrial design graduate is often a result of an over-emphasis on aesthetic training, and their lack of having enough 'real world' problems to solve while in school.

Perhaps no place is this more evident than in the way design faculty require students to frame and approach a project. In selecting and developing advanced project frameworks for students to learn within, design faculty are faced with many application components to integrate: human factors, fostering empathy, business strategy, social implications, ecological impact, just to name a few. In addition, while the structures of institutional education function well in teaching and evaluating the individual student on individual projects, the reality of work in the industrial design profession is both highly collaborative and interdisciplinary in nature.

> *"Collaboration is integral to our profession: 'user-centered' by definition requires teamwork between the designer and the user. Beyond dialog, it's multilog! For us to create successful products and services, it's necessary to tap a whole ecology of experts."* [2]

Since, design faculty seek to establish interdisciplinary projects for students to experience and cultivate viewpoints beyond their own, and when the topic of study is focused or in a specialized field, they are faced with the challenge of how to establish collaborations that will enable this. While locating experts that are willing to share and contribute to a design collaboration may be achievable through university and external networks, time availabilities within course schedules and academic calendars rarely fit the daily work life of a busy professional, making synchronization difficult.

2 Designer Collaboration Roles…Three Models

Problems in syncing collaborative projects have presented faculty in the School of Industrial and Graphic Design at Auburn University with a two-pronged challenge: to charging students to design for someone other than themselves, while providing the specialized experts and expertise required to make their projects substantive and successful. In order to fulfill these requirements for an advanced-level studio project, three distinct collaboration models have been employed across third and fourth year product design studios categorized by the role of the designer within the collaboration: *Designer as Integrated Team Member, Designer as Explorer/Observer, and Designer as Human-Centered Phase.* While describing each of these models by the role of the designer may seem too 'designer-centric,' it is intended to establish for the design educator a general orientation and approach to each project.

2.1 Merging Viewpoints: Designer as Integrated Team Member

In an ideal innovation collaboration, each stakeholder plays an integral part throughout the entire project (problem identification, ideation, development, refinement) in order to create a more holistic and meaningful design solution. The process of designing *with* users rather than 'for' is key for understanding and ultimately innovating for experience [3]. This approach lends itself well to design projects that require personal experience insights in identifying needs and aspirations and concept refinement from someone who would directly benefit from the design.

Designing assistive technology (AT) solutions for people who live each day with a disability poses a unique challenge both to students and faculty. For students, adequately defining the design criteria and problem identification for the project is nearly impossible without close interaction and input from someone who is directly affected. Designing AT can be a particularly unfamiliar and uncomfortable proposition, forcing students to define their entire project around another person's needs that are different from their own. Yet, it is precisely to this end that these projects are so valuable in teaching how to design.

> *"Those of us involved in bringing new products and services to market reduce people to the roles they play in the design development process. We see and think of them as 'consumers' when they shop, as 'customers' when they buy and as 'users' when they interact with the objects of their desire. Rarely do we integrate the roles that people play to come to see them in a holistic manner. Rarely do we take the time to get to know them as people."* [4]

For faculty, organizing and orienting a project to include individuals within a narrow population is also a hurdle, since locating and recruiting people based solely on their disability is awkward, and seemingly undermines the intent of the project. Every effort must be made to maintain the dignity and well-being of people who are involved. This team connection to the user is required to develop an empathetic mindset, not only for developing a solution, but also for generating it (Fig. 1).

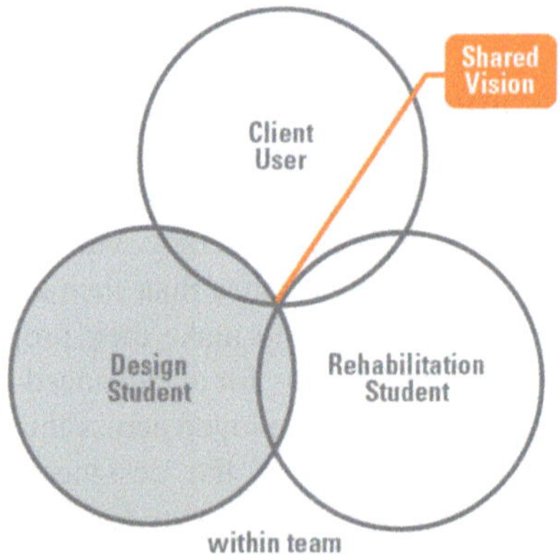

Fig. 1. Integrated Collaborations establish equal footing with designers, experts and users each contributing with each other throughout the project.

In order to better explore how to organize and structure interdisciplinary collaborations, faculty from industrial design and rehabilitation merged their two classes in a six-week studio project designed to create AT solutions. Faculty from rehabilitation drew from their established connections with people living with disabilities, several of whom were enrolled in the class and eager to participate in this project. These participants became the central member of each design team, were termed the 'client user,' and designed with the designer and rehabilitation students throughout the entire project [5]. Each team member (designer, rehab, and client user) was responsible collectively for the outcomes produced and were on equal footing within the team.

Partnering and merging classes with a rehabilitation faculty member allowed for design faculty and students to work directly with both a knowledgeable expert, and a client user from start to finish, all while avoiding potentially awkward participant recruitment. This also enhanced the design student's understanding of considerations within the rehabilitation profession while providing the entire team with a more empathetic approach to design. This high-level personal investment and creativity from each team member created a cohesive, shared vision among the group, ultimately leading to solutions that addressed the client user's personal needs in everyday life. This shared vision allowed design students to develop a more thoughtful concept, rehab students to experience the role design can play in changing lives, and client users to be empowered and enabled by designing through their own input and creativity (Fig. 2).

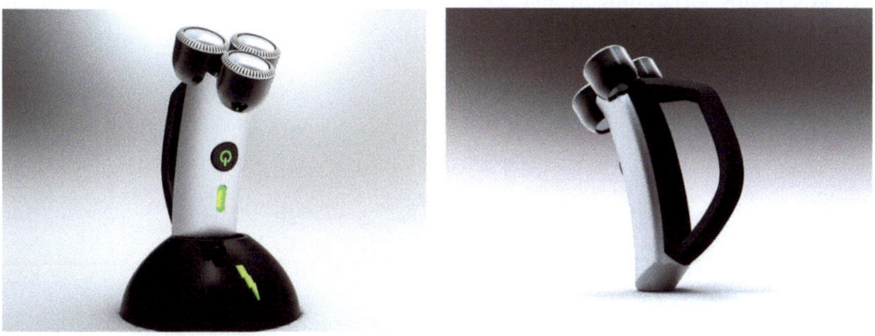

Fig. 2. Assistive technology design solutions (low-grip electric shaver) often answered ordinary personal needs, due to the shared vision established through an Integrated Collaboration with rehab experts and client users.

While this integrated collaboration model represents the ideal, it also brings with it specific challenges. Merging the two class times proved to be difficult based on the demands of two different curricula, and more work spaces were required to adequately house all of the students for effective team tasks. As with most interdisciplinary collaborations, there were initial questions by students struggling to define their roles and responsibilities. Synchronizing team research outside of class time also proved to be problematic, since much of the research stage required in-depth study with the client user in their home or workplace. While this model may create the most inclusive and empathetic team, and possibly circumvent a solution defined too narrowly, it also has the tendency to be inefficient and inflexible to manage during the process.

2.2 Gatekeeper Access: Designer as Explorer/Observer

In some cases, having the both the time and stakeholder participation required for an integrated collaboration is not feasible within a semester undergraduate project. Efforts required to align a large number of students and their schedules across different disciplines can be problematic. And yet, design faculty are still faced with how to dissociate students from their tendencies toward introspective design. In some of these cases, partnering with a 'context gate-keeper' can allow access to vital research environments that would otherwise be off limits, in-effect giving designers an opportunity to work *alongside* a user group.

Generating new product solutions for the commercial construction industry shares many of the same challenges as that of designing Assistive Technology. Despite being one of the largest economic engines in the United States, [6] lack of innovation and investment in industrial products has left the construction industry largely underserved, with many workers simply adapting tools to problems they face. For most design students and faculty, the ecosystem within a large-scale construction zone is foreign with an extreme environment, and busy tradespeople on time-critical and often hazardous tasks (Fig. 3).

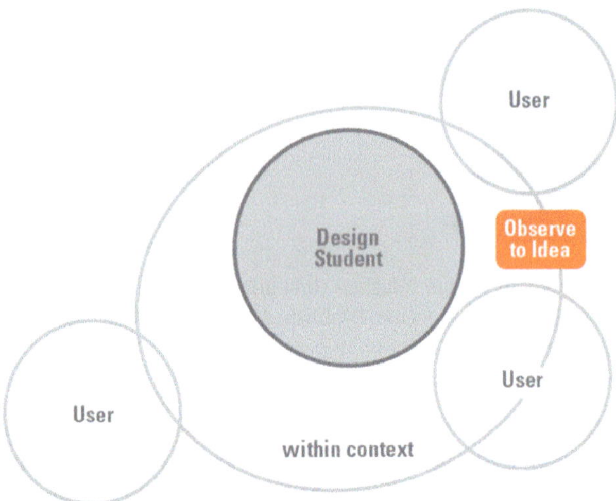

Fig. 3. Explorer/Observer Collaborations rely on a partner gatekeeper to establish access to expert knowledge and specialized contexts, while working alongside users.

In orienting the design student as an 'explorer/observer' within a collaboration, faculty from industrial design partnered with faculty from building science and a corporate sponsor to better understand opportunities for innovation within the construction zone. Small student teams (all designers) were tasked with researching a specific area of a jobsite, identifying potential problems for construction workers, and then individually generating concepts that could solve those problems. The corporate sponsor provided a product category context (laser measuring and visualization) in which students could frame their research. Perhaps most important to the success of this project, however was the faculty partnership with the school of building science who maintained a network of contractors with whom they were regularly able to schedule site visits. Through this, building science faculty were able to schedule in-the-field observation for the design students to study the work dynamics of people on the commercial jobsite. This immersion within this restricted context gave design students the ability to observe and interview construction workers and their practices in real-time. Though this 'outside observer' orientation did not involve the workers as team-members, to the extent of an integrated collaboration, students were able to both see problems and potential design opportunities with 'novice eyes,' all while experiencing the 'real life' context.

> *"...user-centered design strives to enhance the lives of stakeholder and to discover surprising solutions. Searching for unmet human wants and needs opens up the outcomes of the design process to include experiences, systems, and services as well as physical things."* [7] (p. 21)

Following several observation/interview visits, the design students and faculty worked to generate appropriate solutions within the area of laser measuring and visualization. Some concepts generated were even evaluated by construction workers on repeat visits during the middle phase of the project. Building science faculty in turn, became the 'ambassadors of good design' for the jobsite, helping reinforce their innovation reputation among commercial builders (Fig. 4).

Fig. 4. Laser visualization design solutions (cut template gypsum projector) for construction workers were informed by Explorer/Observer Collaborations, and relied on a partner gatekeeper establishing access to expert knowledge and specialized contexts alongside users.

Designer as explorer/observer collaborations are very effective in giving students a viewpoint of themselves as not being the center of a design problem. As an outsider visiting a specialized context, students often ask questions that a construction worker or manager would not. This has the potential for discovering unforeseen problems that can be opportunities for design. These projects also reinforce to design students their necessity to understand the experience of place, rather than simply search for an informative video. Gatekeepers benefit from growing stronger innovation-based relationships with their industry partners, and the context-specific concepts generated benefit the corporate sponsor.

One of the challenges with working through a gatekeeper access model is that the designer is always the outsider. Since the *modus operandi* of work zones are time deadlines for specific tasks, this 'outsider status' can manifest itself in being hurried through an observation or interview, or being given the 'you don't belong here' treatment. Despite this drawback, this still serves as effective in reinforcing to the design students their non-centric role. A strong collaborative relationship with a gatekeeper partner is essential to making this project successful.

2.3 Revision Two: Designer as Human-Centered Phase

Despite the value of integrated collaborations and gatekeeper partnerships in advancing design projects, some fields require even greater expert analysis and research than can be realistically accomplished within a design studio. In these instances, the context may not be practically accessed by design students and faculty, either because of restrictions or physical distance. In addition, the subject matter being investigated to develop a thoughtful and useful solution may be highly technical or regulated. In these cases, it may be effective to establish a collaboration that allows design students to work *through* an expert group's innovation project.

Despite the overwhelmingly negative moniker given to 'silo thinking,' there are certain projects requiring knowledge of specialized contexts and expertise that can benefit from employing design through a phased approach. Though not real-time collaboration, the efforts of a phased project, in which industrial design is employed as a 'revision two,' can in fact, still be more effective than allowing students to generate ideas devoid of context or expertise.

Within the medical device design industry, industrial design is typically employed to address the human factors components of a product. Often the design criteria in new product development is based upon regulatory, the effectiveness and proven track record of predicate devices or processes. In addition, designers involved at the initiation of a project in this sector often have extensive experience in healthcare requirements. This is certainly not the case with undergraduate design students who can find themselves struggling to determine a useful direction in this specialized field (Fig. 5).

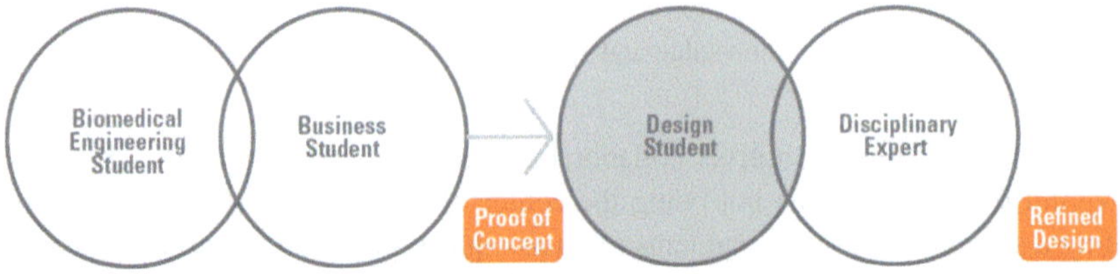

Fig. 5. Human-Centered Phase Collaborations require an initial project phase accomplished by experts on a design revision project can be worked through.

In planning a design studio focusing on medical devices, faculty understood from the outset that an expert partner was required not only to give input on the concept development, but also to frame the project for the design students. With this in mind, design faculty reached out to biomedical engineering faculty at the University of Alabama Birmingham (UAB), a research university that houses a world-class research hospital and extensive expertise in developing solutions for healthcare. Although located two hours apart, each university program complemented the knowledge and abilities of the other. As part of UAB's existing Master of Engineering program, biomedical engineering students and business students are teamed together to research and generate a potential device design. This project requires students to work extensively with biomedical engineering faculty, as well as nurses and physicians within the UAB healthcare system. Each concept generated on this project culminates in a proof-of-concept prototype and a business proposition [8].

While the initial concepts generated by the UAB team represent an idea framed by medical and business experts with a preliminary prototype, they typically lacked the integration, refinement, and attention to user-centered details often understood to be a completed product. Due to conflicts in curricular schedules, and distances, fully integrating and synchronizing the design studio with the UAB project was not viable. Instead, each student within Auburn's design studio selected a previously presented UAB project to revisit from a human-centered perspective. This involved design students and faculty contacting and the members of the original UAB team, as well as the network of medical professionals originally researched as part of the project. Once the framework of the project was understood, design students were free to 'rethink' each aspect of the original concept from the user perspective (aesthetics, intuitive, ease of use, etc.). The functional prototype, though often crude in use and appearance, became an ideal configuration platform for design students to redesign and reconsider an innovative revision. In addition, the design students were able to provide a more realistic version of the product through photo-realistic renderings in their final presentation. While this process effectively ceded the problem identification and collaborative design thinking phase to the UAB team, it provided design students an 'expert informed placeholder,' while allowing them to focus on improving (essentially completing) the established design direction (Fig. 6).

The strength of this 'design phase' approach is its logistical flexibility. Since, expert knowledge is 'cooked in' to the project prior to the involvement of industrial design, the requirement of close coordination of group meetings and locations is minimized. In addition, the follow-up visits for expert advice and specialized access to medical professionals by design students was less intrusive and more focused on the aspects of improving a specific design. Auburn design students gained an advanced design project that answered a clearly identified, specialized need, UAB faculty and teams aided in the advancement and potential commercialization of their original concepts, and the contributing experts and specialists witnessed a realistic embodiment of their input.

As one might expect, the challenges with this model tend to center around this division of labor, namely the tendency to compare the two phases of the project as 'us versus them.' While this dynamic did not adversely affect the project due to UAB students having completed their phase, this mindset has the potential of being detrimental to the collaboration if additional development phases were pursued [9]. In the

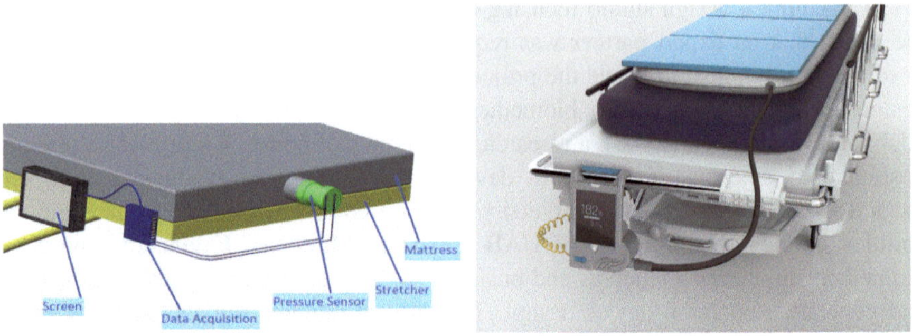

Fig. 6. The development of appropriate medical device solutions (recumbent patient weighing system) benefited from Human-Centered Phase Collaborations that required an initial expert project phase revisited by designers. Initial biomedical engineering/business concept to left, with human-centered design phase revision to right.

case of the UAB projects, the intellectual property had already been reviewed and protected prior to Auburn students beginning the revisit phase. This meant that any improvements made by the Auburn design students would have remained the rights of the original team, unless they fundamentally changed the concept. While this is of some concern, the benefits for Auburn students to have a strong portfolio project on a highly-specialized topic outweighed this potential conflict.

3 Designing Through a Diverse Set of Collaborations

While these three models could accurately be termed as three degrees of collaboration, it should be noted that both teaching and practicing design rarely afford an ideal concurrent collaborative process. In reality, design collaboration occurs at many different levels and phases of a project within product development. Some projects allow a designer to be a contributing team member from the outset, while others may require more external or intermittent involvement. Design faculty can also benefit greatly in their own research agendas through the relationships and partnerships established in developing different approaches and degrees of teaching collaborations. Ensuring that all collaborating stakeholders share in the work and the credit for the outcomes is key to maintaining positive outcomes.

Providing design students with experiences to create a shared vision through a fully integrated collaboration, explore design as a novice observer, or improve an innovation through a design revisit, demonstrates to them the varied role design can play. This is the reality of the profession, and preparing students through a variety of these levels of collaboration better aligns them to the work they will be required to accomplish following graduation.

References

1. New Report Shows Colleges How to Bridge the Gap between the Liberal Arts and the Work Force. https://www.insidehighered.com/news/2018/11/13/new-report-shows-colleges-how-bridge-gap-between-liberal-arts-and-work-force. Accessed 15 Jan 2019
2. Viemeister, T.: Collaboration is Beautility. Innovation: Quarterly of the Industrial Designers Society of America (Winter 2017), pp. 8–9 (2017)
3. Sanders, E.B.-N.: From user-centered to participatory design approaches. In: Frascara, J. (ed.) Design and the Social Sciences, pp. 1–8. Taylor & Francis, New York (2002)
4. Sanders, E.B.-N.: Generative tools for CoDesigning. In: Scrivener, S.A.R., Ball, L.J., Woodcock, A. (eds.) Collaborative Design: Proceedings for CoDesigning, pp. 3–12. Springer, London (2000)
5. Tillman, S., Duncan, J.C.: Guiding interdisciplinary design collaborations in the studio classroom. Des. Principles Pract. Int. J. **3**(4), 271–283 (2009). Common Ground Publishing
6. Construction's Contribution to U.S. Economy Highest in Seven Years. Associated Builders and Contractors - National Office. https://www.abc.org/News-Media/Construction-Economics/State-by-State-Construction-Economics/entryid/9801/constructions-contribution-to-u-s-economy-highest-in-seven-years. Accessed 13 Jan 2019
7. Lupton, E.: Beautiful Users. Princeton Architectural Press and Cooper Hewitt, New York (2014)
8. Eberhardt, A.W.: A project course sequence in innovation and commercialization. J. Biomed. Eng. vol. 139, no. 7, June 2017
9. Haas, M., Mortensen, M.: The Secrets of Great Teamwork. Harv. Bus. Rev. **94**, 70–76 (2016)

Design Embracing Information Science, Intention and Experience

Should the Environment Be a Human Factor?

Stan Ruecker(✉), Claudia Grisales, John A. Capozzo, Juan Alfonso De La Rosa, Gerry Derksen, Priscilla Boff Ferronato, Thais Cristina Sehn, and Cliff Shin

University of Illinois, Urbana, USA
{sruecker, capozzo2, juanad2, gderksen, pf4, thecliff}@illinois.edu, cagrisalesb@gmail.com, crisehn@hotmail.com

Abstract. For many decades, we recognized four human factors: physical, cognitive, social, and cultural. Approximately at the turn of the millennium, we increasingly began to recognize, based on a variety of evidence from psychology, physiology, neurophilosophy, and other fields, that affect should be added as a new category. To understand the human, the argument went, we were missing an important component if we left out emotion. That perspective has been sufficiently well accepted that the dedicated society and conference for the topic, Design and Emotion, was able to declare success, and held its last event in Amsterdam, then closed its doors in 2016.

In this paper, we argue that we should recognize that another aspect of being human is the environment itself. As ecological psychologist Gibson famously said, "Do not ask what is in your head. Ask instead what your head is in" [1]. In order to more completely understand the human factors that are relevant to a particular design, we should be systematically studying the surrounds. These will include everything from the many microscopic symbiotes that make up the human system, to the infrastructure designed and developed by people, to the natural support provided by the planet. Although not typically recognized as part of being human, it is unreasonable to think of people as somehow being apart from the environment that they inhabit. The tendency to separate the two into distinct categories has arguably resulted in some of the worst effects of human activity, as has become increasingly recognized by the people interested in post-human design (e.g. [2–5]).

Keywords: Human factors · Environment · Ecology · Post-humanism

1 Introduction

As the human race has sometimes been construed as *homo faber*, it is reasonable to predict that there will always be human generative activity. We transform the environments where we live. In some instances unfortunately, only some of those transformations have currently been deliberate. However we may decide to proceed as a species, it would be advisable to change that policy, so that our interventions that effect the entire planet are more intentional and less a side effect of other intentions. As with the other human factors, it is possible to understand that good design is going to

C. S. Shin (Ed.): AHFE 2019, AISC 968, pp. 233–239, 2020.
https://doi.org/10.1007/978-3-030-20470-9_25

accommodate all of them in the typical course of the design process; however, just as it was advisable to codify to a certain extent their study, to make those accommodations more predictable and uniform, so it may be useful to begin to think of the environment as another item on the checklist of important factors.

In addition to the practical issues of working within a particular setting, there is the human reaction to the environment to consider. For example, it is not unusual to encounter nostalgic stories of how a particular development transformed the natural landscape. Nostalgia lies over the perception of a trade-off, between what was necessary to achieve the future and what had to be sacrificed—that which apparently belonged to an older world. Stories about rows of trees that had to be chopped down in order to broaden a road, or memories of natural landscapes now disfigured by plastic waste are part of people's narratives all over the world. Because of the intricate networks that support life on the planet the environment is a fundamental part of what being human means, but also, because of the effect that landscapes have in our memories and perceptions, the environment is closely related to our situated experience of humanity.

For the purposes of this paper, we will define nature as those events, processes, materials, and organics on the planet that occur without human intervention. Although it is true that human beings are a natural part of the world, to admit human intervention as a natural phenomenon just shifts the terms until the discussion falls entirely off the table. By the term "environment," we mean the conditions under which people live, including everything from the weather to the virus that is making the rounds to the tulip table in the living room.

2 Three Models of the Future

In one possible future, we have relied entirely on human ingenuity. We call this the principle of human against nature, or "nature is what comes between me and my latte." Are the pollinating insects being wiped out, possibly as a consequence of our using pesticides to kill other kinds of insects that interfere with our crops or transmit human diseases? That is fine–or so this logic goes–because we will just create tiny drones who can act as pollinators, or we will engineer new plants that do not require pollination. Are the trees dying because climate change has introduced a widespread blight that winter would otherwise have killed? No problem: we can genetically engineer new trees that do not have the same weaknesses. Eventually, the entire planet will consist of the designed environment, with no natural environment remaining.

An alternative future, admittedly unlikely, has us change our behavior entirely, choosing to limit forms of human ingenuity that interfere with the natural world. At the present moment, there are arguably no places where this is the case, but there are still wildlife areas where the interventions have largely been accidental, through the introduction of atmospheric lead, radioactive particles, microplastics, and so on, rather than deliberate, as through culling herds of wildlife or burning off forest in order to plant crops. We call this version of the future the hands-off model, where the human agenda is to stand aside and let nature takes its course.

A third possibility constitutes the range in between, which is the world we currently inhabit. We routinely extract resources and make design decisions that destroy the natural world, but not the entire natural world, and we live in hope that the destruction will not have devastating consequences on a large enough scale to trouble the entire human race. This is the future of human stewardship, where our ability to influence the environment is unquestionable, and what remains to be determined is what balance we wish to strike.

3 Using the Environment as a Human Factor

Independent of the version of the future that we are interested in seeing, the use of the environment as a human factor takes the same possible forms. Like the other human factors, it is necessary under the conditions of any given design to take into account the relevant aspects that need to be accommodated. In the case of the environment, these fall into the following categories. Within each category, there is the possibility of the natural and the artificial

- other people (natural or AI), whether individually or in groups
- creatures who are not people
- inanimate or static objects
- kinetic processes (moving without the application of additional forces)
- dynamic processes (accelerating or decelerating under force)

3.1 Other People (Natural or AI), Whether Individually or in Groups

As often as not, other people constitute part of the human environment. They may be involved with the intended design or not involved. If involved, they may be in a supportive role or may represent obstacles [6]. However, the current human factors already include other people and their influence, both at the interpersonal and cultural level, so it is not necessary to consider natural people as a new addition to the checklist.

On the other hand, as artificial intelligence increases in sophistication, it will not be long, and is perhaps already the case in some limited ways, that the environment contains "artificial" people who will need to be accommodated in future design decisions. These might vary from individual systems used, for example, in decision-making in large organizations, as is already done to make the first cut in many human resources departments, to aggregate communities of AIs that together manage more complex situations, including other parts of the environment.

3.2 Creatures Who Are Not People

Animals, plants, and other kingdoms of organic creatures such as bacteria, fungi, and archaea constitute a large part of what people who are interested in design for the post-human are considering. The argument is that, although there were many benefits to realizing that designers were not necessarily the best representatives of the eventual users of a given design, and that other people should therefore be a subject of study, that in fact this focus has meant in practical terms that the rest of the world was either marginalized, or in some cases left out entirely from consideration during the design process.

Including other creatures in the equation is important both in the sense of individuals and in the collective. In fact, there is an argument to be made that the ecosystem as a whole is a sub-component that deserves to be treated separately, given that the systems interactions of the parts are complex enough to be largely unpredictable.

On the artificial side of this category, we have robots. In the same way that we might consider artificial intelligence as a kind of human agency, we can arguably understand robots as a sort of artificial creatures. If they are governed by sufficiently robust AI, they might alternatively be shifted into the category of the artificial human.

3.3 Inanimate or Static Objects

The inorganic world, whether natural or artificial, constitutes the next sub-category to consider. Volumes could be written on the inanimate role in design alone. In some respects, the static environment has largely been the province of designers, who use the inorganic as raw materials for their work. A distinction might be made here in terms of whether the objects being used or impacted are first-generation or natural objects and materials, or whether they are second-generation or more recycled materials.

3.4 Kinetic and Repetitive Processes (Moving Without the Application of Additional Forces)

There are a variety of changes to the environment that can be treated as, or may actually be, kinetic, in the sense that their occurrence is not the result of the immediate application of force. That is to say, there is movement involved, and even change, but locally the situation is not a dynamic one as it would be if there accelerations or decelerations involved. The daily cycle, for instance, from day to night, might fall under this category, while something like weather can be understood as being more dynamic.

3.5 Dynamic Processes (Accelerating or Decelerating Under Force)

If we consider some natural forces as being cyclical, like the change of seasons, then there are also natural forces that occur more randomly, such as the eruption of volcanoes, hurricanes (Fig. 1), tornadoes, and so on. Many design activities already consider the relationship between these events and the humans they influence, but other design projects that might benefit from a more systematic evaluation of these possibilities could benefit by having them included as a standard part of the design process.

From the perspective of human memories of the environment, the dramatic quality of many of the dynamic processes is such that they tend to make a lasting impact on how people remember the world. Anyone who has experienced extreme cold or heat, high winds, unusual ocean waves, and so on, will tend to remember those events for the rest of their lives.

Fig. 1. Should we consider the hurricane as a human factor? Certainly it is a mistake to ignore it…

4 But Is It Arrogance?

To say that animals and robots and the weather are human factors is to propose that everything is in some way human. Clearly this is not the case: a badger, for example, is not a human being, and neither is a bauxite mine. However, from the design perspective, the question is one of the constraints for any given problem: what is in, and what is out, in terms of the factors that must be considered. To date, with the human factors at the center, other things, like ducks and the wetlands that support them, have typically been on the periphery, easy to set aside or to forget because they are not necessarily considered as part of the formal process. Yet in some practical ways and in some ways associated with human thought and sentiment, there are connections that should not be ignored even if thinking only in terms of the exclusively human. In that respect, the part of the Venn diagram that overlaps between the human and the environment is clearly significant, and others might argue that since the actual overlap is complete between the human and the environment (Fig. 2) that contains the human, there is no room to argue.

5 Example of a Pen

Taking the design of a pen as an example, let us see how the addition of the environment might make a difference. To begin with the conventional human factors, the **physical** attributes of the pen are such that it needs to fit comfortably in the human grip

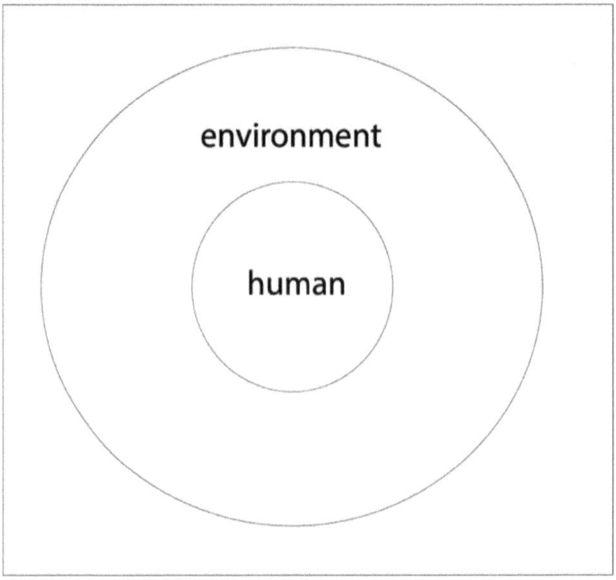

Fig. 2. Humans are contained within the environment.

and conform to writing on the surface of a piece of paper. **Cognitively**, it should be clear how to operate: if the tip retracts, then that mechanism should be in an affordance that is straightforward; if the ink can be replenished, the process should also be clear. **Emotionally**, the feel of the pen should be satisfying in the hand, and it should have an appearance that is pleasing to the eye. The smell of the materials may also have a role to play, so that the ink is either neutral or fragrant, perhaps across a range of choice. **Interpersonally**, the pen should potentially have a role to play in the relationships between people: perhaps it is of sufficiently luxury quality that it can serve as a gift. **Culturally**, the pen has served a wide range of functions from the basic utilitarian to use as a classroom object to being an item of conspicuous consumption.

Now to extend to the **environmental** factor. In terms of its accessibility to artificial humans and robots, it may be necessary to modify the physical attributes somewhat so that it is easy to grasp and use. In terms of the inanimate, we bring in all the considerations of sustainability of materials: perhaps the ink, for example, is sourced from organics that are readily produced and used, and the shaft is either from a recycled material or another renewable source. For the cyclic considerations, most current pens are really only useful when the lighting conditions are appropriate – for this new pen, we might want to consider a kind of ink that is also visible in the dark. Dynamic events are somewhat harder to imagine and plan for, but imagine the pen is in the pocket of a firefighter: will it withstand the heat? Or maybe it is being carried in the coat of someone working in the arctic in winter: will the ink continue to flow at –minus 40 degrees Celsius? Finally, in consideration of other creatures, what are the effects of the cradle-to-cradle life cycle of the pen? Does it have a net positive, negative, or neutral effect in terms of toxicity, especially when multiplied by the hundreds of millions or more?

6 Conclusions

Just as the other human factors are a means of encouraging consistency and completeness in dealing with a design project, so the addition of the environment as a human factor may encourage a new and more comprehensive way of thinking about the human in relation to the rest of the world. Although some design projects have already tended to include elements from the environment, a more systematic approach could be beneficial both to the design outcomes and to the non-human actors who are effected by design.

References

1. Gibson, J.J.: The Ecological Approach to Visual Perception. Houghton-Mifflin, Boston (1979)
2. Forlano, L.: Decentering the human in the design of collaborative cities. Des. Issues **32**(3), 42–54 (2016). https://doi.org/10.1162/desi_a_00398
3. Stengers, I.: Including nonhumans in political theory: opening Pandora's box? In: Braun, B., Whatmore, S.J. (Eds.) Political Matter: Technoscience, Democracy, and Public Life, pp. 3–34 (2010)
4. Tironi, M., Hermansen, P., Neira, J.: The prototype as a cosmopolitical place: ethnographic design practice and research at the National Zoo in Santiago, Chile. In: Proceedings of DRS 2016, Design Research Society 50th Anniversary Conference, 27–30 June 2016, Brighton, UK (2016)
5. Nippert-Eng, C., Dominski, J., Grier, F., Hornor, J., Limb, E., Limb, S., Martinez, M.: Gorillas Up Close. Henry Holt and Company, New York (2016)
6. Ruecker, S., Radzikowska, M., Sinclair, S.: Visual Interface Design for Digital Cultural Heritage: A Guide to Rich-Prospect Browsing. Ashgate Publishing, Farnham (2011)

Quantitative Research on the Relationship Between Design Elements and Kansei Image of Electric Vehicle Styling

Mingzhu Li[1(✉)], Canqun He[2], Zhangping Lu[3], and Liqing Huang[1]

[1] School of Arts, Jiangsu University, Xuefu Road 301, Zhenjiang 212013, China
{limingzhu, hlq}@ujs.edu.cn

[2] College of Mechanical and Electrical Engineering, Hohai University, Jinling N.Road 200, Changzhou 213022, China
hecq@163.com

[3] School of Mechanical Engineering, Jiangsu University, Xuefu Road 301, Zhenjiang 212013, China
lzping@ujs.edu.cn

Abstract. In order to meet the needs of users' emotional experience, quantitative research is carried out on the relationship between product form design elements and Kansei image. The whole research process consists of five steps. Firstly, product samples and Kansei words are collected and selected. Secondly, a new tool is proposed to measure Kansei image, called "Image Circumplex". Thirdly, the form design elements are extracted comprehensively based on Morphological Analysis and Protocol Analysis. Fourthly, Quantification Theory Type I is applied to build the relationship between design elements and Kansei words, and the mathematical model is constructed to support product form design. Finally, the proposed approach is presented and illustrated using a case study of electric vehicle design. The quantitative results of the relationship between form design elements and Kansei image can provide effective support for product form design, evaluation and decision-making.

Keywords: Electric vehicle styling · Kansei image · Form design elements · Image circumplex · Protocol Analysis · Quantification Theory Type I

1 Introduction

With the increasingly growth of material and cultural life, consumers are no longer simply pursuing a product's physical function; they pay more attention to higher emotional needs, such as personalization, the sense of pleasure and emotional experience. Emotional factor has become one of the most important factors in determining their purchase action. Emotional consumption is developing into a dominant form of consumption instead of rational consumption. As an important medium between designers and consumers, Kansei image fully conveys consumers' emotional needs, and can provide a strong support for emotional design.

C. S. Shin (Ed.): AHFE 2019, AISC 968, pp. 240–250, 2020.
https://doi.org/10.1007/978-3-030-20470-9_26

The automobile is a great invention for human beings, which has greatly changed the way of travelling and improved the quality of our lives. Automotive styling design is a complex creative activity involving many subjects such as aerodynamics, ergonomics, material science and aesthetics. It integrates art and engineering technology perfectly through ingenious design. However, the environmental problems caused by automobiles can not be ignored. In recent years, as environmental awareness has grown, people put forward the requirements of clean energy and lightweight for automobiles. The research and development of electric vehicles has been paid more and more attention by governments and major automotive enterprises. The mechanical layout of traditional automobiles may be fundamentally overturned in new energy vehicles, which brings more possibilities for the overall shape design of the body, such as the disappearing grille, non-exposed tires and new man-machine interaction [1]. Therefore, it is important to clear the relationship between Kansei image and EV styling. It is helpful to understand the process of EV styling design, to grasp the direction of EV styling design and to make design evaluation and decision efficiently.

2 Method

2.1 Image Circumplex

Osgood [2] developed a method of measuring the emotional content of a word more objectively, called "Semantic Differential Method". A pair of Kansei words, such as "soft-hard", are located at the left and right ends of the Likert scale, with 5 or 7 points in the middle. SD method has been widely used in Kansei engineering and emotional product design. However, it has some shortcomings. (1) SD method measures people's reactions to stimulus words and concepts in terms of ratings on bipolar scales defined with contrasting adjectives at each end, but for some samples both ends might be selected. (2) SD method assumes that the survey participants' understandings on Kansei words are consistent, so image boards are often used to get the consistent or similar understandings. (3) SD method requires participants to score the product samples on a scale for each Kansei word, but some words might be not suitable to describe all the samples.

On the basis of Russell's circumplex model of emotion [3] and "Emotion Compass" [4], Image Circumplex is presented to improve these issues, and a method based on Image Circumplex is developed to measure Kansei image. It combines SD method and image board. Kansei words are arranged in a circular way. The upper part is the words describing the ideal product, and the lower part is the opposites. The corresponding image boards are distributed around Kansei words. Each Kansei word is measured by a 5-point scale. Participants are allowed to skip some words or choose the words at both ends. In Fig. 1, the black point means that the participant can feel the sense of Kansei Word 02, and the score is 3 points.

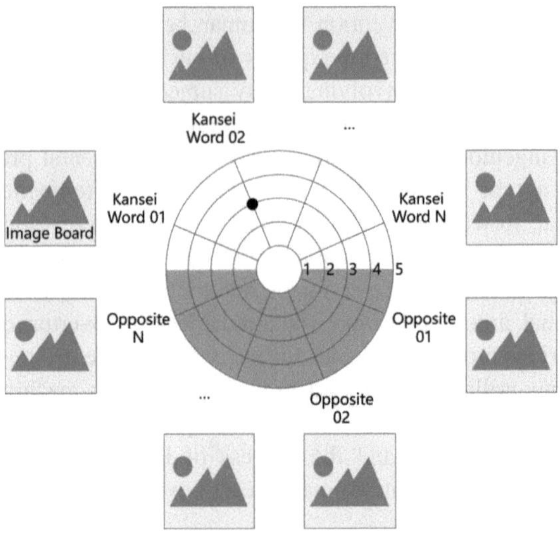

Fig. 1. Image Circumplex

2.2 Morphological Analysis and Protocol Analysis

The theory of Morphological Analysis [5] is widely used in many domains, including product design, manufacturing, and architectural design. In the field of Kansei engineering, it is usually applied to extract the design elements. The main purpose is to break up the product form into different features and the corresponding design elements. Then new forms are created by the replacement and reorganization of these design elements.

Protocol Analysis [6], also known as thinking aloud, is an experimental method of psychological research by analyzing participants' verbal reports. It includes concurrent verbal protocols and retrospective verbal protocols. Protocol Analysis can make the implicit information explicit through verbal reports, so that the design information can be acquired by analyzing these words.

In this research, the form design elements are extracted comprehensively based on Morphological Analysis and Protocol Analysis. Compared with the single method of Morphological Analysis, this method has the two advantages. One is that the design elements include not only the local features, but also the overall features that have a key impact on Kansei image, such as the fading surfaces on the front face; and the other is that the relevant design elements are chosen for each Kansei word, rather than the same.

2.3 Quantification Theory Type I

Quantification Theory is a branch of multivariate statistical analysis, which was developed by Hayashi in the 1950s. According to the different purposes of the research, it can be divided into four categories: Quantification Theory Type I, II, III and IV [7]. Quantification Theory Type I, mainly solves the problem of predicting dependent variables when independent variables are qualitative variables and dependent variables are quantitative variables. Linear regression is used to establish a mathematical model, so that the inherent relationship between things can be revealed. With the help of Quantification Theory Type I method, some results have been achieved [8–10]. The relationship between Kansei image and product form elements is established, and the mathematical model is constructed to obtain the contribution scores of various design elements on Kansei image.

3 Case Study

A case study on electric vehicle design is used to illustrate the proposed method. The whole research process consists of five steps shown in Fig. 2.

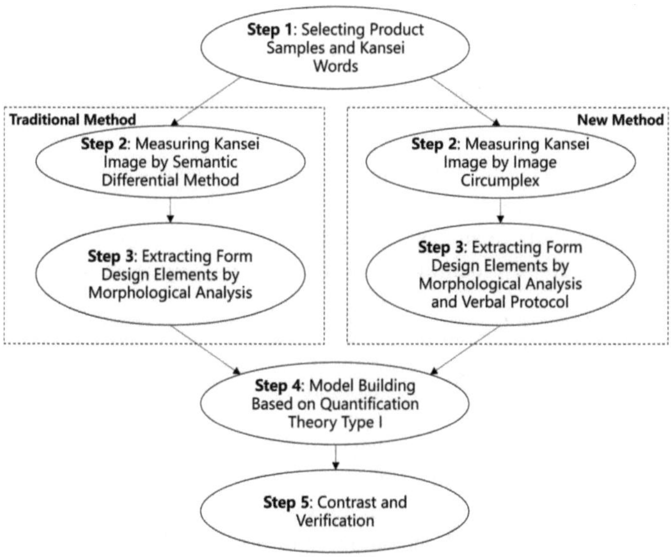

Fig. 2. Research process

3.1 Selecting Product Samples and Kansei Words

By December 2018, a total of 172 new energy vehicles of 80 brands, which are sold in the Chinese market, were collected through major automobile websites. After expert interviews, questionnaire survey and cluster analysis, 22 EV samples were finally obtained. In order to avoid the influence of brand, color and other factors on the images of EV styling, the white front face pictures of the corresponding cars were selected and adjusted as representative samples, shown in Fig. 3.

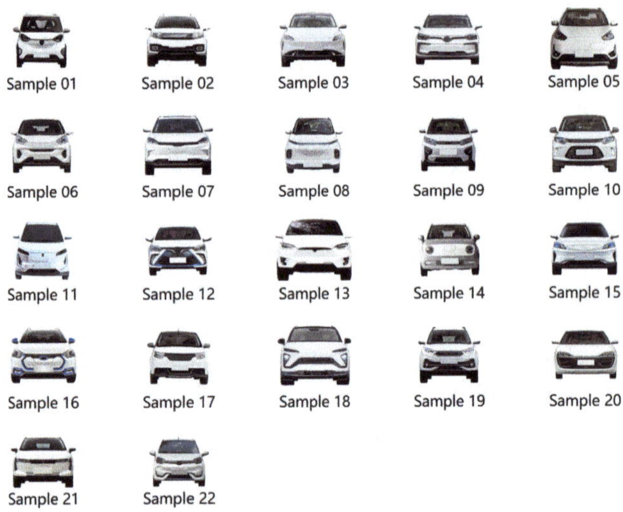

Fig. 3. Representative samples

By the way of looking up automobile magazines, advertisements and articles [11, 12], 65 pairs of Kansei words, suitable for describing the images of EV styling, were collected. Through questionnaire survey, factor analysis and expert revision, 4 pairs of Kansei words were finally identified: soft-hard, technological-retro, simple-complex and light-stable.

3.2 Measuring Kansei Image

Traditional Method. To assess the images of the products, 22 representative samples and 4 pairs of Kansei words were used to establish a 7-point scale based on Semantic Differential Method. A questionnaire survey was distributed on the network platform. A total of 26 data were collected, including 12 males and 14 females, who are users or potential users of automobiles. The average values of Kansei evaluation are shown in Table 1.

Table 1. Average Values of Kansei Evaluation (Traditional Method)

	Soft-hard	Technological-retro	Simple-complex	Light-stable
S 01	−0.77	0.50	0.65	−0.27
S 02	1.00	−0.27	−0.04	1.77
S 03	0.31	−0.38	−1.15	−0.31
S 04	0.85	−0.35	0.35	1.00
S 05	−1.50	0.15	−0.92	−1.23
S 06	−0.69	−0.19	0.12	−0.96

(*continued*)

Table 1. (*continued*)

	Soft-hard	Technological-retro	Simple-complex	Light-stable
S 07	0.46	−0.77	−0.69	0.85
S 08	−1.77	−0.96	−0.35	−1.15
S 09	0.88	−0.27	0.92	−0.04
S 10	0.69	−0.58	0.35	0.96
S 11	−0.12	−0.50	−0.65	−1.46
S 12	0.96	−0.73	2.08	0.81
S 13	0.27	−1.08	−1.31	0.54
S 14	−2.04	1.77	0.42	−0.92
S 15	−0.27	−1.08	−0.81	0.81
S 16	−0.08	−0.42	1.08	1.04
S 17	−0.85	0.50	−0.62	−1.23
S 18	−0.73	−1.00	0.00	1.04
S 19	1.00	0.23	1.15	0.77
S 20	−0.88	−0.81	−0.12	0.15
S 21	0.77	0.23	0.81	1.46
S 22	−0.85	−0.08	−0.08	−0.88

New Method. Then 14 of the above-mentioned participants, including 6 males and 8 females, were re-investigated by using Image Circumplex shown in Fig. 4. They were asked to give a score on Image Circumplex according to their subjective feelings. They can skip some words, or choose the words at both ends. After that, they were interviewed and asked to explain the reasons for the scoring.

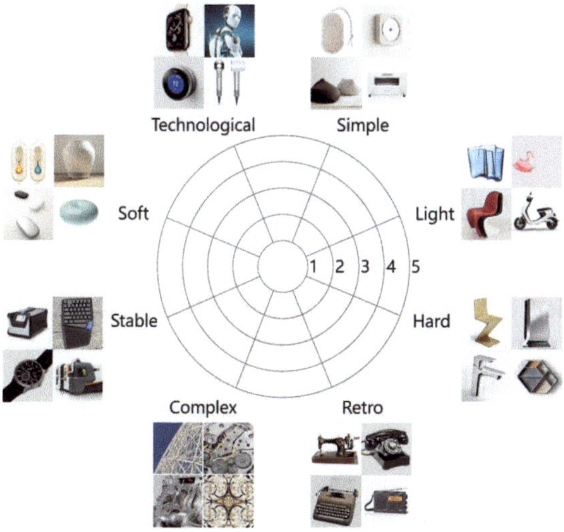

Fig. 4. Image Circumplex for EV styling

The words that more than one third of the participants selected were considered valid for the sample. The average values of these data were calculated. If both ends of the words are valid for a sample, delete the sample. Finally, we got the average values of Kansei evaluation shown in Table 2. The sign "/" means that sample should be deleted for this pair of Kansei words.

Table 2. Average values of Kansei evaluation (new method)

	Soft-hard	Technological-retro	Simple-complex	Light-stable
S 01	−3.00	−3.17	3.00	−3.75
S 02	3.78	−2.44	−3.20	3.36
S 03	−2.83	−3.80	−3.92	3.29
S 04	/	/	3.63	3.18
S 05	−3.33	−2.86	−2.75	/
S 06	/	/	2.88	−3.14
S 07	3.13	−3.20	−3.25	3.55
S 08	−3.91	/	/	−3.44
S 09	2.89	/	3.60	/
S 10	/	−2.40	/	4.00
S 11	−3.57	−3.60	−3.60	−3.75
S 12	/	−3.50	4.69	3.14
S 13	3.40	−3.90	−4.00	/
S 14	−4.00	4.75	3.40	−3.67
S 15	3.14	−3.69	−3.00	3.38
S 16	−2.43	−2.75	3.67	3.14
S 17	−2.00	−1.80	−3.20	−3.50
S 18	−2.43	−3.67	3.00	3.73
S 19	3.10	/	3.33	3.88
S 20	−2.78	−3.13	−2.91	/
S 21	3.00	−2.88	/	3.67
S 22	−2.71	/	−2.29	−3.11

3.3 Extracting Form Design Elements

Traditional Method. Morphological Analysis was used to extract the form design elements. In the field of automobile design, Sun [13] concluded the front face of the car into five design features: engine cover, headlight, upper grille, lower grille and foglight. However, in the field of electric vehicles, the function of the engine cover is gradually weakening, and the parameterized style of the front face has gradually become a trend. Therefore, we extracted the design features shown in Table 3, and refined them into design elements.

Table 3. Classification of design elements (traditional method)

Design features	Design elements			
Headlight shape (T_1)	Circle (T_{11})	Triangle (T_{12})	Trapezoid (T_{13})	Line (T_{14})
Upper grille shape (T_2)	No (T_{21})	Trapezoid (T_{22})	Other (T_{23})	
Foglight shape (T_3)	Trapezoid (T_{31})	Line (T_{32})	Other (T_{33})	
Shape of lower grille (T_4)	No (T_{41})	Trapezoid (T_{42})	Other (T_{43})	
Parametric decoration (T_5)	No (T_{51})	Yes (T_{52})		

New Method. The verbal reports obtained from the interview in Sect. 3.2 were analyzed and coded. Taking the words "technological-retro" as an example, we got some other targeted design features, such as blue line decoration and fading surface. It was also revealed that the item of Lower Grille Shape is not the key feature for the words "technological-retro". Based on the results of Morphological Analysis and Protocol Analysis, we concluded the design features and design elements of electric vehicles, as shown in Table 4. For the item of Foglight Shape (N_3), there was no sample having the trapezoidal shape, so we deleted this element.

Table 4. Classification of Design Elements for "Technological-Retro" (New Method)

Design features	Design elements			
Headlight shape (N_1)	Circle (N_{11})	Triangle (N_{12})	Trapezoid (N_{13})	Line (N_{14})
Upper grille shape (N_2)	No (N_{21})	Trapezoid (N_{22})	Other (N_{23})	
Foglight shape (N_3)	Line (N_{31})	Other (N_{32})		
Parametric decoration (N_4)	No (N_{41})	Yes (N_{42})		
Blue line decoration (N_5)	No (N_{51})	Yes (N_{52})		
Fading surface (N_6)	No (N_{61})	Yes (N_{62})		

3.4 Model Building

According to Quantification Theory Type I, we took the design features as items, the design elements as categories, and the evaluation values of Kansei evaluation as dependent variables. Assuming that m is the number of items, n is the number of samples, and c_i is the number of the item i, $\delta_k(i,j)(i = 1, 2, \cdots, m;\ j = 1, 2, \cdots, c_i;\ k = 1, 2, \cdots, n)$, which is called the reaction, could be calculated by Formula (1):

$$\delta_k(i,j) = \begin{cases} 1 & \text{for the sample } k, \text{ the data of the item } i \text{ is the catogory } j \\ 0 & \text{otherwise} \end{cases} \tag{1}$$

If there is a linear relationship between dependent variables and the reactions of various items and categories, a mathematical model [7] can be established by Formula (2):

$$y_k = \sum_{i=1}^{m} \sum_{j=1}^{c_i} \delta_k(i,j) b_{ij} + \varepsilon_k \tag{2}$$

where b_{ij} is the category weight, and ε_k is the random variable.

Based on Tables 1, 2, 3, 4 and Formula (1), the corresponding reflection matrices of traditional method and new method were established respectively. Then mathematical models were built by using MATLAB, and the results are shown in Tables 5 and 6.

Table 5. Results of "Technological-Retro" (traditional method)

Items	Categories	Category score	Partial correlation coefficient	Multiple correlation coefficient	Constant
T_1	T_{11}	2.477	0.803	0.830	−0.277
	T_{12}	0.037			
	T_{13}	−0.109			
	T_{14}	−0.320			
T_2	T_{21}	−0.152	0.401		
	T_{22}	0.288			
	T_{23}	−0.025			
T_3	T_{31}	−0.387	0.467		
	T_{32}	−0.293			
	T_{33}	0.143			
T_4	T_{41}	−0.135	0.296		
	T_{42}	0.089			
	T_{43}	−0.169			
T_5	T_{51}	−0.105	0.328		
	T_{52}	0.184			

Table 6. Results of "Technological-Retro" (new method)

Items	Categories	Category score	Partial correlation coefficient	Multiple correlation coefficient	Constant
N_1	N_{11}	7.195	0.966	0.981	−2.628
	N_{12}	−0.825			
	N_{13}	−0.301			
	N_{14}	−0.857			
N_2	N_{21}	−0.126	0.420		
	N_{22}	0.395			
	N_{23}	−0.051			
N_3	N_{31}	0.013	0.019		
	N_{32}	−0.008			
N_4	N_{41}	−0.046	0.138		
	N_{42}	0.076			
N_5	N_{51}	0.005	0.020		
	N_{52}	−0.014			
N_6	N_{61}	0.214	0.351		
	N_{62}	−0.167			

3.5 Contrast and Verification

The complex correlation coefficient reflects the accuracy of the model and should generally be greater than 0.85 [14]. Therefore, the mathematical model using the new method is ideal, and is better than the model using the traditional method.

The partial correlation coefficient reflects the contribution of design items to Kansei words. As can be seen from Table 6, the item of Headlight Shape (N_1) contributes most to the words "technological-retro". The linear headlight tends to be technological, while circular headlight is more retro. Secondly, it is the item of Upper Grille Shape (N_2). The trapezoidal upper grille is more traditional and tends to be retro, but the vehicle with no upper grille or irregular upper grille looks like technological. The feeling of "technological-retro" is also influenced by the item of Fading Surface (N_6). The vehicle with fading surfaces is technological.

The quantitative results of the relationship between form design elements and Kansei image can provide effective support for product form design, evaluation and decision-making.

4 Conclusions

Kansei image is an effective way to grasp consumers' emotional needs in the stage of new product development. It enables designers to fully understand consumers' needs and promotes two-way communication between them by externalizing consumers' inner feelings.

In our research, two methods are explored to get the relationship between product form design elements and Kansei image. We take electric vehicle as a case study.

Product samples and Kansei words are collected and selected. "Image Circumplex" is proposed to measure Kansei image in the new method, while SD method is used in the traditional method. Then the design elements are extracted comprehensively by Morphological Analysis and Protocol Analysis. Quantification Theory Type I is applied to build the relationship between design elements and Kansei words, and the mathematical model is constructed to support product form design. Finally, we compared the results of the two methods. It is revealed that Image Circumplex is effective to measure the users' Kansei image, avoiding the shortcomings of the traditional SD method. Combining Protocol Analysis with Morphological Analysis, the form design elements can be extracted more pertinently, so that the mathematical model can be established more accurately. In the later stage, we will focus on the data processing and implementation process of Image Circumplex method, and further standardize and optimize the new method.

Acknowledgments. This work is supported by the Research Project of Philosophy and Social Science Research in Colleges and Universities in Jiangsu Province, China (Grant No. 2018SJA1081). Thanks to Dr. Yang for his help in the MATLAB program.

References

1. Ouyang, B., He, Y.: New energy car design driven by technology, art and users. Packag. Eng. **35**, 44–47 (2014). (in Chinese)
2. Osgood, C.E., Suci, G.J., Tannenbaum, P.H.: The Measurement of Meaning. University of Illinois Press, Urbana (1957)
3. Russell, J.: A circumplex model of affect. J. Pers. Soc. Psychol. **39**, 1161–1178 (1980)
4. Zhou, W.Z., Zhao, J.H.: "Emotion compass": the design method in product styling emotion. Zhuangshi J. **285**, 122–123 (2017). (in Chinese)
5. Zwicky, F.: Morphological astronomy. Observatory **68**, 121–143 (1948)
6. Ericsson, K.A., Simon, H.A.: Protocol Analysis: Verbal Reports as Data. The MIT Press, Cambridge (1993)
7. Dong, W.Q., Zhou, G.Y., Xia, L.X.: Quantification Theory and Its Application. Jilin People's Publishing House, Changchun (1979). (in Chinese)
8. Hsiao, S.W., Chiu, F.Y., Lu, S.H.: Product-form design model based on genetic algorithms. Int. J. Industr. Ergon. **40**, 237–246 (2010)
9. Schütte, S., Eklund, J.: Design of rocker switches for work-vehicles—an application of Kansei engineering. Appl. Ergon. **36**, 557–567 (2005)
10. Li, M.Z., He, C.Q., Lu, Z.P., Yang, L.: Research on car image styling design based on quantification theory type I. J. Mach. Des. **4**, 105–108 (2016). (in Chinese)
11. Deng, Z.: A Research on the Relationship Between Urbanized Small Electric Vehicle' Form Feature and Image. Hunan University, Changsha (2012). (in Chinese)
12. Yuan, F.J.: User Research and Analysis of Electro-mobile Based on Kansei Engineering. Shandong University, Jinan (2012). (in Chinese)
13. Sun, J.: Research on Method of Image-based Produce Form Design. Wuhan University of Technology, Wuhan (2007). (in Chinese)
14. Zeng, G.X., Deng, Z.Y.: Multivariate Analysis (I) - Theoretic Applications. Songgang Books, Taipei (1986). (in Chinese)

Research on Modeling Design of Modern Agricultural Machinery Based on Soft System Methodology

Huabin Wang(✉) and Baoping Xu

School of Design, South China University of Technology, Guangzhou 510006, China
hbwang@scut.edu.cn

Abstract. In order to meet the requirements of domestic users for the increasingly improved comprehensive quality of modern agricultural machinery and improve the current situation of the insufficient perceptual function of modern agricultural machinery in China, an innovative design method of modern agricultural machinery based on Soft System Methodology (SSM) was proposed by using SSM and product design theory. Firstly, through the application of SSM, the seven logical steps of the SSM are: the stage 1 no structural problem scenario; Stage 2: The situation in which the problem is expressed: a description of the status of the bad structure; the root definition of the Stage 3 related system: Initially clarify and improve various factors related to the status quo and their relationship; Stage 4 related system conceptual model construction: use structural model or language model to describe the current status of the system; Stage 5 conceptual model and problem perception comparison: discuss and find the gap and difference between problem perception and conceptual model; Stage 6 Finding expected and feasible change: Identifying reform paths or options that are in line with the intent and feasibility of the decision maker; Stage 7 action to improve the problem scenario and use SSM to solve the new problem scenario based on changes after the action, summarize it into four application ideas, stage 1: perceive the problem situation, and clarify the relevant system root definition; Stage 2: Establish relevant conceptual models; Stage 3: Compare conceptual models and current status; Stage 4: In-depth study and verification of the rationality and feasibility of conceptual models and implementation of proposed reforms. Then use this idea to analyze the soft problems of modeling aesthetic feeling in modern agricultural machinery modeling design layer by layer, and divide the soft problem system into three subsystems: color beauty, form beauty and craft beauty. Then investigation and analysis of domestic and foreign agricultural machinery design situation and describe the current situation of the modern agricultural machinery modelling aesthetic problems, and then through the six elements "CATWOE" form clear problems of modern agricultural machinery modelling aesthetic feeling, and combining the theory of ergonomics, aesthetics, color theory, an analysis of the various product modeling design, the concept of modern agricultural machinery design model is established. Finally, the crawler self-propelled rotary tiller based on this conceptual model is compared with the current product by fuzzy evaluation method. A conceptual model of modern agricultural machinery innovation design based on SSM is proposed, and innovative design of crawler self-propelled rotary tiller. SSM is a human

C. S. Shin (Ed.): AHFE 2019, AISC 968, pp. 251–262, 2020.
https://doi.org/10.1007/978-3-030-20470-9_27

activity system as the core of system theory, and the product modeling design is built around the user as the center of design activities, use of human-centered common between them, SSM with product modeling design related theory, to design practice, crawler self-propelled rotary cultivator and a conceptual model of modern agricultural machinery innovation design is verified and the feasibility and rationality of and provide important reference for the related agricultural machinery product design.

Keywords: Industrial design · SSM · Product modeling · Conceptual model · Crawler self-propelled rotary tiller

1 Introduction

Wang et al. used SSM theory and product emotional design theory to propose a product emotional model based on SSM, which provides a new design idea for product emotional design [1]; Based on the SSM theory, Lu made an innovative design for the food chain in the community life chain, and proposed the concept of shared refrigerator and community food sharing, and completed the innovative design [2]; Based on the research ideas of SSM theory, Lu proposed the performance evaluation strategy of digital library service, which provided a new improvement thinking for library performance evaluation [3].

Regarding the research on the design of agricultural machinery products, domestic scholars mainly study in three aspects of color, man-machine and shape [4]. In terms of color, Liu and Guo discussed the application and design method of the main color of agricultural machinery and the color of indoor and outdoor environment [5]; In terms of man-machine, Li and Ou have explored the healthy, comfortable and efficient work of agricultural machinery operators in a harmonious human-machine-operating environment system from the perspective of ergonomics; In the research of modeling design, Shen-Ce and others use inductive analysis methods to summarize the design of agricultural machinery products under three brand dimensions [6].

In the above research, there are few researches on the application of SSM in product design, and most of the research on the design of agricultural machinery design is the application of traditional research methods.

2 SSM Theory

The soft system methodology was proposed by Professor CheckIandP of Lancaster University in 1981 [7]. This methodology is known as the Cecland soft system soft system methodology. The soft system methodology is a methodology developed from the hard system approach, the practice of systems engineering. Since the hard system methodology cannot handle bad structural problems, it is difficult to describe them with mathematical models, and often can only be treated with semi-quantitative, semi-qualitative or qualitative methods. The core of SSM is not to seek "optimization" but to "survey, compare" or "learn", to learn ways to improve existing systems from models and status quo. From the logical thinking of SSM, every logical process is inseparable from the human activity system, so the human activity system is the core concept of SSM [8].

3 Application Ideas of SSM

SSM has seven logical steps. Stage 1 no structural problem scenario; Stage 2 expression problem scenario: description of the status quo of the bad structure; stage 3 related system root definition: preliminary identification and improvement of various factors related to the status quo and their relationship; Stage 4 related system conceptual model construction: using structural model or language model to describe the current status of the system; stage 5 conceptual model and problem perception comparison: discuss and find the gap between the problem perception and the conceptual model; Stage 6 looks for expected and feasible changes: Identifying reform paths or options that are in line with the decision maker's intentions and actions; Stage 7 actions to improve the problem scenarios and then use the soft system methodology to solve new problem scenarios based on post-action changes [9]. This logical step can be summarized into four phases of application ideas: Phase 1: Perceived problem scenarios, clear related system root definitions; Phase 2: Establish relevant conceptual models; Phase 3: Compare conceptual models and current status; Phase 4: In-depth study, verification of the rationality and feasibility of the conceptual model and implementation of the proposed reforms.

4 Analysis of Modern Agricultural Machinery Model Design Based on SSM

4.1 Perceive the Problem Situation of Modern Agricultural Machinery Design

1. Domestic Agricultural Machinery Status
   The development of agricultural machinery products in China is mainly from the establishment of New China. After more than 60 years of development and continuous encouragement and promotion of national policies, it has made considerable progress, but the development level of domestic agricultural machinery enterprises is uneven. The following is a horizontal distribution map of domestic agricultural machinery product design (Fig. 1). The survey sample is used to screen 30 domestic large and medium-sized agricultural machinery brands that are representative (according to the agricultural machinery network). According to the survey and analysis, In general, the development of domestic modern agricultural machinery product design innovation is still in its infancy.
2. Status of Foreign Agricultural Machinery
   The development of the economy, the improvement of people's living standards, and the level of aesthetics are also constantly improving [10]. The combination of technology and aesthetics has become one of the characteristics of the times. The

Fig. 1. Domestic agricultural machinery enterprise design level map

development level and accumulation of foreign agricultural machinery is much higher than domestic, mainly represented by the United States, Japan, and Europe. Their agricultural level is relatively high, and the design level of agricultural machinery products is also at the world's leading level. A new generation of farmers, the demand for agricultural machinery products has increased from the simplest functional requirements to the aesthetic potential psychological needs. The design of foreign agricultural machinery in product modeling largely meets the functional and psychological needs of modern new generation farmers [11]. Therefore, it can be summarized that the problem scene of modern agricultural machinery design is that the design of foreign agricultural machinery is much higher than that of domestic ones, and how to improve the aesthetics of domestic modern agricultural machinery.

4.2 Clear Root Definition

1. Comparative Analysis of Domestic and Foreign Agricultural Machinery Modeling

The most influential factors in the design of agricultural machinery products by literature investigation and analysis are: feature surface, shape transition mode, detail processing, color. Taking the combine harvester of many agricultural machinery brands at home and abroad as an example, the Table 1 agricultural machinery design analysis table is established, wherein if it is represented by "+", otherwise it is represented by "–". By analyzing the design and processing methods of agricultural machinery products, we can sum up the following shortcomings of domestic agricultural machinery design:

(1) The feature faces are straight, too calm and heavy.
(2) The shape details are too singular and parted, lacking styling details.
(3) The design and processing methods are relatively rigid, rough and messy.

Table 1. Agricultural machinery product shape analysis table with combine harvester as an example.

	Domestic agricultural machinery			Foreign agricultural machinery		
Brand	Bilang	Revo Valley God	Xinyuan	Yanmar	Kubota	New Holland
Product Image						
Right angle transition	+	+	+	-	+	-
Shape rounded transition	-	+	-	+	+	+
Characteristic surface is flat	+	+	+	-	+	+
Feature surface is surface	-	+	-	+	+	+
Detail processing	-	+				
Color transition	-	-	+	+	+	+
Color level	-	+	-	+	+	+

Combined with the above-mentioned domestic and foreign agricultural machine modeling comparison analysis, the six elements of its root definition can be expressed as:

C: The profit of the customer-agricultural model, that is, the agricultural machine user
A: Actor - the actor who performs the system transformation, i.e. the designer
T: transformation—the change of the system, that is, to improve the aesthetic sense of modern agricultural machinery and improve the overall quality of domestic agricultural machinery.
W: Word - the world view of the implementation of the conversion process.

(1) entrepreneurs do not pay enough attention to the design of agricultural machinery.
(2) foreign agricultural machinery because the shape aesthetic is much higher than domestic agricultural machinery, the general grade is higher than domestic.

O—The owner of the system, namely the designer, the entrepreneur, the user.

E—Environmental constraints imposed on the implementation of conversion activities, i.e. the cost and process issues of improving the aesthetics of agricultural machinery.

Therefore, the root of improving the aesthetics of agricultural machinery is defined as the innovative design method of agricultural machinery modeling aesthetics. It can mainly improve the color, craft and form of agricultural machinery modeling from color processing, shape transition processing, feature surface and detail processing.

5 Establishing a Conceptual Model of Modern Agricultural Machinery Modeling Innovation Design

According to the above-mentioned root definition, that is, to improve the aesthetics of modern agricultural machinery modeling, the SSM theory is used to sort out the conceptual model diagram of the modern agricultural machinery modeling innovation design, as shown in Fig. 2.

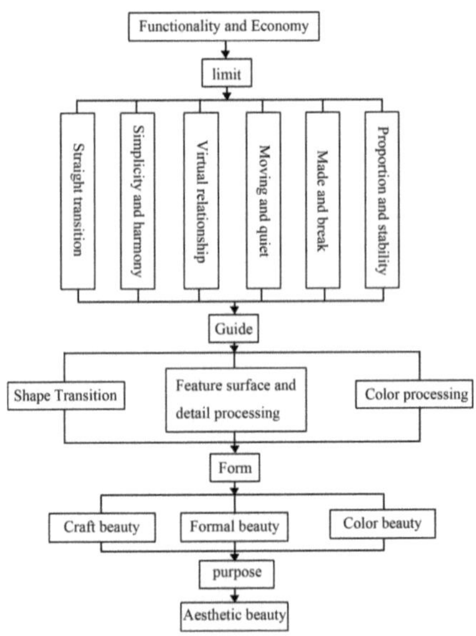

Fig. 2. Modern agricultural machinery modeling innovation design conceptual model

5.1 Ways to Enhance the Aesthetics of the Shape

1. Craft Beauty
   (1) Curved and Straight Body Transition
   The design of agricultural machinery products is mainly composed of sheet metal and pipe parts made of sheet metal. The sheet metal outer cover is the main appearance feature of agricultural machinery products. From the

perspective of sheet metal processing, the morphological characteristics of mechanical equipment can be divided into Planefeature and Bendfeature [12]. The transition of the shape of the body largely determines the feeling of the overall shape. The transition of the surface and surface of the agricultural product is generally achieved by continuous gradual changes in lines, faces, bodies and colors. The way of transition can be divided into two types: straight face transition and surface transition. Face-to-face transitions give people a hard, sharp, and inaccessible feeling, but they also have a more certain and clear outline, reflecting the durable and robust functionality of agricultural machinery products; Curved surface transition gives a person downy, fruity, amiable sense, but too use curved surface transition to be able to cause agricultural machinery product too downy, do not accord with the hale and hearty, durable function that agricultural machinery itself needs. Therefore, in the modeling design of modern agricultural machinery, the roundness transition with small curvature of the surface or the combination of right-angle transition and curved surface transition, in order to achieve the affinity of agricultural machinery products, and do not destroy the agricultural machinery itself is relatively clear contour line, highlighting the strong, durable, strong sense of mechanical products. In the overall or part of the design, taking into account the overall front, back, left and right, up and down echo, part of the shape of the shape of the color, the details of the echo, to achieve the overall shape of the symmetry and harmony. Japan's Kubota's design in this regard is very appropriate, as shown in Fig. 2:

Fig. 3. Kubota agricultural machinery

2. Formal Beauty
   (1) Virtual Relationship

The contrast between virtual and real in the modeling design of agricultural machinery can make the agricultural machinery products more vivid and have a sense of layers. Through the contrast between the virtual part, that is, the transparent part or the exposed part of the mechanical parts, and the closed and thick part of the real part of the shell, the modeling of agricultural machinery products can have a sense of breathing. In the detail processing of part of shape and surface modeling, the punching and hollowing processing of small piece of face shape is carried out, as shown in Fig. 3, or the punching and hollowing processing of large piece of face shape is carried out, and then the gauze is attached for the half-hollowing processing, as shown in Fig. 4, forming the contrast between actual and imaginary parts. "Virtual" gives a sense of lightness, "real" gives a sense of heaviness. Modelling is dealt with when pass "empty" processing can cause the feeling of light wave, too real can appear depressing, bulky. At the same time also need to consider the stability of agricultural machinery products, on the "virtual" under the "real" feeling, so appears stable and yet lightweight agricultural machinery.

Fig. 4. On the left is the John Deere tractor, and on the right is the Yanmar tractor

(2) The Relationship between Standing and Breaking
Establishing and breaking is a contradiction. It is not broken or not. It is said that without breaking the original definition, it is impossible to re-establish a new definition. In the application of agricultural machinery modeling, on the one hand, it breaks the traditional solidification modeling of agricultural machinery, and carries out innovative design on the basis of inheriting the original excellent modeling genes; on the other hand, it is in order to deal with the details of the design of agricultural machinery products. The styling elements are combined or divided into color separation, block division, deformation of the surface, hollowing, decorative belt, etc., as shown in Table 2, breaking the monotonous integrity of the entire large surface, thus creating a new, dynamic, and layered shape.

Table 2. Analysis of the way of "breaking".

"Broken" Way	Modeling Elements	Block Plane Segmentation	Color Separation	Fracture Change	Hollow
Product Picture					
Details Section	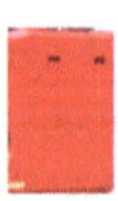				
Analysis	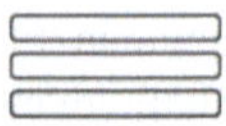	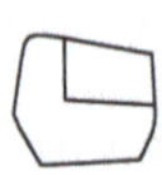	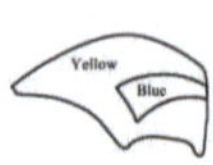	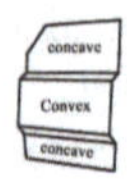	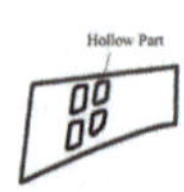

(3) "Moving" and "Quiet" Combination

The styling design is not an arbitrary stacking of styling elements, but an orderly arrangement and combination. In the shape design of agricultural machinery products, especially the shape details of the shape need to follow the aesthetics of the design, such as the use of hollowing, stamping and other methods for detailed processing, it will involve the rhythm and rhythm of the arrangement. Rhythm and rhythm are a kind of periodic movement of things. Regular, organized, and changing repetitive movements are a design way of modeling changes. The shape and detail design of agricultural machinery products often adopts design elements such as points, lines and faces to arrange rhythmically and rhythmically, so that the details of the shape of the body shape are unified and orderly, and there is no lack of sense of change. The main styling features of agricultural machinery products tend to be relatively simple, and can be enriched by the rhythmic changes in various ways. The change in rhythm actually gives us a visual sense of a "moving" trend, that is, adding a "moving" feeling to a more dull, "quiet" shape, and combining dynamics and statics. The entire styling surface will become energetic, fluid, and relaxed, which will be beneficial to the user's boring work.

3. Color Beauty

(1) Coordination Ratio and Stability Center of Gravity

Any large-scale machine will bring some fear to people, so the structure and shape of the machine itself must give people a feeling of safety. When the agricultural machinery products are designed in the shape, the focus of the overall shape is also very important. Grasping the center of gravity of the overall model is a sense of stability. Although the actual mechanical design, the agricultural machine will not fall, or the center of gravity is unstable, but when the user uses it, the user should be given a stable sense of security. For example, considering the shape of the crawler type rotary tiller, it is necessary to adjust the visual center of gravity of the shape to a position close to the middle of the crawler, and try to lower the feeling of the center of gravity to increase the sense of stability. On the one hand, the stability of the agricultural machine can be adjusted by the proportional relationship of the shape. On the other hand, the stability of the agricultural machine is enhanced mainly by the light and heavy feeling of color, the degree of shading of different materials and the thickness of the processed surface texture [13].

6 Innovative Design of Crawler Self-propelled Rotary Tiller

Through the above-mentioned design method for improving the aesthetic shape of the agricultural machine, the design example of the above-mentioned innovative design model applied to the crawler self-propelled rotary tiller is specifically illustrated here, as shown in Fig. 5. The specific design and treatment is based on the design of the rear compartment of the crawler self-propelled rotary tiller. As shown in Fig. 5e, the overall design of the rear compartment is designed according to the shape of the bottom crawler to form an overall harmony; As shown in Fig. 5b, using the relationship between motion and static, the top of the rear compartment adopts two-part neatly arranged wedge-shaped symmetry treatment to increase the richness and symmetry of the top shape; As shown in Fig. 5d, using the virtual-real relationship and the vertical and broken treatment method, the side of the rear compartment uses the side block shape to perform the semi-transparent "virtual" treatment of the gauze-type hollowing, which will not make the side too "virtual", and the rest is closed. The "real" surface forms a virtual and solid relationship, and the designed decorative belt is used to "break" the surface to increase the breathability and richness of the side. The treatment method of proportionality and stability is adopted, and in the overall stability sense processing, in order to meet the feeling of light and heavy on the visual sense, as shown in Fig. 5, the bottom most crawler belt is painted in black, and the rear compartment part is painted in orange. The operating room is lightly colored in white to help ease the operator's operating pressure, and other parts are presented in their own color without losing the richness of their overall shape.

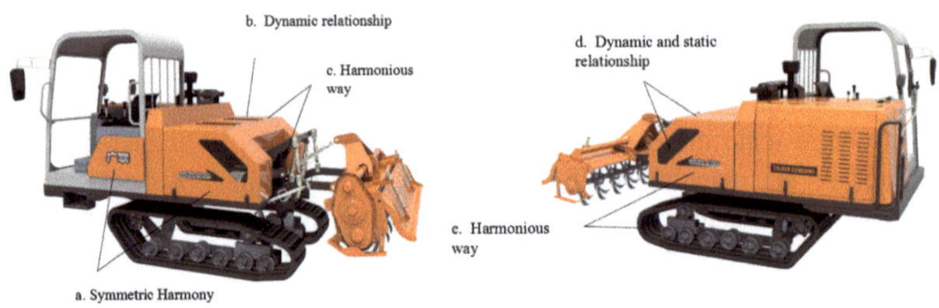

Fig. 5. On the left is the John Deere tractor, and on the right is the Yanmar tractor

7 Comparison of the Current Status of Crawler Self-propelled Rotary Tiller with Conceptual Model

7.1 Comparison

Table 3. The current situation comparison plan renderings

	Plan,1:Domestic status plan	Plan,2:Foreign status	Plan,3:Design example
	Word Group	New Holland	Guan Lian
Product Picture			

Fifty designers with relevant design experience were gathered to evaluate the three schemes in Table 3 above. According to the previous research, feature surface treatment and color hierarchy had a greater impact on the aesthetic feeling of modern agricultural machinery. Therefore, the weight distribution was 0. 3. Color level 0. 3. Details 0. 2. Shape transition 0.2 is directly given to obtain the weight matrix:

A = [0.3, 0.3, 0.2, 0.2]

(1) The evaluation score of scheme 1 can be obtained according to the fuzzy evaluation model:

$$B = AR = \begin{bmatrix} 0.3 & 0.3 & 0.2 & 0.2 \end{bmatrix} \begin{bmatrix} 0 & 0.1 & 0.4 & 0.5 \\ 0.05 & 0.1 & 0.35 & 0.5 \\ 0 & 0.15 & 0.45 & 0.4 \\ 0.04 & 0.16 & 0.5 & 0.3 \end{bmatrix}$$

$$= \begin{bmatrix} 0.023000 & 0.122 & 0.415 & 0.368 \end{bmatrix}$$

The comprehensive aesthetic score of the model 1 is B • VT = 32.2

(2) The evaluation score of scheme 2 can be obtained according to the fuzzy evaluation model:

$$B = AR = \begin{bmatrix} 0.3 & 0.3 & 0.2 & 0.2 \end{bmatrix} \begin{bmatrix} 0.3 & 0.35 & 0.25 & 0.1 \\ 0.25 & 0.4 & 0.32 & 0.03 \\ 0.25 & 0.3 & 0.4 & 0.05 \\ 0.37 & 0.3 & 0.25 & 0.08 \end{bmatrix}$$

$$= \begin{bmatrix} 0.474 & 0.306000 & 0.176 & 0.044000 \end{bmatrix}$$

The comprehensive aesthetic score of the model 1 is B • VT = 79.15.

(3) The evaluation score of scheme 3 can be obtained according to the fuzzy evaluation model:

$$B = AR = \begin{bmatrix} 0.3 & 0.3 & 0.2 & 0.2 \end{bmatrix} \begin{bmatrix} 0.5 & 0.3 & 0.15 & 0.05 \\ 0.5 & 0.2 & 0.25 & 0.05 \\ 0.4 & 0.45 & 0.15 & 0 \\ 0.47 & 0.33 & 0.13 & 0.07 \end{bmatrix}$$

$$= \begin{bmatrix} 0.289 & 0.345 & 0.301000 & 0.065000 \end{bmatrix}$$

The comprehensive aesthetic score of the model 1 is B • VT = 69.825.

By comparing the scores of the three schemes of S, it can be intuitively seen that scheme 3 is much higher than scheme 1 and close to scheme 3 after applying the innovative design model of modern agricultural machinery. It can be seen that applying this conceptual model to the modeling design of modern agricultural machinery has greatly improved the modeling aesthetic feeling of agricultural machinery.

8 Conclusion

By SSM theory combined with product modeling design related theory, perception of modern agricultural machinery modelling of situation both at home and abroad, and according to the description of the root definition for six essence, and then puts forward the corresponding concept model of innovation design, and finally the crawler self-propelled rotary cultivator and design practice, comparing the design practice and the status quo, to verify the rationality of modern agricultural machinery innovation design concept model and the feasibility, and provide important reference for the related agricultural machinery product design.

References

1. Wang, H., Xue-Yan, Z.: Emotional design of products based on crypt's soft system methodology. Packag. Eng. **39**(22), 288–293 (2018). (in Chinese)
2. Lu, C.: The design research of food sharing community based on soft system methodology. South China Univ. Technol. (2018). (in Chinese)
3. Lu, Y.: Research on performance evaluation strategy of digital library service based on SSM. Inf. Stud. Theory Pract. **39**(6), 92–95 (2012). (in Chinese)
4. Liu, J.-L., Mei, G., Xin-Xin, Y.: Using design on agricultural machinery's color. J. Agric. Mech. Res. **7**, 105–109 (2006). (in Chinese)
5. Li, J., Ou, Y., Jiang, Y., et al.: Discussion on the theory of ergonomics in the design of agricultural machinery products. J. Agric. Mech. Res. **1**, 50–51 (2002). (in Chinese)
6. Wang, S.-C., De-Shun, L., Xing, W.: Agricultural machinery modeling design based on brand experience. Packag. Eng. **38**(2), 105–109 (2017). (in Chinese)
7. Jianmei, Y.: On the soft systems methodology of Checkland. J. Syst. Dialect. **3**, 86–91 (1994). (in Chinese)
8. Wheeler, F.P., Checkland, P.: Systems thinking, systems practice: includes a 30-year retrospective (1999)
9. Jianmei, Y.: A thinking about soft systems methodology. Syst. Eng. Theory Pract. **8**, 91–95 (1998). (in Chinese)
10. Miles, L.D.: Techniques of Value Analysis and Engineering, New York (1972)
11. Liu, L.: Empirical analysis on the styling image of agricultural machinery equipment of new-style professional farmers in Jiangsu Province. J. Agric. Mech. **36**(7), 117–123 (2017). (in Chinese)
12. Yan, Z.: The application of form constitution elements on industrial design of numerical control machine tool. Mod. Manuf. Eng. **4**, 49–53 (2013). (in Chinese)
13. Min, W., Aibing, Y., Yi, Z.: Method of model design for machine tools. Mach. Manuf. Eng. (09), 100–103 (2005). (in Chinese)

Research-Driven Design

A Critical Usability Problem-Solving Case of MazeCube Through Design Exploration Based on Scientific Experiments

Jihyeon Yang[1], Hwisu Jeon[2], Muhammad Tufail[1], and KwanMyung Kim[1(✉)]

[1] Graduate School of Creative Design Engineering, Ulsan National Institute of Science and Technology, Ulsan, Republic of Korea
{yjh5220, tufail, kmyung}@unist.ac.kr
[2] School of Design and Human Engineering, Ulsan National Institute of Science and Technology, Ulsan, Republic of Korea
jhs8891@unist.ac.kr

Abstract. This study presents a design development case in which a critical design problem raised in a commercialization process was solved through design exploration based on scientific experiments. A critical usability problem-solving case of MazeCube is described. MazeCube is a toy designed for commercialization via testing and experiment without compromising the functionality and aesthetics of the design. During the usability test with working prototypes, a critical usability problem was found. As such, several working principles were considered to solve the problem. The purpose was to select the appropriate one to adopt and conducted a series of experiments to verify and to determine the design specification of the implementation. This study demonstrates how usability testing and design exploration can be incorporated into a design process to improve the finished design more user-friendly. This study concludes that the case can be considered as an example of research-driven design.

Keywords: Usability · Product design · Commercialization

1 Introduction

The design of a product is constantly revised through making and testing several prototypes until a concept design is implemented as a commercialized product. In this process, designers continuously improve design elements, such as aesthetics, usability, functionality, and performance to fulfill the product requirements [1]. Designers undergo several decision-making processes based on their intuition, reasoning, and results of experiments to reduce the number of raising design problems [2]. The present study introduces a design development case in which a critical design problem raised in a commercialization process was solved through design exploration based on scientific experiments by which improved design requirements were derived. The case is a critical usability problem-solving of MazeCube that is designed by our team. MazeCube is a toy designed for commercialization via testing and experiment without

C. S. Shin (Ed.): AHFE 2019, AISC 968, pp. 265–273, 2020.
https://doi.org/10.1007/978-3-030-20470-9_28

compromising the functionality and aesthetics of the design. It has three dimensional maze units on Rubik's cube structure [3]. Unlike Rubik's cube that is designed to match colors of each side, the proposed MazeCube design has two holes on its surface to move in and exit a ball alongside several tunnels under the surfaces forming dynamic mazes by rotating the cube. As shown in Fig. 1, users put the ball into the entrance hole first, then move a ball through a 3-dimensional dynamic maze creating their own paths and rotating the toy to exit ball from another hole.

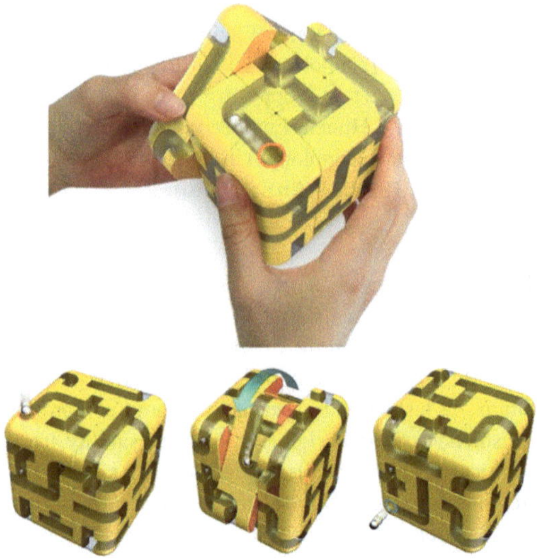

Fig. 1. Images of MazeCube

During the usability test with working prototypes, our design team found that the user unintentionally missed the running ball at the entrance and exit holes. Users could not catch the ball while pulling out from the exit hole. As a result, the ball misplaced and the user had to find the ball manually. This was regarded as a critical usability problem that has a negative impact on the usability of MazeCube. To solve this, several working principles were investigated to select an appropriate one that will be adopted in the later development stage. Therefore, a series of experiments were conducted to verify the idea and to determine the design specification of the adoption.

At first, our design team came up with potential solutions, and then selected one idea according to a design criterion: aesthetics, reliability, functionality, and perfection. As a result, the selected method was to adopt magnets to catch the running ball. This idea was justified and detailed through a series of experiments. Various running routes of the ball and the position of magnets were designed by considering the influence of magnetic force to the running ball. Furthermore, the optimal specification was determined by manipulating the size and the number of magnets, the relative position of the magnets and the ball, and the shape of the magnetic field formed by the arrangement of magnets. The final defined specification was applied into following prototypes, which had received a positive evaluation at the user subjective responses.

2 A Critical Usability Problem of MazeCube

We arranged a usability test with a working prototype of MazeCube. Subsequently, we recruited 67 voluntary participants for to test the prototype at an open-exhibition. They were asked to rate the prototype for its completeness on 1 to 10 rating scale. Of which, 20.2% responses rated 10 score, 16.1% as 9 score, 29.8% as 8 score, and 23.8% responses were 7 score. Overall, most of the responses were above 7 score (M = 8.35, S.D = 1.45), indicating a positive sign of completeness of the prototype. However, most users had concerns raised in the subsequent interviews. These concerns were weight of the prototype, durability and a ball dropping and misplacing.

In fact, the prototype was a design mock-up manufactured for implementing the appearance and basic functions. The sub parts were filled with solid materials, which were planned to reduce later in the mass production. We expect that the final weight of MazeCube at mass production will be one-fifth of the designed mock-up's weight (See Fig. 2). Moreover, the problem of poor reliability and durability is the result of materials used and the fastening method. This problem will be automatically solved with plastic injection and appropriate assembling method at mass production stage. The most critical problem is the ball dropping and misplacing. Here, the ball accidently drops, rolls on the floor, and misplaces, which is a big hinder for the users.

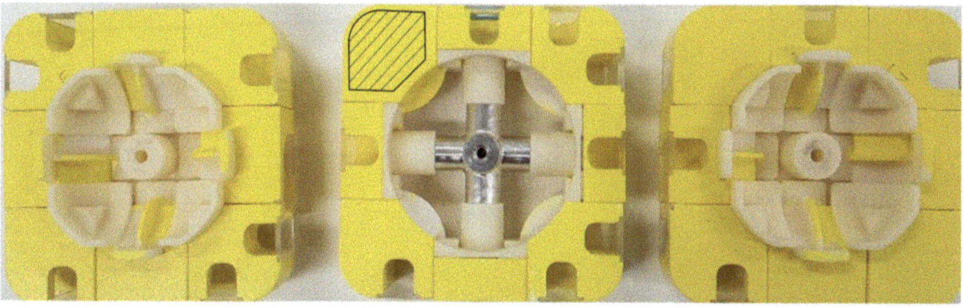

Fig. 2. Inside of the prototype – the hatched area is an example of redundant material.

We define a critical usability problem as a usability problem in which the overall quality of a product does not increase without resolving it. In our experiments, the critical usability problem is the ball dropping and misplacing. Therefore, a new design that grabs the ball and does not drop out of it is required.

3 Solution Exploration

3.1 Investigation of Potential Solutions

In order to solve the critical problem identified in the earlier stage, our design team came up with possible ideas with quick and rough sketches. We decided to exclude one of the ideas that blocks the holes permanently, because it did not reflect user needs that they desire to pull out the ball. There were five possible ideas to block the running ball (Fig. 3):

(1) Using a rubber cap: We can stop the running ball and pull out it by closing and opening a rubber cap.
(2) Rotating cover: We can change the hole size with a rotating cover.
(3) Sliding cover: We can change the hole size with a sliding cover.
(4) Using a device like a magnet at hidden place near the holes to make the ball stopover.
(5) Digging a spot in the path to keep the ball static.

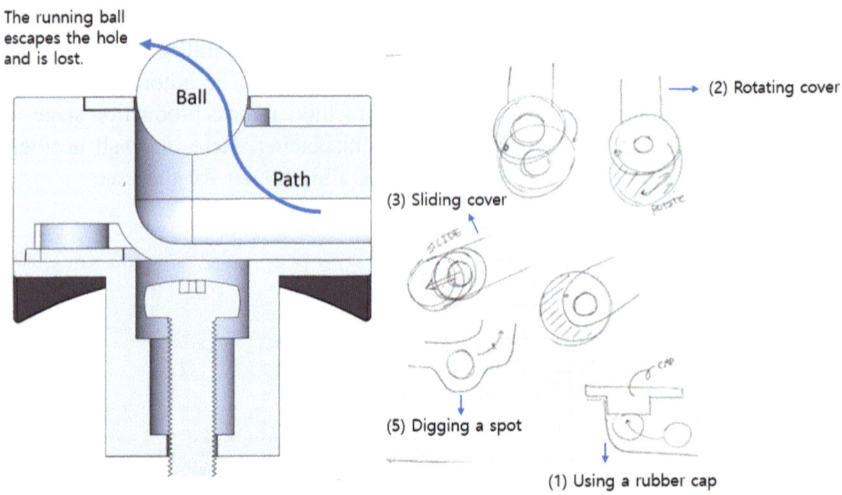

Fig. 3. Origin state of the problematic part (left) and idea sketches for problem solving (right)

Our design team evaluated the disadvantages of each idea in the process of ideation that help to developed as a selection criterion for ideas. The method of blocking with the rubber caps would cause to misplace the rubber cap like the ball. By making rotary covers and sliding covers on surface of the holes may also look aesthetically unpleasant. While the method of digging a site in the route seemed to be difficult to produce and magnet devices if use may raise production costs. Therefore, we synthesized these opinions to establish four criteria: (a) aesthetics, whether the method do not hurt the appearance design (b) reliability, the product can be produced and durable with the method (c) functionality, whether the method can solve the critical problem that we defined, that is, whether the method is problem-solving, and (d) perfection, whether there is no any other problem that the method occurs. We defined these four criteria as design requirements.

As shown in Table 1, we created a matrix and evaluated each idea by expert judges consisting of three professional designers. They discussed and marked whether each idea meets or fails the criteria. For example, the methods to make rotary covers or sliding covers affect the appearance of the product as a guideline must be created along the place where the cover moves over the holes. Similarly, these methods along with making rubber cap are not durable because a user may need to frequently touch and move the rubber cap or covers. The method of digging a site in route method cannot always catch the ball, thus cannot be a way to problem-solving.

Table 1. Result of evaluation by expert judges

	Aesthetics	Reliability	Functionality	Perfection
Blocking with rubber caps	Δ	X	O	X
Making rotary covers	X	X	O	O
Making sliding covers	X	X	O	O
Digging a site in route	O	Δ	X	O
Using magnets	O	O	O	Δ

Our design team decided to choose a solution that meets all four criteria. The idea that best fits the design requirements was to place magnet parts. This is because the hidden magnet does not affect the appearance. It can be reflected in the production phase by making a space where the magnet can be placed. This solution does not cause any problem while keeping original design other than the unit cost of the finished product. Although the solution was appropriate in the evaluation phase but other conditions have to be considered. It should not only catch the ball but also start the ball effectively. For this purpose, further refined conditions of the critical usability problem in MazeCube were considered: (1) when the ball comes out, it should not be dropped off the exit, (2) when a play begins, the ball must starts efficiently overwhelming the magnetic force by simple tapping. To this end, we conducted a series of exploration and experiments to find best condition of arranging magnets and determining adjacent features.

3.2 Experiments Design

Users should be able to start the ball at the beginning and should not miss the ball when it arriving at the exit. As such, we conducted two experiments. The first experiment was conducted to find optimal magnetic field to hold the ball. The second experiment was done to design new path to initialize the ball effectively. In order to test for making an optimal magnetic field around the entrance and exit, the position and number of magnets were considered. To initialize the ball successfully with a finger tapping, we designed a new path where the distance between the ball of the magnet is abruptly increased by a small horizontal movement of the ball.

Experiment 1: Magnet arrangement to hold the ball effectively.
Process. The first condition is to catch the ball. The purpose of the experiment was to apply the existing magnets certainly and use small-sized magnets that could be fit into actual products because the intention was to develop the product for commercialization. We finally stated our experiment by arranging ∅ 3 × 6 mm ND magnets in various ways. As shown in Fig. 4, a set of experiments was prepared to minimize the number and position of the magnets. Thus, we designed test beds with a 3D CAD program, and made test prototype using an FDM 3D printer. A metal ball with a diameter of 9 mm was also used (See Fig. 4).

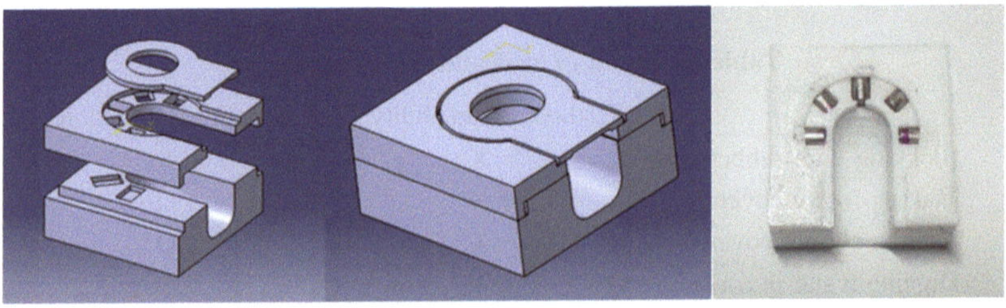

Fig. 4. 3D CAD modeling of one of test bed (left and middle), and test bed made by 3D printer (right)

The experiment was performed by rolling the ball beside the path with the number and position of the magnets being different as shown in 4 cases: (a) using 5 magnets, (b) using 3 magnets, (c) using 2 magnets, and (d) using 1 magnet (See Fig. 5).

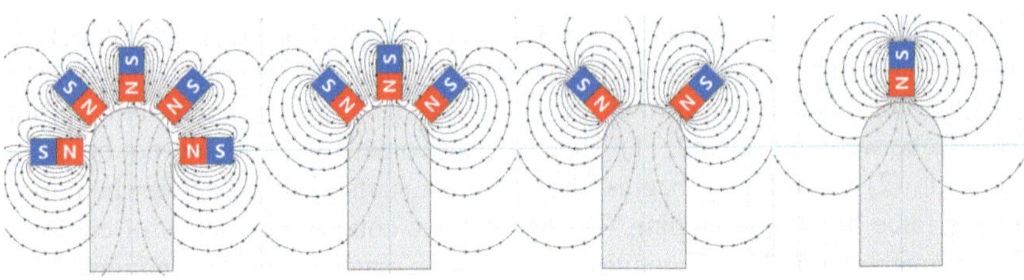

Fig. 5. The magnets field depends on the position and number of magnets in test beds

Since the magnet has a polarity, a pushing force between the same poles, and a pulling force between opposite poles, arranging the poles of several magnets is an important variable. It affects a range of the magnetic field, which is an important factor whether the ball can be hold. The purpose of the experiment was to find the condition for catching the ball effectively, and they were all arranged in the same pole direction to form a wide range of magnetic field. Thus, we drove the ball using the test beds and observed what we conditions were relatively effective (See Fig. 6).

Result. If the number of magnets is one, the ball that rolls straight toward the magnet is effectively fixed. However, the ball escapes when it comes from a different angle.

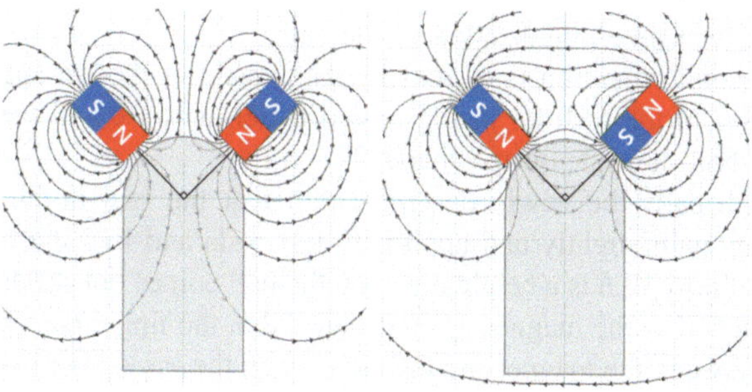

Fig. 6. Different magnetic field due to the direction of the pole

The larger the number of magnets, the wider the range of the magnetic field, thus better to hold the ball. As the number of magnets increases, the production cost of the product and the manufacturing time increases. Thus, a solution was required in which the magnetic field becomes wide enough and the number of magnets required is small. Therefore, our team decided to develop the location using two magnets with the same polar arrangement accordingly. Here, one magnet was used but the magnetic shape and size were changed to form a magnetic field and magnetic force which are similar with the test bed. The use of large magnets will create a magnetic field that can hold the ball effectively even with a single magnet. However, a curved magnet should be invested because it is unusual one in the market (See Fig. 7).

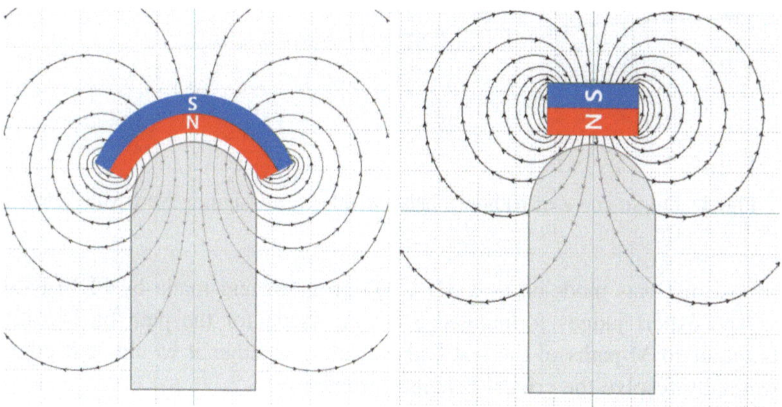

Fig. 7. Other ways of using one magnet similar magnetic field

Hence, a rectangular magnet was picked that was available on the market. Moreover, we found a positive effect similar with employing the initial two magnets in the same direction at 90°. Through this process, a condition for holding the ball was found. Alongside, we can reduce the manufacturing time by using one magnet that can produce the desired effect in the manufacturing process as well as there is no need to manage arranging the pole of the magnet.

Experiment 2: Stress-free initialization of the ball.

Process. The second condition is to start the ball easily at the entrance and exit. We changed the shape of the path for the design condition because a distance between the magnet and the ball depends on the shape.

Let New-path calls the newly designed path that the ball initializes easily. We changed the New-path slightly and applied to test beds and find the optimal position and shape accordingly. If h is a height at which the ball comes out of MazeCube in case if the ball is attached to the magnet, f is a distance that the finger can push the ball and d is a distance between the magnet and the ball. When the same f is pressed, the d in the final New-path becomes larger than d in the initial design.

$$d'_{New-path} > d'_{Origin-path} \quad (1)$$

That is, when a game is started in the newly designed New-path, the ball can quickly roll away from the area of magnetic field (See Fig. 8).

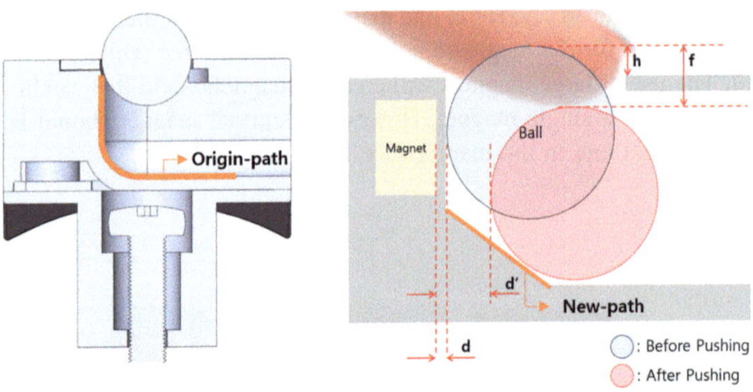

Fig. 8. Origin curve of path(left) and variables for designing New-path(right)

The test bed was modeled as a 3D CAD program and made by FDM-based 3D printer. The expert judges formulated a design shape for the path by investigating various modified M-paths of the test beds to fulfill whether it would start effectively and eventually resolve the critical usability problem.

Result. The distance between a magnet and a ball was adjusted by varying the curve of the path in the second test. The longer the distance between the two is the more the ball out of range of the magnetic field, and the ball can start again effectively. Therefore, when users press the ball with their finger, the ball should move away from the magnetic field and users can start to play a game.

As we changed the curvature of M-path sharply started from the magnet, the ball cannot only be caught effectively at the entrance and exit holes but also easily moved by lightly pressing it when plays begin. Particularly, it can greatly improve user requirements with a small difference that does not massively modify the manufacturing design of the product, and it can save time and cost for design modification.

4 Discussion and Conclusions

In this study, we defined a critical usability problem of MazeCube based on user feedbacks. We then evaluated various potential solutions from four perspectives as described above, and performed problem-solving through testing. Thus, a solution was needed to be applied at the actual manufacturing stage. Meanwhile we avoided to harm the appearance design of the product. Therefore, possible solution ideas were evaluated by design experts based on aesthetics, reliability, functionality, and perfection. As a result, a method using a hidden magnet was evaluated the highest, and then two additional tests were conducted to derive a specific modification plan. We identified optimal magnitude and shape of the magnetic field by changing the position and number of magnets. In addition, it was possible to solve the problem, increase usability, and fulfill user requirements by changing the curve of path designed to induce changes in distance between ball and magnet. In this design process, we solved the critical usability problem of which is yet to be manufactured and produced.

Industrial designers rarely consider implementation of their designs because they perceive it as an engineer's job [4]. Furthermore, it is believed that industrial designers' job is more intuitive and subjective whereas engineering designers' jobs are scientific and systematic. Design process cannot be described with a single isolated phase; it requires integration of both expertise and concepts from industrial design and engineering design [1]. The present study revealed how usability testing and design exploration can be incorporated into a design process to improve the design more user-friendly as well as robust, especially when a concept design undergoes commercialization process. This can be considered as an example of research-driven design.

Acknowledgments. The work is supported by the Promotion of Special Design-Technology Convergence Graduate School of Korea Institute of Design Promotion with a grant from the Ministry of Trade, Industry & Energy of the Republic of Korea (N0001436).

References

1. Kim, K.M., Lee, K.P.: Collaborative product design processes of industrial design and engineering design in consumer product companies. Des. Stud. **46**, 226–260 (2016)
2. Lawson, B.: How Designers Think, vol. 3. Routledge, Abingdon (2006)
3. Rubik, E.: Rubik's Cubic Compendium, vol. 3. Oxford University Press, Oxford (1987)
4. Kim, K., Lee, K.: Industrial designers and engineering designers; causes of conflicts, resolving strategies, and perceived image of each other. In: Proceedings of DRS 2014 (2014)

Project Pilot Run: A Sewn Collaboration

Adam Feld(✉)

The School of Architecture and Design, The University of Louisiana at Lafayette, 421 E. Lewis St., Lafayette, LA 70503, USA
ajf4151@louisiana.edu

Abstract. Flatland BMX is performed on flat surfaces without ramps or jumps resulting in less wear and tear on the bike than is common to extreme BMX. There are many niché products but in the world of flatland BMX, an appropriate air travel bicycle bag is missing. It costs upwards of $150, each way, to check a bicycle on a plane. In this sponsored project, 21 students designed the means of safely and cost-effectively transporting a Flatland BMX bicycle by air. Students worked with a local professional rider and consulted with a sewn goods designer both of whom assisted students in the development of their designs and manufacturing strategies. At the end of the project, a select group of students was chosen to deliver a working prototype to the professional rider. This pedagogical paper follows the progress of the students from the initial pitch through the final presentation of the product.

Keywords: Sewn goods · Air travel · BMX flatland · Sponsored project

1 Introduction

In Design there are many conceptual ideas taught. Students need: to understand the elements and principles of design, to understand how to extract ideas from their head and put those ideas onto paper, to translate those ideas from paper to three-dimensional form, to incorporate the user, to understand manufacturing methodologies, and they need to incorporate all of that information into a format so others can understand. While students are learning these conceptual ideas, they approach them from a personal standpoint. In the industry, an Industrial Designer designs for people not only themselves. By learning all of those conceptual ideas, in school, they develop an innate understanding of how people live and behave. One way to strengthen that understanding is to provide students with the ability to design for others in a project. The project for this paper followed the externally sponsored project to develop a new travel case for a sponsored BMX rider. Goals for this project were to provide students the opportunity to develop a sewn good product, deepen their level of understanding on how sewn good products are made, connect with local industry, and work with a global company. Through three interactions with the sponsored rider and two interactions with a local sewn goods business owner, each students developed 73 sketches from ideation exploration through digital sketch render, three physical models from ¼ scale process to one to one scale final, a vector pattern for sewn goods, and a final presentation poster. Produced material was used to host a contest where the best four projects, the

C. S. Shin (Ed.): AHFE 2019, AISC 968, pp. 274–282, 2020.
https://doi.org/10.1007/978-3-030-20470-9_29

students who design the four, were then hired for a summer internship and to complete a prototype of the best project in the class.

2 Primary Source

Before the project began, students were tasked with delving into the world of Red Bull. The first assignment was to understand Red Bull the company, inspiration boards, as well as the brand, brand analysis, and how it is applied to products. In addition, the students needed to understand TSA and FAA regulations on travel and luggage. On the official project kickoff, students had boards prepared to show to Terry Adams, the sponsored BMX rider. At the start, Terry showed the students his current scenario and the products that he deals with on a daily basis. He brought a bike bag that was provided by the company as well as what he actually used. The bag he actually used was a thin, nylon bag that will fit over a set of golf clubs. This bag was produced by a bicycle company but did nothing to suit Terry's needs. Students also learned the cost of air travel with a bike which is upwards of one hundred and fifty dollars each way but select items, one of which being golf clubs, were free to check (Fig. 1).

Fig. 1. Terry Adams displaying his bike in his bike bag. Photo: Zach Breaux

After his current bike bags were covered, the students and Terry discussed the anatomy of his bike, how he will prepare the bike it for packing, how he protects certain places on it and the need for that protection, and ultimately, how the bike is his livelihood but he is not guaranteed that it will arrive at his destination in working order. One part of the bike that needed to remain in pristine order was a part called the head tube. This part is where the forks connect to the handlebars. The head tube has the logo another of his sponsors who design bikes for him. An important part of the bike, for sure, and it was protected with his wife's old worn out sneaker turned gardening shoe. Getting the opportunity to interact with the clientele on that level gave students insight as to the power of having a primary source of information. Afterward, students documented the bike parts for scale and drew them on a long role of paper that was hung on the studio wall for the remainder of the project.

With the first meeting with Terry complete, students reflected on the lessons learned as well as reviewed their image and brand boards and then began their visualizations for the project. The process for these visualizations is to develop a large amount ideation visualizations, fifty or more, which are quick freehand sketches usually three to five on a page and containing appropriate line weights, values, and minimal surface shading. Students then take a small sample of the ideation visualizations and refine them. These first level refined sketches are a single per page and contain basic orthographic views in the upper right corner as well as have notes or callouts. The first level of refined sketches were prepared for a critique from a local sewn goods business owner, Patti Dunn. Patti came to studio and critiqued the students on the viability of their designs and what she saw as innovative. It was at this time that she also gave a lecture on how to develop a pattern for a sewn good product. Students then worked with the professor to further refine their concepts and select five for the next stage of visualizations that would be digitally hand rendered in Adobe Photoshop, these were referred to as sketch renders. Sketch renders take a digitized paper sketch and add value to add dimension in a raster software application. On a sketch page, a sketch render appears as a more complete concept direction and this is a way to show completeness of an idea for clients. Typically this Industrial Design Program will have students develop three sketch renders for multiple concepts to show a client. They are cheaper than developing digital 3D models or physical prototypes. For the purposes of this project, and to give Terry a larger amount of options to choose from, the students developed five sketch renders for feedback. To get feedback the students set up a conference call with Terry, who was sitting in a hotel lobby somewhere waiting for a cab to take him to an airport to fly to his next event. Terry had a print out of all of the students work, 20 pages of sketch renders, and gave each student two to three minutes to give them the direction he would like for their project.

3 Refinement

Since 2D sketching can only take a designer so far and because their Industrial Design Program stresses the importance of making, students were assigned to begin translating their idea into three dimensions. This was done through an orthographic drawing. The orthographic drawing forced students to accept real world proportions, dimensions, and how sketch elements would appear on a 3D form. To assist with the orthographic drawings, students found an digital 3D model of a BMX bicycle and converted it to vector line work in the three views, front, side, and top. These vector views could then be manipulated in Adobe Illustrator, where the students were working in a one to one scale. With the drawing complete, students began their process, ¼ scale models made out of craft paper. Craft paper is a great material to use for process models for sewn goods. The paper allows for slight structure and can be cut similar to how some fabrics would be cut. In addition, it is a good material to be able to take notes on or mark up. Students can develop patterns in this material without the need to sew. One might ask at this point, isn't' this a project about sewn goods? Wouldn't sewing be necessary for these models? The answer is, yes, sewing is a very large part of sewn goods projects, however, the understanding of how two-dimensional parts and pieces (the pattern) can be combined, on a seam tolerance edge, to create a three-dimensional product was paramount. Pattern making can be easily done with craft paper and it gave the students a comfort level, which allowed them to explore with the material (Fig. 2).

Fig. 2. Patti Dunn critiquing students sketches and model

These models were derived from the orthographic drawing which were in turn derived from the sketch render. What that means is that someone could review all of the work and see the progression from initial concept through the process model. The ¼ models were prepared for a second critique with Patti Dunn who came to the studio and critiqued sketches as well as the ¼ scale models. Her feedback was based on how to development of a pattern for sewing from their model as well as how each individual project could be manufactured. The ¼ scale models were a perfect representation of the project direction but with the addition of the sketches, she could see how the students made their decisions and could also sample from different sketch ideas and incorporate those into the model for the next round of building which was the refined model.

The refined model heavily focused on the developed pattern. That developed pattern was printed out and then placed on the material and cut in the same manner that a person would use a pattern to make any sewn good product. It was at this time there the students began to realize what worked and what did not work. To make the refined model, the pattern had to be correct and many students needed multiple iterations to correct their designs. Continually refining the pattern was not an exercise in punishment or even perfection rather a need to have the confidence that when they created their final prototype for the semester, the pattern was resolved and they would not have to modify their design at the last minute. It was at this time that the students began preparations for the final presentation and documentation for the project. This was done through their prototype as well as a poster that documented the entire semester.

4 Final Presentation/Contest

The final presentation occurred on the last day of the semester. This was great in two ways, one- it was the last day and students could sleep after and two- being the last day, there was a general air of excitement for not only the project but also the last RedBull

Fig. 3. Terry Adams and student Victoria Roux discussing the final project. Photo: Zach Breaux

visit. Students set up a gallery in an open space in the building and the final was orchestrated as a gallery opening. RedBull came as well as Terry Adams. At this time, Terry went to all 20 students and listened to their pitch and reviewed their models. While that was taking place, the professor was also reviewing the work for the semester. After an epic review by Terry, he gave a demonstration of his flatland BMX profession in the building. To many of the students, they had only seen a few videos of his work but to see it in person was truly a unique experience. In a great way, the students shared their profession with Terry and then he had the ability to do the same for them. Once the demonstration was complete and all of the RedBull was drank, the students and the professor left the gallery to allow Terry the opportunity to privately review the work and select the students who would continue the project. Four students, with the best designs would be selected for as an internship over the summer to complete a prototype but with the limited budget, only one design would be fully actualized (Figs. 3 and 4).

Fig. 4. Contest winner and prototype direction Student Sam Riehl. Photo: Zach Breaux

5 Summer Work

With the intensity of the semester over and a little break after finals, the four students and one faculty member got to work on the final prototype. The students and faculty member went into the summer thinking it would be a quick turn around but when then began to review the design and discovered that the selected project was perfect at ¼ scale but if it would be developed at 100% scale, it would be too large. To discover that, the students developed a volumetric model from laminated pink foam insulation and cut to the overall dimensions. When the group reflected on the overall scale, the determination was to find places to trim. The trimming led to a reworking and restructuring of the pattern as well as new renders to make sure the product would have the same qualities of the chosen design.

While the students and the faculty member were perfecting the pattern, they then needed to consider materials. To develop the prototype, there would need to be a full-scale model made prior to working with the final materials. Sourcing materials was an interesting experience for the group as many of them had not had the opportunity to do that in the past. In the process of sourcing materials, students had the opportunity to connect with another local cobbler who assisted in which leather to purchase and how to manipulate that material. Patti Dunn who assisted in helping the group on what materials to purchase for a travel bag such as ballistic nylon. If anyone has worked at a university you know how quick the procurement process can be and the group needed the materials expeditiously. Because of staff, at the university, the materials were able to get in on time. The final prototype would be built in Tchoup Industries, Patti Dunn's company. Since we were not professional sewers, Patti worked with the professor, at the beginning of the project, to rent her facility rather than purchasing our own sewing machines.

With the redeveloped pattern and all of the materials, the group drove to Tchoup Industries, located in New Orleans, LA. They were met by Red Bull and the day began. This would turn out to be an epic day. The first thing that had to be completed was the organization of the patter. This was chaotic as there were so many parts and pieces (Fig. 5).

Fig. 5. Pattern organization at Tchoup Industries

Once the pattern was organized, everyone went to work developing the initial mockup for verification of the pattern. There were on-the-fly changes made to the pattern when developing the mockup. Once the mockup was complete, the group began pinning the pattern to the final material and began the final prototype. Since the students did not need to know how to sew, during the semester project, this was a great opportunity for the students to see the process and ask questions. At the end of the day the rider, Terry Adams showed up to accept the final bag (Fig. 6).

Fig. 6. Terry Adams receiving his final product

6 Reflections

Sponsored projects are a great opportunity for faculty, students, and the school. For the faculty member running the project, it was a great opportunity to learn a new part of the profession. Faculty cannot know everything and sponsored projects are a way to allow faculty to collaborate with industry to learn new methods and techniques as well as continue to remain current in their skill set. It also allowed the students the chance to drink RedBull and do some work and drink RedBull some more. Actually, the projects assigned allowed students to invest in the company through their image and brand analyses and by doing so, they were able to speak with authority on their designs. The many visits from the client and local professional reminded students that there is a much larger and real design world out there. It was also a great opportunity for the students to see Patti Dunn who began her own business and could speak to the students about potentially doing the same for themselves in the future. Entrepreneurial spirit runs through and through in Industrial Designers and for many, that is the future they wish for themselves. That exposure prompted many discussions on how to create a start up. Opportunities for the university were that through this project, they gained recognition and exposure. At the end of the project, it was determined that taking on the responsibility of completing the final prototype, having not done so in the past, may have been an over reaching goal. However, the product was completed and put into the field and got rave reviews demonstrating to the students that they can accomplish all of their goals if they work for it. Things that could have been seen as set backs or roadblocks were successfully overcome. This lesson was so important for the students and the faculty member as it is always easy to say that you cannot do something but when you don't have that option, you get it complete. In fact, this project was successful enough that there is another project set to run Spring 2019.

Photo Credits

Figure 1, 3, 4, Zach Breaux Red Bull
Figure 2, 5, 6, Adam Feld- ULID

Professionals and Consultants

Terry Adams- Professional BMX Rider
Zach Breaux- Red Bull USA
Patti Dunn- Owner/Operator Tchoup Industries. LLC

Students

Gabriel Agudelo, Hunter Ayo, Christina Broussard, Krystle Brown, Christopher Denais, Alexus Dunbar, Corey Dupuis, Tyler Gagnard, Andrew Keeler, Kelsey Kenney, Dillon Landry, Chloe Lauland, Corey Louque, Erica Moore, Brian Nguyen, Sirikhwan Pamornkul, Tyler Rabel, Hannah Rossler, Victoria Roux, Ciara Sanders, and Sam Rhiel

Discomfort with Low-back Pain Relief Exercise Training for Older Adult Women

Muhammad Tufail[1], Sangjin Park[1], Haebin Lee[1], Yang Gyu Moon[2], and KwanMyung Kim[1](✉)

[1] Graduate School of Creative Design Engineering, Ulsan National Institute of Science and Technology, Ulsan, Republic of Korea
{tufail, superjin, geamongc, kmyung}@unist.ac.kr
[2] The Balance-Korea Inc., Nam-gu, Ulsan, Republic of Korea
anybaro@naver.com

Abstract. Low-back pain is a common health problem in older adult women. Due to various health problems, they cannot continue the use of medications for low-back pain. Recently, specific exercise training are considered as more promising and widely used interventions for low-back pain. However, participation in exercise training remains low among older adult women. The main reason is that these training cause discomfort to other parts of the body due to which they quit the training in its early stage. The purpose of this study is to examine the effect of specific exercise training on low-back pain intensity and overall body parts discomfort. A group-based training for muscles around the spine was arranged in a martial art club. An empirical study with 8 older adult women was conducted for the duration of five weeks. The findings showed a significant difference in the pain intensity and body parts discomfort over the specified duration. Body parts discomfort score was gradually and relatively decreased along with the pain intensity score. The findings revealed that the pain intensity and overall all body parts discomfort remain unchanged in the first and second weeks of the exercise training. However, the pain intensity and body parts discomfort was influenced in the later weeks. This study suggests that exercise training should be strategized to encourage devotion, enable group-based exercise activities, and educate self-efficacy for exercise training.

Keywords: Body parts discomfort · Low-back pain · Exercise training

1 Introduction

Low-back pain is a common health problem in most older adult women. Due to various health problems, they cannot continue the use of painkillers and related medications for low-back pain [1]. Older adult women are informed that exercise trainings can improve physical and mental functions, keep them flexible and independent, and minimize medications that has negative effects as well as reverse some effects of chronic diseases [2]. As such, many older adult women prefer exercise interventions and spinal manipulative therapies to relieve low-back pain intensity. Recently, specific exercise trainings for the manipulation of muscles around the spine are considered as more promising and widely used interventions for low-back pain [3]. Besides, the

C. S. Shin (Ed.): AHFE 2019, AISC 968, pp. 283–288, 2020.
https://doi.org/10.1007/978-3-030-20470-9_30

participation rate in exercise trainings remains low among older adult women. The reason is that these trainings cause discomfort to other parts of the body due to which many older adult women quit exercise trainings in its early stage.

The purpose of this study is to examine the effect of a specific exercise training on low-back pain intensity and overall body parts discomfort that is caused by the duration of the exercise training. To achieve this, a group-based specific exercise training for muscles around the spine was arranged. Subsequently, an empirical study with 8 older adult women was conducted for the duration of five weeks. Two scales were used to measure low-back pain intensity and body parts discomfort. The pain intensity and body part discomfort was measured subsequently from Monday to Friday for five weeks except Saturday and Sunday. The findings showed a significant difference in pain intensity and body parts discomfort score over the duration of five weeks. Body parts discomfort score was gradually and relatively decreased along with the pain intensity score. The findings revealed that the pain intensity and overall all body parts discomfort remain unchanged in the first and second week of the exercise training. However, the pain intensity and body parts discomfort was significantly influenced in the later weeks. This study suggests that older adult women with low-back pain should be encouraged to increase their physical activities if influenced by clinicians, therapists, trainers, enabling group-based activities and educating self-efficacy for exercise trainings.

2 Methods

2.1 Subjects

In total, eight older adult women (age range, 70–80) participated in the group-based exercise training program for the duration of five weeks. These subjects were recruited from an old care center located in Ulsan City of South Korea. Each subject was suffering from low-back pain for a mean 2.3 and 2.6 years, respectively. Before the training session, subjects were fully educated about the exercise training and the subsequent subjective measures. Low-back pain intensity and overall body parts discomfort was measured at the eand of each exercise session in five consecutive weeks. The subjects' anthropometric characteristics are shown in Table 1.

Table 1. Subjects' anthropometric characteristics (n = 8)

Mean age (years)	Height (cm)	Body mass (kg)	BMI (kg/m^2)
73.5 ± 3.2	151.01 ± 4.84	53.46 ± 5.09	23.42 ± 1.96

Before participating in the exercise training, subjects were asked to provide informed consent that was approved from our facility's institutional review board. The training was conducted in a martial art training club with a trainer, health physician and two researchers. The arrangement is shown in Fig. 1 as an example.

Fig. 1. Exercise training arrangement

2.2 Apparatus and Exercise Training

A 15-steps exercise-training program along with a toes-up inclined slope platform with 15–20° slope angle for muscles of the spine, legs, and hips was presented to the subjects. This program was adapted from the study of Tufail et al. [4] in which the focus was made on inducing exercise practice, and influencing muscle stretching that can ultimately decrease low-back pain. The training was performed Monday through Friday for five consecutive weeks. During this period, the pain intensity and discomfort was measured after each exercise session. For this purpose, a post evaluation Wong-Baker Faces Pain Rating Scale[1] with anchors "No Hurt = 0" to "Hurts Worst = 10", and body parts discomfort scale with anchors ("Not uncomfortable = 1" to "Extremely uncomfortable = 5") devised by Corlett and Bishop [5] was used. Body parts discomfort scale divides body into 9 parts. These are head and neck, shoulder, arm, middle back, low back, buttock, thigh, knee, and leg and foot.

The training program focuses on specific muscle stretching that were previously studied by Rainoldi et al. [6], Nam et al. [7], Ginn and Halaki [8], Lovell et al. [9], and Callaghan et al. [10]. These muscles are gastrocnemius, the thickest part at the back of legs and below the hips, Latissimus Dorsi, the later trunk flat muscles located 4 cm below the shoulder bone, Adductor Magnus, hip muscles between the lateral trunk and the legs, and Lumbar Erector Spinae muscles located on each side of the vertebral column and extend alongside the lumbar spine. The training program for these muscles is shown in Fig. 2.

[1] For Wong–Baker Faces Pain Rating Scale: http://wongbakerfaces.org/.

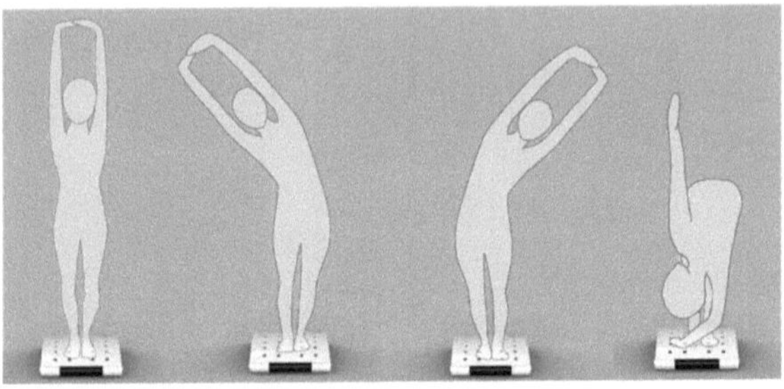

Fig. 2. Muscle training exercise on a toes-up inclined slope platform

2.3 Design and Procedure

This study includes one independent variable: Duration of exercise with five levels: 1st week training (W1), 2nd week training (W2), 3rd week training (W3), 4th week training (W4), and 5th week training (W5). The dependent variables are low-back pain intensity (LBP-I), and overall body parts discomfort (OBP-D) on subjective basis. These subjective response measures are the ratings of pain intensity and discomfort that are the feelings about pain and discomfort. The purpose was to observe change in pain intensity and overall body parts discomfort against the duration of the exercise training.

For the statistical analyses, a mixed design repeated measure analysis of variance (ANOVA) was used to determine whether any change in low-back pain intensity was due to the duration of the exercise training. Similarly, the same procedure was applied to measure the overall body parts discomfort against the duration of the training. Each subject performed the exercise in a group-based training using a toes-up inclined slope platform. An exercise guide was present for the entire training session. Each subject required 10–15 min to complete the everyday training.

3 Results

The basic descriptive statistics of the LBP-I and OBP-D are presented in Table 2. The score of these variables were analyzed using a repeated measure design ANOVA at a significance level of 0.05. The analyses compared the effect of the duration of exercise training on the mean values of LBP-I and OBP-D. Follow up post hoc comparison was conducted to explore a significant main effect of each level of exercise duration (W1 to W5) on LBP-I and OBP-D.

Table 2. Means and standard deviations of LBP-I and OBP-D (n = 8)

Duration	LBP-I	OBP-D
W1	6.25 (1.48)	3.30 (0.595)
W2	5.83 (1.78)	3.13 (0.651)
W3	4.45 (1.35)	2.72 (0.411)
W4	4 (1.06)	2.54 (0.442)
W5	3.62 (0.49)	2.06 (0.487)

Mean LBP-I and OBP-D across the levels of exercise duration

A repeated measures ANOVA with a Greenhouse-Geisser correction determined that mean LBP-I differed statistically significantly between the levels of exercise duration (F (2.98, 68.54) = 19.01, $p < 0.05$, ηp2 = 0.453). Post-hoc comparisons revealed that LBP-I was decreased at W3 (M = 4.45, S.D = 1.35), W4 (M = 4, S. D = 1.06), and W5 (M = 3.62, S.D = 0.49) when compared with W1, and W2. As shown in Fig. 3, LBP-I score decreases gradually when the exercise training is continued because there is a significant difference between the mean LBP-I at W1 (M = 6.25, S.D = 1.48) and W5 (M = 3.62, S.D = 0.49).

A repeated measures ANOVA determined that mean OBP-D differed statistically significantly between the levels of exercise duration (F (4, 92) = 23.52, p < 0.05, ηp2 = 0.506). Follow-up post hoc comparisons showed a statistical difference between the mean OBP-D at W1 (M = 3.30, S.D = 0.59) and W3 (M = 2.72, S.D = 0.41) and W5 (M = 2.06, S.D = 0.48). These results show that the OBP-D associated with the exercise training is gradually decreased against the levels of the exercise duration. The results also show that the score of OBP-D score at W5 (M = 2.06, S.D = 0.48) is less than the score at W1 (M = 3.30, S.D = 0.59), indicating a decline in OBP-D.

4 Discussion and Conclusions

This study focuses on overall body parts discomfort caused by the exercise training for the muscles around the spine during the five weeks duration. The back muscles are mainly Latissimus Dorsi (LD) and Lumbar Erector Spinae (LES) muscles. LD is located 4 cm below the inferior angle of Scapula that connects humerus with clavicle. LES is located on the right and left side of the spine at lumbar and vertebral column. Studies have considered these muscles for spine manipulative therapies and other exercise activities, such as standing on inclined sloped surface. Nelson-Wong and Callaghan [11] reported that the overall back pain of participants was decreased by 59.4% with the use of an inclined slope surface with an angle of 16 degrees. Gallaghar et al. [12] showed that a sloped standing platform during a 2-h occupational standing simulation decreased low-back pain in participants. In the present study, the exercise training has a small but significant benefit to minimize the subjective LBP-I and OBP-D in older adult women. However, this study is based on subjective scales, which may have lack of consistency, validation, and clinical effectiveness. Low-back pain is associated with troubled muscle activation patterns and weaknesses, and a sudden increase in back muscle fatigue caused by the exercise training [13]. At this juncture, future studies should be continued to examine muscles associated with pain and discomfort. However, some studies reported that exercise trainings for stretching and strengthening of the back muscles are more effective than other types of physical therapies but of low quality with different outcome measures [14, 15]. These studies further demonstrated that exercise therapies that include supervised muscle stretching or strengthening might improve low-back pain.

This study concludes that exercise trainings should be strategized to encourage devotion in the training and assess the exercise model characteristics including each individual muscle study and types for each individual patient. In particular, each muscle should be studied meticulously to examine the effect of the exercise training on the specific muscle. This can be achieved through the recording of the electrical activity using electromyography (EMG) of muscle tissues where the muscle contraction and activation patterns with the exercise duration can be explored. Furthermore, EMG activities can expose nerve abnormalities, abnormal muscle functioning or problems with nerve-to-muscle signal transmission.

Acknowledgments. The work was supported by the Promotion of Special Design-Technology Convergence Graduate School of Korea Institute of Design Promotion (N0001436) with a grant from Ministry of Trade, Industry & Energy of the Republic of Korea. This work was also supported by the UNIST funded product development project (1.170097.01).

References

1. Coole, C., et al.: What concerns workers with low back pain? Findings of a qualitative study of patients referred for rehabilitation. J. Occup. Rehabil. **20**, 472–480 (2010)
2. Nelson, M.E., et al.: Physical activity and public health in older adults: recommendation from the American College of Sports Medicine and the American Heart Association (2007)
3. Chodzko-Zajko, W.J., et al.: Exercise and physical activity for older adults (2009)
4. Tufail, M., Kim, M., Park, S., Moon, Y.G., Kim, K.M.: An ergonomic based postural control and balance body exercise design for young martial art trainees. In: Chung, W.J., Shin, C.S. (eds.) AHFE 2018. AISC, vol. 790, pp. 143–150. Springer, Cham (2019). https://doi.org/10.1007/978-3-319-94601-6_16
5. Corlett, E.N., Bishop, R.P.: A technique for assessing postural discomfort. Ergonomics **19**, 175–182 (1976)
6. Rainoldi, A., et al.: A method for positioning electrodes during surface EMG recordings in lower limb muscles. J. Neurosci. Methods **134**, 37–43 (2004)
7. Nam, H., et al.: The effect of the base of support on anticipatory postural adjustment and postural stability. J. Korean Phy. Ther. **29**(3), 135–141 (2017)
8. Ginn, K.A., Halaki, M.: Do surface electrode recordings validly represent latissimus dorsi activation patterns during shoulder tasks? J. Electromyogr. Kinesiol. **25**, 8–13 (2015)
9. Lovell, G.A., et al.: EMG of the hip adductor muscles in six clinical examination tests. Phys. Ther. Sport. **13**, 134–140 (2012)
10. Callaghan, J.P., et al.: The relationship between lumbar spine load and muscle activity during extensor exercises. Phys. Ther. **78**, 8–18 (1998)
11. Nelson-Wong, E., Callaghan, J.P.: The impact of a sloped surface on low back pain during prolonged standing work: a biomechanical analysis. Appl. Ergon. **41**, 787–795 (2010)
12. Gallagher, K.M., et al.: The influence of a seated break on prolonged standing induced low back pain development. Ergonomics **57**, 555–562 (2014)
13. Searle, A., et al.: Exercise interventions for the treatment of chronic low back pain: a systematic review and meta-analysis of randomised controlled trials (2015)
14. Lizier, D.T., et al.: Exercises for treatment of nonspecific low back pain (2012)
15. Hayden, J.A., et al.: Meta-analysis: exercise therapy for nonspecific low back pain (2005)

From Digital to Analogue: An Interdisciplinary Case Study to Enhance the Communication of Design Products Through Physical Constraints

Simge Esin Orhun(✉)

Department of Communication Design, Özyegin University, 34794 İstanbul, Turkey
simge.orhun@ozyegin.edu.tr

Abstract. Emerging technologies have the potential to make everyday life easier with an increasing number of benefits. However, it is argued that digital products are developed from a techno-centric perspective rather than being human-centered. With the complicated structures they have, they may also cause dissatisfaction and difficulty in their use, due to the unfamiliarity to the mental models of the users. In the area of human-computer interaction, many learning and outcomes have been achieved. This research aims to explore how this knowledge can be benefited for the design of products that are controlled by analogue means. Within the scope of the research, 5-week workshop was planned and applied to Communication Design senior students, who had many design experience with digital media. The outcomes revealed that, when user-centered design method and metaphorical thinking are integrated in the design process, the products' cognitive, physical, sensory and functional affordance improved.

Keywords: User-centered · Analogue · Metaphorical thinking

1 Introduction

In recent years, advanced displaying technologies have changed human beings' interaction with information from physical mechanisms to direct manipulation of screen-based icons. This affected the form of physical artefacts, user interfaces and interaction modes, which caused a contextual change mainly in terms of the insufficiency of physical interaction and visual feedback from digital surfaces, the change in the proximity between the user and the displayed information due to zooming or pausing, the inconsistent flow of experience and lacking variables and the process of repetition due to displaying in a conservative manner [1]. As the technology gives the possibility to implement the smallest item into any tool, we become more distinct from our natural actions and operations that were gained earlier in our life time. However, successful interaction with products, tools and technologies depend on usable designs that accommodates the needs of potential users through their ergonomics, thus affordance [2–4]. Besides, the area of cognitive science points how human beings relate

C. S. Shin (Ed.): AHFE 2019, AISC 968, pp. 289–301, 2020.
https://doi.org/10.1007/978-3-030-20470-9_31

themselves through spatial means and identifies the tendency to represent their mental models by referring to spatial metaphors based on orientation, movement and space-based conception, that is lacking with digital surfaces and interfaces [5]. Although technology is suspected to cause negative outcomes on human beings in terms of cognitive aspects, introducing spatiality for the design of new products has a huge potential to enhance the communication experience and to support memory, perception and reasoning.

Today, the emerging technologies like VR, X-box and AR requires natural user interfaces supported gestures that refer to spatial communication and spatial interaction [6]. As mobile and internet of things are bringing physical and digital worlds together, we need to go beyond screen-based design approaches and clarify "the nature of information" which involves creating new architectures and user scenarios [7]. Researches indicate that user experience is a driving force for all types of interface designs- whether be graphic, solid or natural [8–10]. Yet, user experience is closely related with cognition, memory and perception [2]. This research aims to explore the transformational effects on product design regarding form, content and interaction modes, when the knowledge of user experience for screens is adapted to physical mechanisms.

In order to reach the aim of the research, 5-week workshop, which intends to explore the communication of design products that refers to basic metaphors of daily life and natural way of doing things was planned and applied to the students of Communication Design Program. This discipline has a strong focus in developing interactions for varying media by addressing human needs, emotions, thoughts, and actions on the basis of ergonomics. Senior students, who have a strong background for digital design, were introduced to the idea of making use of physical constraints for designing interactions, that has been implied as vital for the design processes by [11]. The design problem was based on displaying/exhibiting COD students' works at the department's corridor, which were traditionally hung on walls. The brief guided the students to design by making use of user-centered design method and metaphoric thinking. They were expected to build displaying mechanisms in one to one scale that were interacted by analogue means. This brief was developed with the purpose of analyzing basic physical actions through everyday behaviors and how they were constructed into meaningful interactions through the stages of user-centered design and metaphorical thinking in parallel with the spatial and structural constraints of the real world.

This workshop has been applied 6 times between 2010–2019 in two different universities in Communication Design programs. The analysis 45 projects have shown that working with physical aspects in the design process has brought refinement to the mechanisms, consistency in system design and originality of user scenarios. The interaction modes obtained by the works can be classified into 2 groups: manipulating flat and layers of surfaces and self-standing spatial structures. Moreover, the designed mechanisms were evaluated to what extent usability is achieved through provided cognitive, physical, ambient and layout ergonomics and 85% were able to match the criteria. With the outcomes of this study, this interdisciplinary research attempted to fulfill the gap in the ergonomics of products by developing alternative modes of interactivity, which in turn, improved user-experience in multiple ways.

2 Background

The Interaction design is a creation of dialogue between a user and a product, service or system. Structuring dialogue is both an important and difficult task, which requires a reactionary and visionary understanding to create a natural and fluid communication between user and products. On that sense, it is also related with cognition, memory and perception [2]. This section investigates user-experience and cognition related issues concerning product designs.

2.1 Ergonomics of Products in the Frame of User Centered Design

The technological developments expand the range of applicable areas for interactivity. Alternative interaction modes that include bodily involvement, spatial and more natural and real-life gestures to manipulate the content become the focus of the researchers. While the interactions as well as products get more complicated, obtaining rich and efficient and satisfactory user experience solutions at the same time becomes challenging [4]. The difficulty lies in structuring the dialogue, which requires a reactionary and visionary understanding to create a natural and fluid communication between user and products, which is directly connected with user-centered design (UCD).

These products need to be designed to assist the users in order for the user to understand the product. Moreover, the required information needs to be visible to show how the interaction occurs. In this respect, the importance of ergonomics arises in establishing the preliminary features that determine how these technologies are perceived [3]. When the ergonomics of the products are provided correctly, the users easily understand the possible actions and functions of the products, which is also denoted by the term affordance. With the diversity of possibilities, they provide, affordances also allow users to determine the appropriate mode of action even in new situations [3]. By focusing on the usability of design, he suggested some fundamental aspects on how a product need to be designed such as: specifying what actions take place at any time, providing alternative actions and their consequences as visible as possible, allowing to evaluate the current system by providing a natural mapping between intentions, required actions, and presented information products [12].

UCD, which was originated by Donald Norman in 1980s, tries to optimize the design around how users can, want, or need to use the product, rather than forcing them to change their behavior to accommodate the product [13]. From the point of view of Norman [13] systems have to be designed to serve users, rather than to be used by a specific technology or a component of programming. So, the users' needs dominate the design of the interface and in turn, the need of interface is closely connected with the design of the rest of the system.

The term 'User Centered Design' is often used when we apply ergonomics principles to product design. The field of ergonomics has emphasized the value of utilizing users' existing knowledge and behavior so that when they are faced with a new product, they will be able to accommodate its use with their previous experience. Various usefulness of the products is defined as affordances [14]. Researches on affordance concepts also include approaches emphasizing the environment, whereas a specific situation, a certain context and a particular location driving aspects in that sense. People's perceptions vary according to this contextual arrangement and, inherently, the physical environment affects the users' behaviors where the interaction and communication take place [15].

Referred under the area of ergonomics, Hartson [16, 17] classified the affordances as cognitive affordance, sensory affordance, physical affordance, and functional affordance, defined as below [16, 17]:

1. **Cognitive Affordance** plays an essential role which facilitates thinking, understanding and learning a design product. Preliminary information about how to use an object is provided before the object is used. This feature especially helps less experienced users to learn how to use a product. In this context, WYSIWYG condition is also supported by cognitive affordance by giving information about what happens upon acting. Another feature of cognitive affordance is feedback which reassures the user that the action has been successful in terms of performing the functionality of the product.
2. **Physical Affordance** is a design feature that allows users to perform an action physically. This feature is related to the ability of a design product to provide sufficient physical properties and to be easily accessible. Physical affordance enables users to complete a task with less need of cognitive affordance and to be able to perform physical actions quickly while doing so. In this context, with an effective physical affordance, products can be experienced by users easily if the appropriate physical actions are provided as compatible with product's physical environment.
3. **Sensory Affordance** is associated with features such as the noticeability, visibility, legibility and audibility of the products which allows users to perceive products through visual, auditory, haptic, or other sensations. In interaction design, sensory affordance has a complementary role regarding the quality of the user experience by supporting the cognitive and physical affordances.
4. **Functional Affordance** promotes the intentional actions which are purposed in design and is described as an extension of physical affordance. It reinforces the predictability quality of the product by providing a connection between the user's physical actions and the product to activate the system. It is related to the user enablement which helps the users in doing actions to experience the products.

UCD has common objectives with any other approach to design, in that it seeks to create a product that meets the above-mentioned set of requirements. It is our claim that these affordances are valid to maintain the ergonomics of both digital and physical products.

2.2 Metaphorical Thinking to Connect with Human's Cognition

Today we are more accustomed to using screen-based interactive tools to any other mechanical tool. Although most of our versatile interactions that take place between the brain, body and the designed environment are manipulated and shaped by these tools, cognitive scientists stress on the impossibility to separate body from its emotional and performance-based activities [18]. Moreover, in recent years, the developments in technology are recognized to favor interactions for VR, AR and mixed reality.

Researches has started to focus on how to create usable designs when 3rd and 4th dimensions are included in the design process, with the aim of manipulating information more intuitively and naturally. The foundations of intuitive behavior, natural reactions and the cognitive connection has been an important standpoint within the design process of products, in order to maintain the ergonomics for these new technologies and environments.

As referred by Neisser in 1967, cognition is defined as "the processes by which the sensory input is transformed, reduced, elaborated, stored and recovered" [19]. A cognitive structure is a hypothetical construct representing the relationship of concepts in a person's long-term memory [20]. A mental schema, being the basic cognitive structure, is developed by the individual in relation to the bits of data to process information for meaning. Alternatively, cognitive structures include mental structures, mental tools, and patterns of thought. Metaphorical thinking, being one of the mental tools, is essential in human beings, which when practiced consciously can help to develop cognitive structures to help to process information. It is highly denoted that each individual develops his own cognitive structures by means of practice of metaphorical thinking [21].

A metaphor is a figure of speech in which a word or phrase that means one thing is used to describe an object or idea to which it is not literally applicable [5]. Researchers have identified that metaphorical thinking is the ability of an individual to associate the present information with what has already been stored in the memory [22]. Another research report by [5] revealed that the usage of metaphorical thinking develops mind patterns which in turn will influence the cognition of the individuals. Bargh and Barndollar suggested the use of metaphors consciously in language processing would lead to subconscious activation of goals and motivation [23]. Moser's study revealed that metaphorical expressions used in self-concept studies are found to promote cognitive abilities of analogical reasoning [24].

Through metaphorical thinking, a person can conceptualize language or design through time and space. Structural metaphors help to construct one concept in case of another. Orientational metaphors organizes a whole system of concepts with respect to one another by giving spatial orientation like up-down, front-back, on-off, center-periphery, near-far. Likewise, ontological metaphors, rely on our experiences with substances, non-physical things and the environment or the ways we perceive events, activities, emotions or ideas, upon which we treat them as entities. Like in orientational metaphors, ontological metaphors can refer, quantify and also can be used to identify aspects, identify causes or setting goals and motivating actions [5].

Cognitive structures develop in everyone from infancy to old age having neurological capacity to communicate. Hence, creativity requires looking at the world in a different way and trying fresh approaches to problems. So, metaphorical thinking supported by cognitive structures can ease information processing by i. Making connections of present with existing knowledge, ii. Finding Patterns and Relationships, iii. Identifying Rules and iv. Abstracting Generalizable Principles [21, 25].

From above, we understand how metaphorical thinking is important in the design decision of products, as it triggers affordances in many ways. While digital space and coding diminishes the effect of metaphors to a degree, we believe they will be more powerful, when used in the scope of spatial communication.

3 Method

The aim of the study is to explore the alternative mechanisms other than screens for displaying information when the user centered design approaches and the metaphorical thinking that are widely used in screen-based design, is adapted to the design of products that are manipulated by analogue means and build upon physical constraints. In order to reach our aim, 5-week workshop that focuses on spatial interaction and spatial communication was planned. A specific situation, a certain context and a particular environment is decided to be integrated within the brief [15] and working with physical prototypes [10, 26] were the focus of the assignment. The assignment focused on the design and construction of an interactive displaying tool manipulated by analogue means for a definite physical space, which would be used for exposing printed or model student works. The design works would be presented by 1/1-scale prototypes constructed in selected physical spaces. The organization of information, the form of the mechanism and the interface design was expected to be guided by the spatial qualities of the place.

In broad terms, user-centered design is both a design philosophy and a design process. As a philosophy, it makes the needs, wants, and limitations of the end user of a product the priority focus, and as a process it offers designers a range of methods and techniques to ensure this focus is sustained through the stages of design, which includes exploration, creation and evaluation [26] These stages are adapted to this research as follows:

i. **Exploration:** students are asked to identify users' problems, understand users' tasks, characteristics and needs, research ergonomic implications and relevant products.
ii. **Creation:** students developed ideas on the basis of metaphorical thinking, decided on ergonomic guidelines and standards that are relevant to the Project, used feedback to refine the design, and built the prototype.
iii. **Evaluation:** A jury evaluation is made for the general outcomes. Afterwards, testing the products for cognitive, functional, physical and sensory affordance was done and the results are analysed.

In parallel with the aims and method of this research, the product is planned to be a mechanism. In this system, users get in touch with content, users have control over the system or are controlled by the system and the system. In the light of the above-mentioned issues, the design decisions for the product has to be influenced by users' cognitive functions, the semantics and the structure of the content and the characteristic of the system, whereas the system facilitates access to the content. Having implied that, there is synergy between the system, the user and the content which interact two-ways with each other as user-system interaction, system-content interaction and user-content interaction [27]. So, if user-centered approaches and metaphorical thinking can be adapted to user-content-system relationship, the communicative aspects of the products will be enhanced and the products will be affordable.

4 Results and Discussions

This method was applied in the years 2009 fall, 2010 spring, 2010 fall, 2011, spring, 2018 spring, 2019 Fall. With the analysis of the prototypes, all the students lived through the process of exploration, creation and evaluation and managed to solve the expected stages to some extent. At the end, nearly all of them were able to create content, select ta specific place and managed to build a working prototype. Similarly, nearly all of the works yielded the relationship between user-system, user- content and system-content. From Fig. 1, 2, 3, 4 and 5, different projects belonging to different years are introduces, so as to give idea about the works, their initial standpoint and the outcome on the basis of above mentioned relationships.

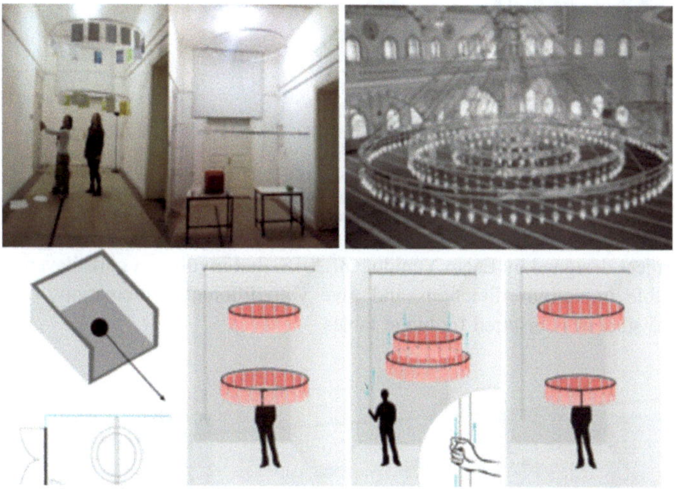

Fig. 1. A project designed by Pınar Eskikan, 2009, that focused on creating a mechanism that can be used for displaying storyboard or animation. The metaphor use is a mosque chandelier and window blind mechanism.

We have recognized that the places selected as working site really affected the outcomes. The project that are set up on the corridor had the tendency to base the design strategy on movable surfaces (Fig. 2). While this solution reminds of the screen experiences, the user-system-content relationship guided for double or multi layered solutions. The projects that are located in open areas are developed like self-standing sculptures, which made use of spatiality at top most level (Fig. 1). The level of interactivity was higher than the first group and the projects were less restricted to develop multilayered surface solutions.

Another interesting outcome was to observe how students overcame the disadvantages brought by screen based interactions and solutions, as indicated in the introduction section. Due to analogue means for interaction, all the projects overcame the insufficiency brought by screen interaction and visual feedback. While the spatial solutions solved the flow of experience without cut in a refined way, the multilayered screens were able to be more successful in integrating proximity to their design.

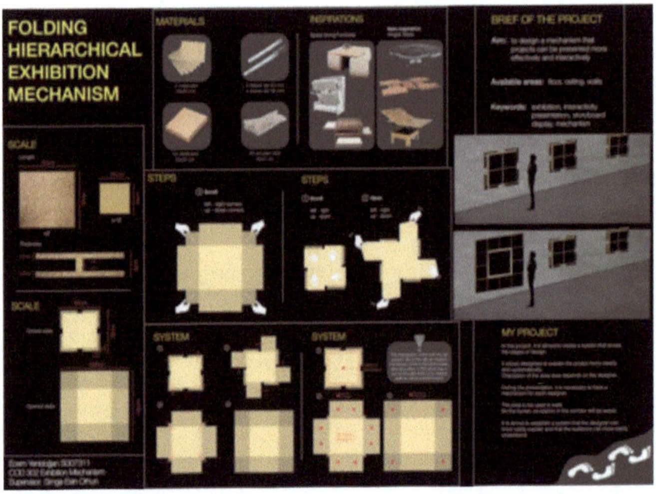

Fig. 2. A project designed by Ecem Yenidoğan, 2018, that involves a mulstilayered mechanism that is openable in various directions and give you different combinations of surfaces for displayin. The student is inspired by hinge tables.

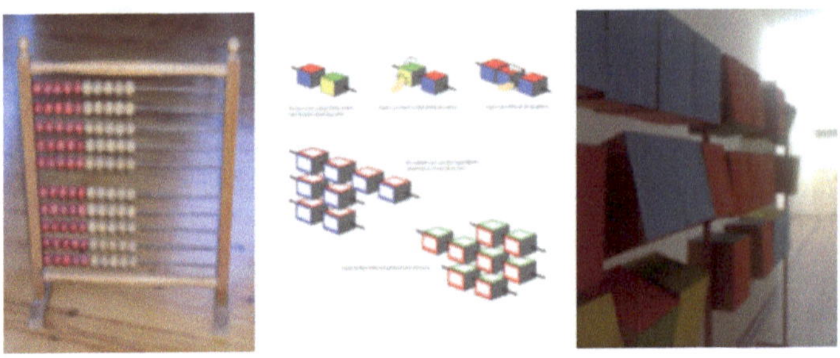

Fig. 3. A project designed by Yiğit Hepsev, 2009, that imitates the idea of abacus, as it can be viewed from both sides when placed in a corridor. He transformed the balls of abacus to cubes. With the aim of using proximity, each face of the cube is a different color so as to indicate the varying compositions to be used when displaying.

From the jury evaluations, making use of metaphorical thinking for the design of such tools was very helpful, as the idea to interact with the mechanism can be perceived and achieved very easily.

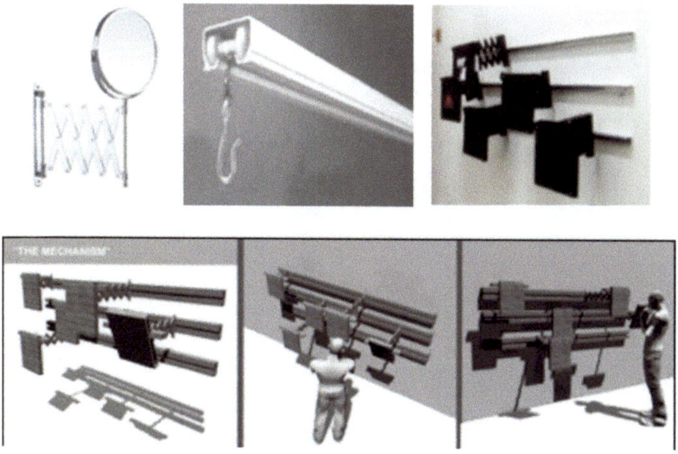

Fig. 4. A project designed by Cansın Bozoglu, 2010, that involves double surface units that can move forward and back, slide left to right. Double side mirror with scissor arms and rails are the guiding elements for the design.

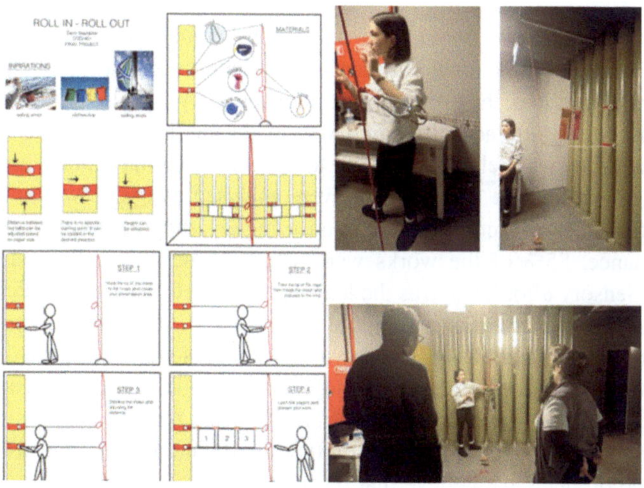

Fig. 5. A project designed by Derin Bayraktar, 2019, made use of her experience in yatching and knotting. She made use of the permanent columns at the end of the coridor to bring out a spatial experience. The free swinging knots are movable up and down, so that you can arrange paper size. The belts can also be placed on different columns, so user can personalize.

The solid user interfaces that will be used to manipulate the product were lacking in some projects. When designing graphic user interfaces, students were so accustomed to place ikons and symbols to direct the user to reach information. With the 3 dimensional prototypes, as the scale changed, they felt like the prototype itself easily defines how to be used (Fig. 6).

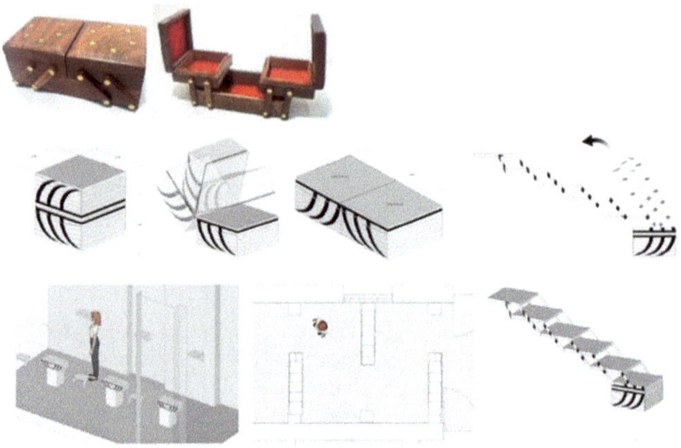

Fig. 6. A project designed by Esra Sağıroğlu, that serves to display 2 different people's works. The design is based on the articulation of a cube by its middle section in parallel with the solid user interface placed on the sides, When the cube is a whole, the name of the project of the student will take place on the top and when the top layer is pulled in one direction, multiple layers of displaying faces connected to each other will appear.

The products were also evaluated for satisfying ergonomics in terms of cognitive, physical, sensory and functional affordances. As the process was planned on the basis of metaphorical thinking, the projects were all inspired by daily activities and objects or well known cultural artefacts. Research was made for the objects that include user-content-system two way relationships and the idea of what the designs are expected to include were formed. The end products were analysed and 95% were able to satisfy cognitive affordance. As cognitive affordance hugely support the physical and functional affordance, 85% of the works were agreed to be physically and functionally ergonomic. Sensory affordance was the least achieved because Communication Design students did not have any prior experience with material, structure and prototyping. So the durability of the materials and the gravitational forces were unexpected components of the project. Still, this experience was so important in this process because today, as technology gets smaller and can be integrated into many objects, sensory inconsistency of the products were neglected. Based on the investigations and analysis of the ergonomics of the products we believe that our mental models are formed on the basis of our earlier experiences and activities and designers have to depend and invest on them for the new product, form and interface designs.

5 Conclusion

The last 20 years of academic researchers in the area of interaction design invested mostly in the area of human computer interaction, where digital communication took place on screen based interactions with the input devices like mouse, tablet, keyboard. However, in time, technological tools are suspected to transform the perceptual and behavioural mechanisms of human beings. In order to overcome these problems caused by digital technologies different methods for designing were developed, such as human-centred design, design thinking method, participatory design … etc. While researches still go on for maintaining the ergonomics of screen based interactions, the emerging technologies like AR, VR, MR call for the growing need for intuitive and natural interaction modes. This research claims that the foundations of ergonomics and affordance used in digital interaction design can be adapted to spatial interaction design in order to obtain intuitive and natural interaction modes and mechanisms.

The researches and the examples show us that many of our experiences and activities are metaphorical in nature; it is possible to understand how our conceptual system is constructed by metaphors. The similarities we perceive match with the categories of our conceptual system or our recent experiences, which can be beneficial for the design of refined and pure experiences. Meanwhile, human-centered design is recognized to be an important source for this aim because it focuses on the theory that puts forward the users' needs and limitations and also provides a range of methods for the design process to the designers in parallel with its focus.

On the basis of these issues, a workshop which aims the design and building of displaying mechanism prototypes that will be controlled through analogue means by the user, was planned. This workshop has been applied 6 times in years between 2009–2019 with the same focus in different universities. The outcomes have shown us that first of all the designers avoided the problems of screen based interactions and graphic user interfaces by providing spatial solutions in various ways. Self-standing mechanisms or multiple layered surfaces were the formal solutions for these mechanisms, however, the metaphorical thinking approach triggered the use of natural actions and instinctive behavior that improved the ergonomics of the products. Moreover, the products were evaluated if they maintain cognitive, functional, sensory and physical affordances. Cognitive affordance was obtained in 95% of the works. Similarly, functional and physical affordances were also achieved in around 85% of the works. The problems raised due to not having enough knowledge for materials and construction. The least with 70% was obtained for sensory affordance because the solid user interfaces for the products were not very well thought and visible and not efficient enough.

To conclude, we came to a thinking that the solution will come through changing the present design approach, from an "artifact-centric" model, towards a "user-centric" model, that focuses on personalization and adaptability and can be transformed according to the customized needs of each. We believe this strategy will be beneficial to prevent the undesirable effects and bring positive outcomes for the design of new modes of interactions and user interface designs of the future technologies.

References

1. Dawes, B.: Analog in, Digital out. New Riders, Berkeley (2007)
2. Kolko, J.: Thoughts on Interaction Design. Brown Bear LLC (2007)
3. Norman, A.D.: The Design of Everyday Things. Basic Books, Inc., New York (2002)

4. Yantaç, A.E.: A method for teaching affordance for user experience design in interactive media design education. In: Marcus, A. (ed.) DUXU 2013. LNCS, vol. 8012, pp. 630–638. Springer, Heidelberg (2013). https://doi.org/10.1007/978-3-642-39229-0_67
5. Lakof, G.: Methaphors We Live By. University of Chicago Press, Chicago (2003)
6. Ünlüer, A., Özcan, O.: Mime based creative drama implementations for learning gestural shadow mapping. Digit. Creativity **24**, 275–290 (2013)
7. Morville, P.: Intertwingled: Information Changes Everything. Semantic Studios, Ann Arbor (2014)
8. Laurel, B.: Interface as mimesis. In: User Centered System Design: New Perspectives on Human–Computer Interaction, pp. 67–85. Lawrence Erlbaum, Hillsdale (1986)
9. Nardi, B., O'Day, V.L.: Information Ecologies: Using Technology with Heart. MIT Press, Cambridge (1999)
10. Özcan, O.: Cultures, the traditional shadow play, and interactive media design. Des. Issues **18**, 18–26 (2002)
11. Faiola, A., Matei, S.A.: Enhancing human–computer interaction design education: teaching affordance design for emerging mobile devices. Int. J. Technol. Des. Educ. **20**, 239–254 (2009)
12. Norman, D.A.: The Psychology of Everyday Things. Basic Books, New York (1988)
13. Norman, D.A., Draper, S.W.: User Centered System Design: New Perspectives on Human-Computer Interaction. Lawrence Erlbaum, Hillsdale (1986)
14. McGrenere, J., Ho, W.: Affordances: Clarifying and evolving a concept. In: Proceedings of Graphic Interface 2000, Montreal, pp. 179–186 (2000)
15. Schmidt, A.: Interactive context-aware systems interacting with ambient intelligence. In: Riva, G., Riva, G. (eds.) Ambient Intelligence: The Evolution of Technology, Communication and Cognition Towards the Future of Human-Computer Interaction, pp. 159–178. IOS Press, Amsterdam (2005)
16. Hartson, H.R.: Cognitive, physical, sensory, and functional affordances in interaction design. Behav. Inf. Technol. **22**, 315–338 (2003)
17. Hartson, R., Pyla, P.S.: The UX Book: Process and Guidelines for Ensuring a Quality User Experience. Morgan Kaufmann, Amsterdam (2012)
18. Harrison, S., Art, C., Sengers, P., Tatar, D.: The 3 Paradigms of HCI. In: Proceedings of the Conference on Human Factors in Computing Systems (CHI2007), San Jose (2007)
19. Dade-Robertson, M.: The Architecture of Information. Routledge, New York (2011)
20. Shavelson, R.J.: Methods for examining representations of a subject-matter structure in a students memory. J. Res. Sci. Teach. **11**, 231–249 (1974)
21. Navaneedhan, C.G., Kamalanabhan, T.J.: Is metaphorical thinking related to development of Cognitive structures among learners? World Sci. News **52**, 1–13 (2016)
22. Siegler, R.S., Shipley, C.: Variation, selection, and cognitive change. In: Simon, T.J., Halford, G.S. (eds.) Developing Cognitive Competence: New Approaches to Process Modeling, pp. 31–76. Lawrence Erlbaum Associates Inc., Hillsdale (1995)
23. Bargh, J.A., Barndollar, K.: Automaticity in action: the unconscious as repository of chronic goals and motives. In: Gollwitzer, P.M., Bargh, J.A. (eds.) The Psychology of Action: Linking Cognition and Motivation to Behavior, pp. 457–481. Guilford Press, New York (1996)
24. Moser, E.I.: Impaired spatial learning after saturation of long-term potentiation. Science **28**, 2038–2042 (1998)
25. Garner, B.: Getting to Got It!—Helping Struggling Students Learn How to Learn. Cambridge University Press, Cambridge (2000)

26. Chartered Institute of Ergonomics and Human Factors (CIEHF): User Centered Design, A Practical Guide for Teachers of Design and Technology (2017). https://www.ergonomics.org.uk/Public/Resources/Publications/User_Centred_Design/Public/Resources/Publications/UCD.aspx?hkey=2138360f-c44a-4c20-b500-1efb4f04e43a
27. Toms, E.G.: Information interaction: providing a framework for information architecture. J. Am. Soc. Inf. Sci. Technol. **53**, 855–862 (2002)

Innovative Product Design for Hand-muscle Weakness Therapy in the Elderly People

Parichart Rattanapol(✉)

Department of Industrial Product Design, Faculty of Industrial Technology, Phranakhon Rajabhat University, Bangkok, Thailand
parichart@pnru.ac.th

Abstract. Currently, the structure of population all over the world seems to become more of aging society. Elderly people in Thailand are rapidly increasing in number as well and this affects many aspects of the country, including economics and society. More than 40% of health degrading problems make the elderly become dependent on others because they are unable to help themselves. Weakening of hand muscle from aging is caused by the damage of muscles or tendons directly related to the movement of hands and fingers. The weakening happens gradually; sometimes it can be temporary symptom and sometimes it can become more severe. Designing products for occupational therapy of hand muscle exercising in elderly people originates from the idea of decelerating the weakening of the hand muscle to enable the elderly to live and do day-to-day activities more conveniently so they may have better quality of life. The process of designing consists of: interviewing about the needs of 10 elderly persons, and using the inspiration from the shape of Siamese fighting fish to specify the size and the usage of the product. This type of fish has strong fighting instinct. It symbolizes obstacles fighting in the past experience of a person. The color and swimming manner of the fish were also used as the inspiration in designing the products according to 3 types of occupational therapy. Then, the products were tested and satisfaction of users was evaluated after 3 product prototypes had been used. Percentage calculation was used in analyzing the data from the questionnaire. As for the analysis of data from satisfaction evaluation form, the researcher used statistical method of finding mean and standard deviation (S.D.). It was found that the product "neck-supporting pillow" received the highest satisfaction ($\bar{x} = 4.6$) (S.D. = 0.69). The product "plier" received second highest satisfaction ($\bar{x} = 4.5$) (S.D. = 0.59). The product "knuckle" received high satisfaction ($\bar{x} = 4.3$) (S.D. = 0.60). Therefore, this designing resulted in the products for hand muscle exercising that can effectively decelerate the weakening of hand muscle in elderly people.

Keywords: Occupational therapy · Occupational therapy product · Elderly people · Hand muscle

C. S. Shin (Ed.): AHFE 2019, AISC 968, pp. 302–315, 2020.
https://doi.org/10.1007/978-3-030-20470-9_32

1 Introduction

In the year 2007, aging of societies became a global phenomenon. The proportion of people who are 60 years old or older has doubled, increasing 11% from the year 2006. It is predicted that the number will increase 22% in the year 2050 [1]. The major factor that causes aging of societies is the increase of the elderly and the decrease of birth rate [2]. This is a problem that occurs all over the world, including developed countries like Japan or South Korea [2]. As for Thailand, this problem is evident in the fact that its population of the elderly has rapidly grown from 5% in 1970 to 10% in 2006. There is a tendency that it may reach 30% in the year 2050 [3]. Office of National Economic and Social Development Council shows that in the next 10 years, or in 2023, the number of Thai people who are 60 years old or above will increase to 14.1 million people or 21% of the whole population. It means that Thailand will become an "aged society." People who are 15 years old or younger will be only 14% of the whole population in the year 2033. Therefore, it can be said in less than 20 years, Thailand will become a "complete aged society" because the people who are 60 years old or older will become a quarter of the whole population [3]. This increase of elderly people results in the needs for changes in environment to make it easier for the elderly people to live. The main problems that this group of people face are concerning age, physical fitness, and lower ability to work, so the overall picture is determined by related factors, such as way of life, society, environment, and economics [4].

Aging is inescapable for everyone; thus, it is necessary to consider the physical limit of the population and increase general knowledge of each elderly person about weight [5], body mass index [6], and muscle strength [7]. When a person ages, the loss of muscle has positive correspondence with the weakening of hands [8]. Human hand is a complex structure and can be adapted to skills that are necessary for daily human life. There are many conditions that can have effects on hands and may lead to abnormality and disability [9]. Symptoms of hand-related diseases might not directly affect the function of the whole body and internal organs, but when hands cannot function properly, quality of human life can drastically go down [10]. Health risks may differ [11] in cooking and washing housewives, computer users, factory workers, and road-drilling workers [12]. Playing musical instrument has risks as well. For instance, drum set is an instrument that needs to be hit with a stick continuously in order to create vibration, and this vibration can cause injuries to hands and arms. Many diseases can develop from minor hand symptoms such as trigger finger, weakening of hand muscle, and carpal tunnel syndrome [11]. Occupational therapy is a method that can restore the muscle, focusing on repeated exercise with an aim to improve physical, mental, and social skills while the patients do their daily activities, work and recreation. It begins with a small movement and gradually increase, depending on the capacity of the patient. This will enable them to have a better life [13, 14].

As mentioned above, it can be seen that aging of societies is a problem that affects many countries around the world. The population of the elderly increase and that results in changes in the condition of society and people's way of life. Elderly people should pay much attention to their physical and mental health, society, and the environment. Some of the most important parts of the body are hands and arms. When an elderly person loses these parts or some damages occur to these parts from daily activities or aging of the body, he or she may be in a state of anxiety and depression, and is unable to live normally like other people. Therefore, the researcher wanted to design a product for hand and arm exercise in order to decelerate the weakening of hand muscle, a symptom commonly found in the elderly, caused by aging and activities. The product was designed to facilitate exercise movement that is suitable with daily life of the elderly to slow down the weakening of hand muscle and enable them to live happily like other in the society.

2 Research Method and Procedure

In designing a product for hand muscle exercise in occupational therapy for elderly people, the main method and procedure are established as stated below:

2.1 Related Information Study

This step is the study of information related to the elderly and the design, such as daily behaviors, symptoms, and needs of elderly population. Existing products and tools from various resources were studied in terms of usage, shape, mechanical system, material, and benefits for the elderly people. All other related theories and literature were also studied to gather data that will support the design of product for hand muscle exercise occupational therapy for the elderly. Materials and production method were studied to ensure that the product was suitable for the body, mind, and capacity of the elderly. Furthermore, academic data and conclusions from brainstorming among the population and sample group, which were 10 male elderly persons whose bodies have aged naturally but are able to exercise or work out.

2.2 Requirements Interviews

The interview conducted with the elderly was about their daily living, current health condition, illness, and exercise habit. The information obtained were translated as much as necessary before product design, so that it could be concluded and use as the guidance for the design. The questions in the interview are shown in Table 1.

Table 1. Interview data analysis.

Questions	Responses
How does hand muscle weakening affect an elderly person's life?	It can be concluded that hand muscle weakening can affect daily life tremendously; for example, it causes pain or ache in hands and arms and make a person have difficulty in holding things or using hand to push the body when getting up. They will not be able to hold on to the walking stick tightly and this causes difficulty in walking as well

(continued)

Table 1. (*continued*)

Questions	Responses
What are the characteristics that product for hand muscle exercise in occupational therapy should have and how much it should cost to be considered an affordable price?	- It should be light and at a size fit for hand holding. It should not be slippery or too soft - Some do not have the financial mean to purchase such product and some can purchase a product at a price not higher than 300 Baht
In regular days, how do elderly people exercise their arms and hands? For how long? What methods do they use?	- Walking slowly, body exercising, hand movement (holding and releasing), arm swinging, squeezing a cloth bag full of beans; exercise period is about 10–30 min
If there is a product for hand and arm exercising that can decelerate muscle weakening and help you to become more flexible in holding and grabbing things in daily life, how do you view such product?	They really want to use such product and believe that it will be very helpful for the elderly
What are the characteristics of product for occupational therapy that you are interested in?	They are interested in a product that is light and small, hand-fitting, not slippery, soft but not too soft, and can be used in hand exercising

The opinions gained from interviewing with the elderly were used by gathering specific data and additional data for designing the product for occupational therapy hand muscle exercise that is effective and matches the needs of users.

2.3 Attribute Data Analysis

After gathering data, opinions, suggestions, and conclusions from documents, theories, and properties with facts, the researcher arranged the data based on the dimension of the problems and the characters of data that are needed for supporting the design of the product and used the method of attribute data analysis in order to clearly show the overall picture. The details are as follows:

(1) Analysis Summary of Appropriateness of Exercise for the Elderly
Elderly people should not do exercise that requires strong movement of the body. They require exercises that have simple and light movement or equipment. They should focus on flexible exercises and avoid those that require long-period of muscle contraction. They should not exercise for too long; only 10–20 min but more frequent.
(2) Analysis Summary of Hand Muscle Weakening Conditions
Most symptoms that occur at the hand muscle mostly come from abnormal conditions of nerves and diseases. Severity of the symptoms depend on each patient. Besides that, there might be comorbidity that is the main cause of hand muscle weakening. This condition is found more commonly in the elderly group than other age groups. Doing some activities repeatedly, stretching and folding the wrist repeatedly, jobs that require stretching of wrist when lifting an object, jobs that require using drilling or digging

tools, jobs that requires twisting movement of the wrist in an abnormal position such as using computer, or using the hand muscle too hard each day, can be major factors that cause the weakening of hand muscle. In addition, this condition can make a person face difficulty in holding objects, particularly small ones, as well as clasping hand and using strength from hand and wrist. Therefore, hand muscle weakening causes problems in daily life.

(3) Analysis Summary on Hand Muscle Weakening Age Range

This condition is usually found in elderly people who are older than 50 years old. The cause can be heredity, environment, medicine, toxin, abnormal immunity reaction, and aging. Furthermore, people who are at risk of hand muscle weakening are those who use hand and arm muscle to labor hard or do the same activities repeatedly are also vulnerable to this condition.

(4) Analysis Summary of Physical, Emotional, and Social Condition of the Elderly

Elderly people have lived their lives for a long time and have been through much experience. Their bodies deteriorate. They begin to have wrinkle skin, grey hair, changed body shape and changed personality. They move more slowly. Their emotions also change; less enthusiasm, more indifference, needing more privacy, and reluctant to depend on others.

(5) Analysis Summary of Each Occupational Therapy Product

After the analysis of each occupational therapy product, it can be concluded that the tool for squeezing and releasing is the best. It has the highest efficiency in healing hand muscle weakening in patients. It works directly to the affected area and its shape can be changed into various appearance so that the patients will not be bored. It is easy and convenient to use so the patients can do the exercise with this product by themselves. In occupational therapy, to gain the highest benefit, one must use many positions to exercise the hand muscle. The ratio is: 40% squeezing-releasing activity and 20% opening-folding. One should not only squeeze their hand while doing occupational therapy because the muscle will contract more in the position of curve fingers.

(6) Analysis Summary on Shapes that Affect Product Design

Since the product for exercising hands and arms in elderly people can be used in 3 movements: squeezing-releasing, opening-closing, and stretching-retracting, it is necessary to be stable enough to withstand the pulling force. Shapes inspired by nature can create relaxation and reduce stress during occupational therapy.

(7) Analysis Summary on Main Materials for Producing Prototype

Pine wood is the best material for producing the prototype. This is because it is a natural material so it is environment-friendly and will not cause pollution. It has good properties such as density of wood, beautiful texture and color, colorable nature, light weight, low shrinkage, and long duration. The surface of pine wood can be easy to decorate and its price is cheap when comparing with other kinds of wood in terms of benefits and capacity. Secondary material is loincloth because it is safe and has proper weight for occupational therapy. It also has beautiful pattern and long durability. The price is cheap and suitable for exercising. The details are shown in Table 2.

Table 2. Analysis of materials used in producing the prototype.

Topics for Consideration	Rubber 	Rubber wood 	Pine Wood 	Cotton 	Loincloth 
Strong and enduring	2	3	3	3	3
High safety	2	4	4	4	4
Suitable weight for occupational therapy	3	4	4	4	4
Beautiful pattern	2	3	3	2	4
Long durability	3	3	4	2	2
Low price	3	3	3	3	3
Easy to produce	2	3	3	3	3
Total	17	23	24	21	23

Remark: 4 = Very good, 3 = Good, 2 = Ok, 1 = Poor

(8) Analysis Summary on Design Theories

Product design and development in the present time is based on some theories that establish style, industrial design process, and method of design. This usually leads to creative products from the past to the present. There is a theory that is still commonly used today in a global level; it is called "Function Follow Form" design. This is a concept of designing by focusing on the beauty of shape and size. Based on this theory, beauty is more important than usage. This kind of design is normally used when the producer wants to upgrade the product to attract another group of consumers who have more purchasing power because of the economic status. The product will be more valuable and will be sold at a higher price. This kind of design concept absorbs a lot of art and beauty and does not depend much on rules or principles. The designer is flexible to design as he or she feels like it. Another type of design is "Form Follow Function", which is a concept that focuses on the usage of the product more than its beauty. This is necessary when the product needs to be used according to its purpose. Most product designs rely on this theory, especially products that are industrially manufactured in mass. The emphasis is on making design that corresponds with the machine, materials, and ensuring that the products are produced in sufficient amount for the consumers.

In order to make it suitable for the emotions, feelings, and psychological condition of the elderly, who normally have repeated behaviors and the same form of daily life, and to make the product suitable with occupational therapy hand muscle exercising, all three types of products are needed: squeezing-releasing, opening-folding, and stretching-retracting. Therefore, the researcher adapted both theories ("Function Follow Form" and "Form Follow Function") in designing the product. Many kinds of arts are used in the design in order to make the product relaxing and stimulating the desire for exercise in elderly people.

3 Finding Concept Design

This step is the establishing of design concept and product design process as well as the details in the design and development of hand muscle exercising product. This includes the final design and the drawing of the blue print for the production of the prototype. Then, after testing the product with the population group, the results were concluded.

Product use satisfaction assessment from the population is shown in Fig. 1.

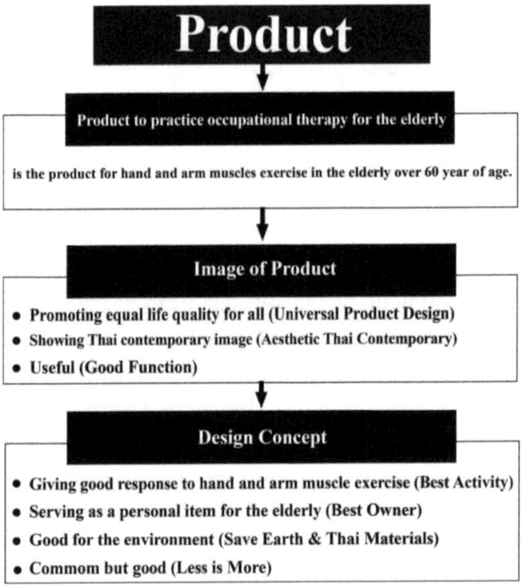

Fig. 1. Idea map.

3.1 Product to Practice Occupational Therapy for the Elderly

These products were designed and produced based on the design concept of the tools for exercising hand and arm muscles to decelerate the weakening of the hand muscle in the elderly who are still healthy or with just the initial stage of health deterioration. The designing of the products was aimed at encouraging the types of exercises that are suitable for the elderly's daily lives. Since the study population or target users of the products was the elderly group; therefore, the concept of product development needed various elements which led to the design that contains product identity, as shown in the details below.

3.2 Image of Product

Products for therapeutic activity of hand muscle exercise in elderly people were designed with the image of product (image concept) in mind as follows:

Promoting Equal Life Quality for All (Universal Product Design)
Since more than 70% of the elderly group had a restriction on the use and limited access to exercise products, which most of them were outdoor type or indoor type with

a limited number, thus, this group of population had limited chance and access to basic necessary facilities. The researcher decided to design the products based on the concept of equal life quality, so they can live the lives happily without being socially discriminated. The designing of these products was considered Universal Product Design that promotes equality for all.

Showing Thai Contemporary Image (Aesthetic Thai Contemporary)

On the average, the age of the study population was over 60 years of age. They were born from the year 1932 to 1957. Thus, the image of this generation is clearly Thai lifestyle, such as, Thai food, Thai outfit, Thai house, Thai recreation, including Thai rural lifestyle. The elderly's lives in that period of time can be considered a lifestyle with a strong bondage with Thai living.

Useful (Good Function)

The products must be able to reflect good image, great function with comprehensive use according to occupational therapy of hand muscle exercising principle in elderly people. Thus, they can respond to the need of the consumer group in this occupational therapy.

3.3 Design Concept

The researcher defined a broad design concept including the usage of supportive, urging and encouraging words that would bring the elderly to take part in this therapeutic activity of hand muscle exercise. Data was collected in various dimensions and resulted in the 4 design concepts based on the needs of the elderly population group (Fig. 2) as follows:

(1) Giving Good Response to Hand and Arm Muscle Exercise (Best Activity)

These products were used in the therapeutic activity of hand and arm muscle in the target group of elderly over 60 years of age. Hand and arm muscle exercise products were designed into a set design that fits 3 types of therapeutic activities of hand muscle, including grasp-release, spread-close and stretch-retract types.

(2) Serving as a Personal Item for the Elderly (Best Owner)

This concept reduces daily behavioral problems in the elderly. The research, the survey and the interview showed that the stated elderly group was not fond of field exercise machines, of a medium to a large size, due to the high temperature when being used in the open space of Thailand and the hot sensation the metal machine released. Moreover, the stated elderly group did not like to socialize in group. Due to their high privacy, they tended to stay alone in their room. Based on this understanding, the researcher wanted to provide an alternative for the elderly to be able to own the exercise machines that are easy to carry around, so they can enjoy occupational therapy of hand and arm muscles in their private space, both indoor and outdoor, in their preferred time.

(3) Good for the Environment (Save Earth & Thai Materials)

The elderly population with the age over 60 years old has been living for more than 60 years. Most of them love and care for the environment. They love to use products made from natural material and Thai material based on the idea of "Made by the Thais, used by the Thais and the Thais flourish". Thus, the researcher came up with the design concept of using natural material and Thai material which are beautiful and safe to use and the elderly felt good from saving the environment and using Thai material.

(4) Common but Good (Less is More)

The researcher offered a good alternative and trend to use products that contain high benefits. They designed these products to be simple, beautiful and attractive in term of initial cost, based on the concept of "Less is More".

Fig. 2. Design concept

Thus, the researcher designed 3 product prototypes for therapeutic activity of hand and arm muscles for the elderly. The design of each product was based on the principle of therapeutic activity for hand and arm muscles in the elderly, including 1 grasp-release exercise product, 1 spread-close exercise product and 1 stretch-retract exercise product. The researcher developed the product prototypes based on the shape and the moving of the colorful tails of Thai fighting fish, including Betta Fish and Half Moon Fish. The above characters of the fighting fish were added into the set design of the products and gave harmonious shape to each piece of products. The researcher also redesigned the shape of the fish body to be round in order to fit into the palm of the elderly and gave a pleasant feeling when being used in therapeutic activities of hand and arm muscles exercise.

3.4 Setting the Design

This step is the setting of design for the 3 products for occupational therapy hand muscle exercising, as shown in Tables 3 and 4.

Table 3. Design development of Product 1.

Exercise Movement	Shape inspired from fighting fish	Details of design	Idea Sketch
Squeezing-releasing	Body and tail 	The plier is for the squeezing- releasing movement of hand. It uses 4-6 mm. of elastic. The focus is to pull the hand open. The main material was wood pallets. The material padded the holder was sponge, with colorful loincloth to make it look like the tail of fighting fish. The product comes with a ring for hanging the product and make it more convenient for use.	

Table 4. Design development of Product 2.

Exercise Movement	Shape inspired from fighting fish	Details of design	Idea Sketch
Opening-folding	The spreading of the fighting fish's tail 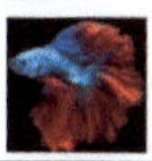	It is a piece of wood shaped like an open hand, similar to the spread of the fish tail. The main material was wood pallet and elastic of 6-mm. thick for placing between the fingers. More elastic can be added just in case one is broken.	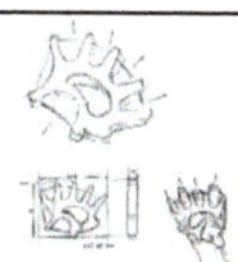
Stretching - retracting	Body of fighting fish 	Neck-supporting pillow was inspired from the body part of fighting fish in its curve swimming position. It has multiple purposes or Three-in-one use. 1. Can be used as additional pillow when sleeping or as neck-suppor-ting pillow when sitting and relaxing. 2. There is a zip for closing and opening, like a package that contains a product. 3. Can be stretched and retract because it has resistance elastic band of 8-10 mm. long, with a hook at both ends of the pillow.	

From the development of the 3 products from the idea sketches, it can be seen that the shape of the products are beautiful and fits the usage purpose. The form of the fighting fish, such as its body, tail, or the spreading of the tail, were adopted in the mind of the designer. Then it was developed into idea sketch. This is to ensure that the product has the perfect shape for the convenience of hand and muscle exercising occupational therapy in the elderly.

After the research designed the product and finished the whole process, the prototype of the product was then tested with the elderly people and the analytical results on satisfaction of the research population and evaluation of the prototype were obtained.

4 Results and Discussion

In this research of designing and developing product for hand muscle exercising occupational therapy for the elderly, the researcher has studied related information and has gone through the whole process of designing and producing the product. It resulted in the production of 3 products for hand and arm muscle exercise, as shown in Fig. 3.

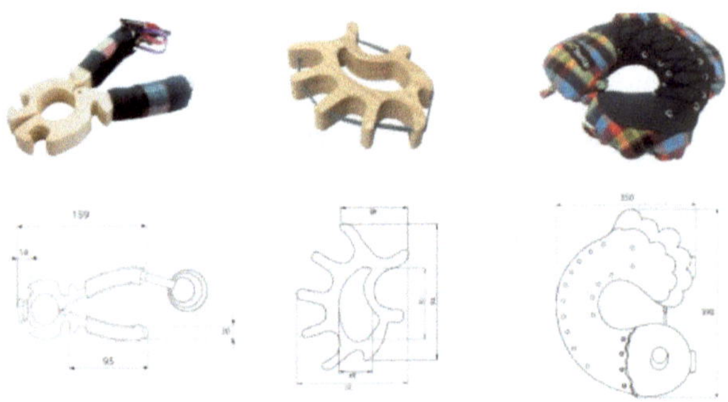

Fig. 3. Product prototypes and their sizes.

The prototypes of the products were used with the population group or 10 elderly people in order to assess their satisfaction with the prototype. The 3 products can be separately described in their usage (Fig. 4).

Product 1 – When using this product, the elderly person should bring the lower hoop toward the upper part, then hang the elastic hoop with the top of the wooden handler, and repeatedly squeeze and release it.

Product 2 – When using this product, the elderly person should put each of their fingers into each groove of the wood, then slowly open their hand and fold it back in order to exercise the opening and folding of the fingers.

Product 3 – The usage can be divided into 3 steps: (1) put the pillow around the neck and pull the elastic string to exercise; (2) put the pillow around the waist and pull the string to exercise the fingers, and (3) Pull open the zip and place the pillow into the bag, the pull the bag over before putting it away.

Ten sets of questionnaires were given to the 10 elderly people in the population group. The table below shows the number, mean, and standard deviation that reflects the satisfaction that the users have toward the products for hand and arm muscle exercising occupational therapy. It can be concluded into the following list of aspects.

Fig. 4. The products being used with target group.

Table 5. Number, Percentage, Mean, and Standard Deviation of the Satisfaction that the Elderly People Have toward the Product Design (N = 10).

Opinions	Plier		Knuckle		Neck Pillow	
	$\bar{x}$	SD.	$\bar{x}$	SD.	$\bar{x}$	SD.
1. Design						
1.1 Shape and form of the products	4.6	0.52	4.4	0.70	4.5	0.71
1.2 Color of the products	4.7	0.48	3.8	0.79	4.8	0.42
1.3 Materials suitable for product users	4.3	0.52	4.3	0.67	4.6	0.52
Total	**4.5**	**0.51**	**4.2**	**0.72**	**4.6**	**0.55**
2. Convenience						
2.1 Convenience in using	4.5	0.71	4.1	0.52	4.7	0.68
2.2 Being easy to use	4.6	0.52	4.1	0.57	4.1	0.74
2.3 Being comfortable for holding	4.3	0.68	4.5	0.52	4.5	0.71
Total	**4.5**	**0.64**	**4.2**	**0.54**	**4.4**	**0.71**
3. Safety						
3.1 Strong and enduring	4.5	0.68	4.6	0.70	4.5	0.53
3.2 Harmless for users	4.3	0.68	4.3	0.48	4.6	0.52
3.3 Made of safe materials	4.6	0.52	4.4	0.52	4.6	0.70

(continued)

Table 5. (*continued*)

Opinions	Plier		Knuckle		Neck Pillow	
	$\bar{x}$	SD.	$\bar{x}$	SD.	$\bar{x}$	SD.
Total	**4.5**	**0.63**	**4.4**	**0.57**	**4.6**	**0.58**
4. Satisfaction with overall material use and products for occupational therapy of the elderly						
4.1 Product is attractive and interesting	4.7	0.68	4.6	0.52	4.7	0.48
4.2 Product is suitable for the elderly	4.7	0.48	4.5	0.53	4.7	0.48
4.3 Product helps the elderly become aware of their need for exercise	4.5	0.53	4.3	0.67	4.6	0.52
Total	**4.6**	**0.56**	**4.5**	**0.57**	**4.6**	**0.50**
Total Products	**4.5**	**0.59**	**4.3**	**0.60**	**4.6**	**0.59**

According to Table 5, the satisfaction with neck-supporting pillow is at the highest level or ($\bar{x}$ = 4.6) (S.D. = 0.59). The satisfaction with the overall material use and products for occupational therapy of the elderly is at the highest level or ($\bar{x}$ = 4.6) (S. D. = 0.50). In design, the satisfaction is also at the highest level or ($\bar{x}$ = 4.6) (S. D. = 0.55). The same with the safety, which also has the satisfaction at the highest level or ($\bar{x}$ = 4.6) (S.D. = 0.58). The product that receives the second highest satisfaction is the plier, with the result ($\bar{x}$ = 4.5) (S.D. = 0.59). Its convenience receives the highest level of satisfaction or ($\bar{x}$ = 4.5) (S.D. = 0.64). Knuckle receives high satisfaction or ($\bar{x}$ = 4.3) (S.D. = 0.60).

5 Conclusion

Products for hand muscle exercise occupational therapy for the elderly are intentionally designed and developed in order to create equality of people in living daily life normally, and reduce the social gap. Today, there are many products for exercising hands and arms. They are sold in medical places and most of them are imported products that are sold in Thailand at high prices. This negative affects the feeling of the people who have less opportunity to access good products. It is possible to produce such products with lower production cost by using natural materials from Thai local intelligence. This kind of products can have multiple purposes and cover the exercises of both arm and hand muscle in the elderly. They can be use anytime as one is convenient because they are small and handy. They also have uniqueness in Thai-style shape, inspired from Thai fighting fish. This also has good meaning in terms of the experience of life of the elderly. The use of products for hand muscle exercise occupational therapy for the elderly shows high effectiveness in reducing or decelerating hand and arm muscle weakening.

Acknowledgments. The author would like to thank Phranakhon Rajabhat University for providing financial means for the research. Further, the author I also want to thank many people from various groups and divisions for their cooperation, especially the staff, the elderly people, and the qualified persons who gave me much useful advice about the research.

References

1. WHO: Global Age-Friendly Cities: A Guide, Geneva (2007)
2. JICA: Facing up to the Problem of Population Aging in Developing Countries, Tokyo (2006)
3. UNFPA: Population Ageing in Thailand: Prognosis and Policy Response, Thailand (2006)
4. National Research Council of Thailand Executive Summary: The Elderly and Aging Society Strategy (2013–2016). http://www.nrct.go.th/
5. Kenny, G.P., Yardley, J.E., Martineau, L., Jay, O.: Physical work capacity in older adults: implications for the aging worker. Am. J. Ind. Med. **51**(8), 610–625 (2008)
6. Corish, C.A., Kennedy, N.P.: Anthropometric measurements from a crosssectional survey of Irish free-living elderly subjects with smoothed centile curves. Br. J. Nutr. **89**(1), 137–145 (2003)
7. Tsubaki, A., Kubo, M., Jigami, H., Takahashi, H.E.: Age-related changes in physical function in community-dwelling people aged 50–79 years. J. Phys. Ther. Sci. **22**(1), 23–27 (2010)
8. Metter, E.J., Talbot, L.A., Schrager, M.A., Conwit, R.: Skeletal muscle strength as a predictor of all-cause mortality in healthy men. J. Gerontol. Ser. A Biol. Sci. Med. Sci. **57**, 359–365 (2002)
9. Williams, M., Temperley, D., Murali, R.: Radiology of the hand. Orthop. Trauma **33**, 45–52 (2019)
10. Phanuphan, S.: Nerve Injuries. Ruenkaew Publising, Bangkok (2013)
11. Piligian, G., Herbert, R., Hearns, M., Dropkin, J., Landsbergis, P., Cherniack, M.: Evaluation and management of chronic work-related musculoskeletal disorders of the distal upper extremity. Am. J. Ind. Med. **37**(1), 75–93 (2000)
12. Natthachai, C.: Rehabilitation Physician (Carpal Tunnel Syndrome), 13 Dec 2011. http://www.ladpraohospital.com/healthKnowledges.asp?id=15. Accessed 20 Nov 2011
13. Kriščiūnas, A., Kerpė, R., Kimtys, A., Mingaila, S., Rapolienė, J., Petruševičienė, D., et al.: Ergoterapija (Occupational therapy). Vitae Litera, Kaunas (2008)
14. Petruševičienė, D., Kriščiūnas, A.: Ligonių, sirgusių galvos smegenų insultu, ergoterapija [Occupational therapy for patients after stroke]. Medicina (Kaunas) **11**, 1065–1070 (2003)

Author Index

C. S. Shin (Ed.): AHFE 2019, AISC 968, pp. 317–318, 2020.
https://doi.org/10.1007/978-3-030-20470-9

MIX
Papier aus verantwortungsvollen Quellen
Paper from responsible sources
FSC® C105338

If you have any concerns about our products,
you can contact us on
ProductSafety@springernature.com

In case Publisher is established outside the EU,
the EU authorized representative is:
Springer Nature Customer Service Center GmbH
Europaplatz 3, 69115 Heidelberg, Germany

Printed by Libri Plureos GmbH
in Hamburg, Germany